MW00475934

STRESSES IN PLATES AND SHELLS

STRESSES IN PLATES AND SHELLS

A. C. Ugural, Ph.D.
Professor and Chairman
Mechanical Engineering Department
Fairleigh Dickinson University

McGraw-Hill Book Company

New York St. Louis San Francisco Auckland Bogotá Hamburg
Johannesburg London Madrid Mexico Montreal New Delhi
Panama Paris São Paulo Singapore Sydney Tokyo Toronto

This book was set in Times Roman.
The editor was Frank J. Cerra;
the production supervisor was Donna Piligra.

STRESSES IN PLATES AND SHELLS

23456789 HDHD 89876543

Library of Congress Cataloging in Publication Data

Ugural, A C
Stresses in plates and shells.

Includes bibliographical references and index.
1. Plates (Engineering) 2. Shells (Engineering)
3. Strains and stresses. I. Title.
TA660.P6U39 624.1'776 80-13927
ISBN 0-07-065730-0

CONTENTS

PREFACE

The subject matter of this text is usually covered in one-semester senior and one-semester graduate level courses dealing with the *analysis of plates and shells*. As sufficient material is provided for a full year of study, the book may stimulate the development of courses in *advanced statics* and *structural analysis*. The coverage presumes a knowledge of elementary mechanics of materials.

The text is intended to serve a twofold purpose: to complement classroom lectures and to accommodate the needs of practicing engineers in the analysis of plate and shell structures. The material presented is applicable to aeronautical, astronautical, chemical, civil, mechanical, and ocean engineering; engineering mechanics; and science curricula.

Emphasis is given computer oriented *numerical techniques* in the solution of problems resisting *analytical approaches*. The reader is helped to realize that a firm grasp of fundamentals is necessary to perform the critical interpretations, so important when computer-based solutions are employed. However, the stress placed upon numerical methods is not intended to deny the merit of *classical analysis* which is given a rather full treatment. The volume attempts to fill what the writer believes is a void in the world of texts.

The book offers a simple, comprehensive, and methodical presentation of the principles of plate and shell theories and their applications to numerous structural elements, including domes, pressure vessels, tanks, and pipes. Theories of failures are employed in predicting the behavior of plates and shells under combined loading. Above all, an effort has been made to provide a visual interpretation of the basic equations and of the means by which loads are resisted in shells, plates, and beams. A balance is presented between the theory necessary to gain insight into the mechanics and the numerical solutions, both so useful in performing stress analysis in a more realistic setting. Throughout the text, the author has attempted to provide the fundamentals of theory and application necessary to prepare students for more advanced study and for professional practice. Development of the physical and mathematical aspects of the subject is deliberately pursued.

The physical significance of the solutions, and practical applications, are given emphasis. With regard to application, often classical engineering examples are used to maintain simplicity and lucidity. The author has made a special effort to illustrate important principles and applications with numerical examples. A variety of problems is provided for solution by the student. The International System of Units (SI) is used.

The expression defining the small lateral deflection of the midplane of a thin plate is formulated two ways. The first utilizes the fundamental assumptions made in the elementary *theory of beams*. The second is based upon the differential equations of equilibrium for the *three-dimensional stress*. The former approach, which requires less mathematical rigor but more physical interpretation, is regarded as more appealing to the engineer, and is equally used in the case of thin shells. Emphasized also are the energy aspects of plate and shell bending and buckling because of the importance of *energy methods* in the solution of many real-life problems and in modern computational techniques. Because of the introductory nature of this book, the classical approaches requiring extensive mathematical background are not treated.

Recent publications dealing with shell theory include analytical presentations generally valid for any shell under any kind of loading. These formulations usually necessitate the employment of tensor notation, vector analysis, and a system of curvilinear coordinates. The theory introduced in this text is a special case of the above. The equations governing shells are developed only to the extent necessary for solving the more usual engineering problems. The finite element method is applied to treat plates of nonuniform thickness and arbitrary shape as well as to represent shells of arbitrary form.

The volume may be divided into two parts. Chapters 1 to 9 contain the fundamental definitions and the analysis of plates. Chapters 10 to 13 deal with shells. The book is organized so that Chap. 1 must be studied first. The remaining chapters may be taken in any sequence except that Chaps. 10 and 11 should be read before Chaps. 12 and 13. Numerous alternatives are possible in making selections from the book for two single-semester courses. The chapters have been arranged in a sequence compatible with an orderly study of the analysis of plates and shells.

This text offers a wide range of fully worked out illustrative examples, approximately 170 problem sets, many of which are drawn from engineering practice, a multitude of formulas and tabulations of plate and shell theory solutions from which direct and practical design calculations can be made, analyses of plates and shells made of isotropic as well as composite materials under ordinary and high temperature loadings, numerical methods amenable to computer solution, and applications of the formulas developed and of the theories of failures to increasingly important structural members.

Thanks are due to the many students who offered constructive suggestions when drafts of this work were used as a text. Dr. S. K. Fenster read the entire manuscript and made many corrections for which the author is most grateful. He is indebted to Dr. B. Lefkowitz, who read Chaps. 7 and 8, and to Mrs. H. Stanek for her most skillful services in the preparation of the manuscript.

A. C. Ugural

LIST OF SYMBOLS

A area, constant
a, b dimensions, outer and inner radii of annular plate
D flexural rigidity $[D = Et^3/12(1 - \nu^2)]$
$[D]$ elasticity matrix
E modulus of elasticity
F resultant external loading on shell element
G modulus of elasticity in shear
g acceleration of gravity (≈ 9.81 m/s^2)
h mesh width, numerical factor
k modulus of elastic foundation, numerical factor, axial load factor for slender members in compression
$[k]$ stiffness matrix of finite element
$[K]$ stiffness matrix of whole structure
m, n integers, numerical factors
M moment per unit distance, moment-sum $[M = (M_x + M_y)/(1 + \nu)]$
M^* thermal moment resultant per unit distance
M_x, M_y bending moments per unit distance on x and y planes
M_{xy} twisting moment per unit distance on x plane
M_r, M_θ radial and tangential moments per unit distance
$M_{r\theta}$ twisting moment per unit distance on radial plane
M_ϕ meridional bending moment per unit distance on parallel plane
M_s meridional bending moment per unit distance on parallel plane of conical shell
$M_{x\theta}$ twisting moment per unit distance on axial plane of cylindrical shell
N normal force per unit distance
N_{cr} critical compressive load per unit distance

N^*	thermal force resultant per unit distance
N_x, N_y	normal forces per unit distance on x and y planes
N_{xy}	shearing force per unit distance on x plane and parallel to y axis
N_r, N_θ	radial and tengential forces per unit distance
N_ϕ	meridional force per unit distance on parallel plane
$N_{\phi\theta}$	shear force per unit distance on parallel plane and perpendicular to meridional plane.
$N_{x\theta}$	shear force per unit distance on axial plane and parallel to y axis of cylindrical shell
N_s	normal force per unit distance on parallel plane of conical shell
p	intensity of distributed transverse load per unit area, pressure
p^*	equivalent transverse load per unit area
P	concentrated force
$\{Q\}$	nodal force matrix of finite element
Q_x, Q_y	shear force per unit distance on x and y planes
Q_r, Q_θ	radial and tangential shear forces per unit distance
Q_θ	shear force per unit distance on plane perpendicular to the axial plane of cylindrical shell
Q_ϕ	meridional shear force per unit distance on parallel plane
R	reactive forces
r	radius
r, θ	polar coordinates
r_x, r_y	radii of curvature of midsurface in xz and yz planes
r_1, r_2	radii of curvature of midsurface in meridional and parallel planes, principal radii of curvature
T	surface forces per unit area, temperature
t	thickness
s	distance measured along generator in conical shell
u, v, w	displacements in x, y, and z directions; axial, tangential, and radial displacements in shell midsurface
U	strain energy
V_x, V_y	effective shear force per unit distance on x and y planes
V_r, V_θ	radial and tangential effective shear forces per unit distance
W	work, weight
x, y, z	distances, rectangular coordinates
α	angle, coefficient of thermal expansion, numerical factor
β	angle, cylinder geometry parameter $[\beta^4 = 3(1 - \nu^2)/a^2t^2]$, numerical factor
γ	shear strain, weight per unit volume or specific weight
$\gamma_{xy}, \gamma_{yz}, \gamma_{zx}$	shear strains in the xy, yz, and zx planes

$\gamma_{r\theta}$	shear strain in the $r\theta$ plane
δ	deflection, finite difference operator, numerical factor, variational symbol
$\{\delta\}$	nodal displacement matrix of finite element
ε	normal strain
$\varepsilon_x, \varepsilon_y, \varepsilon_z$	normal strains in x, y, and z directions
$\varepsilon_r, \varepsilon_\theta$	radial and tangential normal strains
$\varepsilon_\theta, \varepsilon_\phi$	normal strain of the parallel circle and of the meridian
θ	angle, angular nodal displacement
κ	curvature
ν	Poisson's ratio
Π	potential energy
σ	normal stress
$\sigma_x, \sigma_y, \sigma_z$	normal stresses on the x, y, and z planes
σ_r, σ_θ	radial and tangential normal stresses
σ_ϕ	meridional normal stress on parallel plane
$\sigma_1, \sigma_2, \sigma_3$	principal stresses
σ_{cr}	compressive stress at critical load
σ_u	ultimate stress
σ_{yp}	yield stress
τ	shear stress
$\tau_{xy}, \tau_{yz}, \tau_{zx}$	shear stresses on the x, y, and z planes and parallel to the y, z, and x directions
$\tau_{r\theta}$	shear stress on radial plane and parallel to the tangential plane
τ_u	ultimate stress in shear
τ_{yp}	yield stress in shear
ϕ	angle, stress function, numerical factor
χ	change of curvature in shell

CHAPTER
ONE
ELEMENTS OF PLATE-BENDING THEORY

1.1 INTRODUCTION

Plates and shells are initially flat and curved structural elements, respectively, for which the thicknesses are much smaller than the other dimensions. Included among the more familiar examples of plates are table tops, street manhole covers, side panels and roofs of buildings, turbine disks, bulkheads, and tank bottoms. Many practical engineering problems fall into categories "plates in bending" or "shells in bending." In Chaps. 1 to 9, we treat plates, for which it is common to divide the thickness t into equal halves by a plane parallel to the faces. This plane is called the *midplane* of the plate. The plate thickness is measured in a direction normal to the midplane. The flexural properties of a plate depend greatly upon its thickness in comparison with its other dimensions.

Plates may be classified into three groups: *thin* plates with small deflections, *thin* plates with large deflections, and *thick* plates. According to the criterion often applied to define a *thin plate* (for purposes of technical calculations) the ratio of the thickness to the smaller span length should be less than 1/20. With the exception of Chap. 8, we shall consider only small deflections of thin plates, a simplification consistent with the magnitude of deformation commonly found in plate structures. For clarity, however, the deflections and thicknesses of plates will be shown greatly exaggerated on some diagrams. It is assumed, unless otherwise specified, that plate and shell materials are homogeneous and isotropic. A *homogeneous* material displays identical properties throughout. When the properties are the same in all directions, the material is called *isotropic*.

We are concerned with the relationships of external force or moment to strain, stress, and displacement. External forces acting on a plate may be considered as *surface forces* and *body forces*. A surface force is of the concentrated type when its acts at a point. It may also be distributed arbitrarily over a finite area. Body forces act on volumetric elements of the plate. They are attributable to fields such as gravity and magnetism, or in the case of motion, to the inertia of the plate.

To ascertain the distribution of stress and displacement for a plate subject to a given set of forces requires consideration of a number of basic conditions. These pertain to certain physical laws, material properties, geometry, and surface forces. These conditions, stated mathematically in this chapter, are used to solve the bending problems of plates in the chapters to follow. It is often advantageous, where the shape of the plate or loading configuration preclude a theoretical solution or where verification is sought, to employ experimental methods. The approximate numerical and energy approaches (Chap. 5 and Sec. 1.9, respectively), are also efficient for this purpose. For the cases of nonuniformly heated and orthotropic plates, Chaps. 9 and 6, respectively, rederivation of some basic relationships and the governing equation are required.

The first significant treatment of plates occurred in the 1800s. Since then a great many cases of plate-bending problems have been worked out: the fundamental theory (principally by Navier, Kirchhoff, and Lévy) and numerical approaches (by Galerkin and Wahl, and others). The literature related to plate and shell analysis is extensive. A number of *selected references* are supplied at the end of the text.

The determination of plate shape and the selection of a material that is most efficient for resisting a given system of forces under specified conditions of operation is the *design function*. A basic understanding of material behavior and evaluation of the most likely *modes of failure* under anticipated conditions of service are essential for this purpose (Sec. 1.10). The *rational design* of plates and shells relies greatly upon their stress and deformation analysis to which this book is directed.

1.2 GENERAL BEHAVIOR OF PLATES

Consider a load-free plate, shown in Fig. 1.1*a*, in which the xy plane coincides with the midplane and hence the z deflection is zero. The components of displacement at a point, occurring in the x, y, and z directions, are denoted by u, v, and w, respectively. When, due to lateral loading, deformation takes place, the *midsurface* at any point (x_a, y_a) has deflection w (Fig. 1.1*b*). The *fundamental assumptions* of the small-deflection theory of bending or so-called classical theory for isotropic, homogeneous, elastic, thin plates is based on the geometry of deformations. They may be stated as follows:

(1) The deflection of the midsurface is small compared with the thickness of the plate. The slope of the deflected surface is therefore very small and the square of the slope is a negligible quantity in comparison with unity.

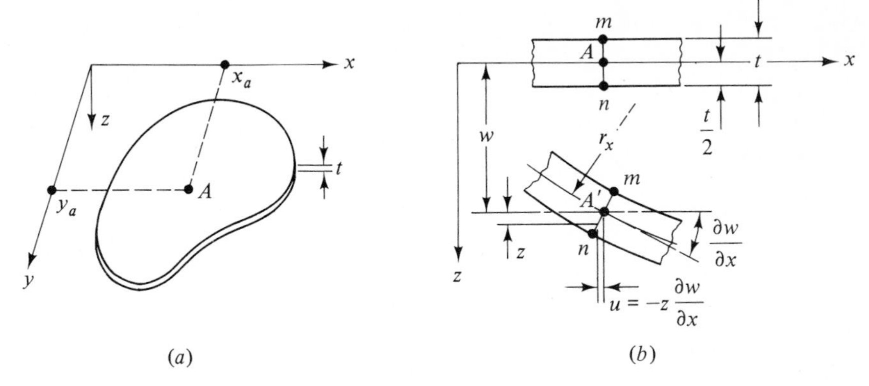

Figure 1.1

(2) The midplane remains unstrained subsequent to bending.
(3) Plane sections initially normal to the midsurface *remain plane* and normal to that surface after the bending. This means that the vertical shear strains γ_{xz} and γ_{yz} are negligible. The deflection of the plate is thus associated principally with bending strains. It is deduced therefore that the normal strain ε_z resulting from transverse loading may also be omitted.
(4) The stress normal to the midplane, σ_z, is small compared with the other stress components and may be neglected. This supposition becomes unreliable in the vicinity of highly concentrated transverse loads.

The above assumptions, known as the *Kirchhoff hypotheses*, are analogous to those associated with the simple *bending theory of beams*. In the vast majority of engineering applications, adequate justification may be found for the simplifications stated with respect to the state of deformation and stress. Because of the resulting decrease in complexity, a three-dimensional plate problem reduces to one involving only two dimensions. Consequently, the governing plate equation can be derived in a concise and straightforward manner.

When the deflections are *not* small, the bending of plates is accompanied by strain in the midplane, and assumptions (1) and (2) are inapplicable. An exception, however, applies when a plate bends into a *developable surface* (e.g., surfaces of cones and cylinders), Sec. 8.2. This type of surface can be bent back to a plane without variation in the distances between any two points on the surface. If the midsurface of a freely or simply supported and loaded plate has a developable form, it remains unstrained even for deflections that are *equal* or *larger* than its thickness but are still small as compared with other dimensions of the plate. Only under this limitation on the deflections will the squares of slopes be small compared with unity; hence the approximate expression used for the curvatures (Sec. 1.3) is sufficiently accurate.

In *thick* plates, the shearing stresses are important, as in short, deep beams. Such plates are treated by means of a more general theory owing to the fact that assumptions (3) and (4) are no longer appropriate.

1.3 STRAIN-CURVATURE RELATIONS

In order to gain insight into the plate-bending problem, consideration is now given to the geometry of deformation. As a consequence of the assumption (3) of the foregoing section, the strain-displacement relations[1] reduce to

$$\varepsilon_x = \frac{\partial u}{\partial x} \qquad \varepsilon_z = \frac{\partial w}{\partial z} = 0$$

$$\varepsilon_y = \frac{\partial v}{\partial y} \qquad \gamma_{xz} = \frac{\partial w}{\partial x} + \frac{\partial u}{\partial z} = 0 \tag{1.1a-f}$$

$$\gamma_{xy} = \frac{\partial u}{\partial y} + \frac{\partial v}{\partial x} \qquad \gamma_{yz} = \frac{\partial w}{\partial y} + \frac{\partial v}{\partial z} = 0$$

where $\gamma_{ij} = \gamma_{ji}(i, j = x, y, z)$. Note that these expressions are also referred to as the *kinematic* relations, treating the *geometry* of strain rather than the matter of cause and effect. Integrating Eq. (1.1*d*), we obtain

$$w = w(x, y) \tag{a}$$

indicating that the lateral deflection does not vary over the plate thickness. In a like manner, integration of the expressions for γ_{xz} and γ_{yz} gives

$$u = -z\frac{\partial w}{\partial x} + u_0(x, y) \qquad v = -z\frac{\partial w}{\partial y} + v_0(x, y) \tag{b}$$

It is clear that $u_0(x, y)$ and $v_0(x, y)$ represent, respectively, the values of u and v on the midplane. Based upon assumption (2) of Sec. 1.2, we conclude that $u_0 = v_0 = 0$. Thus

$$u = -z\frac{\partial w}{\partial x} \qquad v = -z\frac{\partial w}{\partial y} \tag{1.2}$$

The above expression for u is represented in Fig. 1.1*b* at section *mn* passing through arbitrary point $A(x_a, y_a)$. A similar illustration applies for v in the zy plane. We see that Eqs. (1.2) are consistent with assumption (3). Substitution of Eqs. (1.2) into the first three of Eqs. (1.1) yields

$$\varepsilon_x = -z\frac{\partial^2 w}{\partial x^2} \qquad \varepsilon_y = -z\frac{\partial^2 w}{\partial y^2} \qquad \gamma_{xy} = -2z\frac{\partial^2 w}{\partial x\,\partial y} \tag{1.3a}$$

These formulas provide the strains at any point in the plate.

The *curvature* of a plane curve is defined as the *rate* of change of the slope angle of the curve with respect to distance along the curve. Because of assumption (1) of Sec. 1.2, the square of a slope may be regarded as negligible, and the partial derivatives of Eqs. (1.3*a*) represent the curvatures of the plate. Therefore

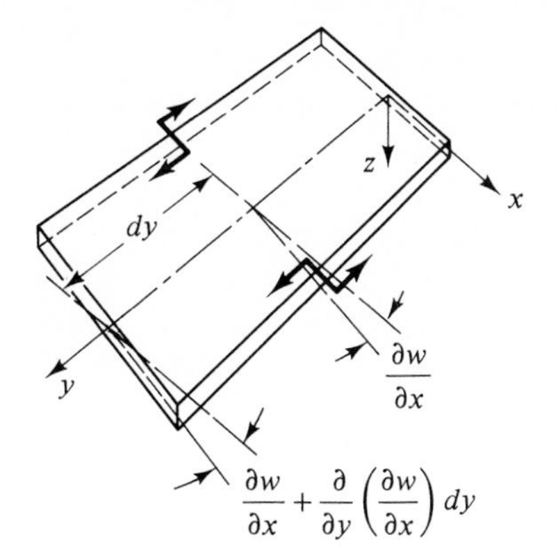

Figure 1.2

the curvatures κ (kappa) at the midsurface in planes *parallel* to the xz (Fig. 1.1*b*), yz, and xy planes are, respectively

$$\begin{aligned} \frac{1}{r_x} &= \frac{\partial}{\partial x}\left(\frac{\partial w}{\partial x}\right) = \kappa_x \\ \frac{1}{r_y} &= \frac{\partial}{\partial y}\left(\frac{\partial w}{\partial y}\right) = \kappa_y \\ \frac{1}{r_{xy}} &= \frac{\partial}{\partial x}\left(\frac{\partial w}{\partial y}\right) = \kappa_{xy} \end{aligned} \tag{1.4}$$

where $\kappa_{xy} = \kappa_{yx}$. Clearly, Eqs. (1.4) are the *rates* at which the slopes vary over the plate. The last of these expressions is also referred to as the *twist* of the midplane with respect to x and y axes. The local twist of a plate element is shown in Fig. 1.2.

The strain-curvature relations, by means of Eqs. (1.3*a*) and (1.4) may be expressed in the form

$$\varepsilon_x = -z\kappa_z \qquad \varepsilon_y = -z\kappa_y \qquad \gamma_{xy} = -2z\kappa_{xy} \tag{1.3b}$$

An examination of Eqs. (1.3) shows that a circle of curvature can be constructed similarly to *Mohr's circle* of strain. The curvatures therefore transform in the same manner as the strains. Figure 1.3 shows a plate element and a circle of

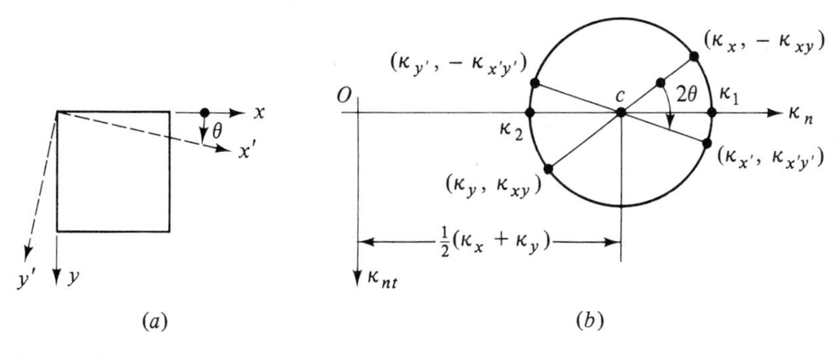

Figure 1.3

curvature in which n and t represent perpendicular directions at a point on the midsurface. The *principal* or maximum and minimum curvatures are indicated by κ_1 and κ_2. The planes associated with these curvatures are called the *principal planes* of curvature. The curvature and the twist of a surface vary with the angle θ, measured in the clockwise direction from the set of axes xy to the $x'y'$ set. It is seen that when the two principal curvatures are the same, Mohr's circle shrinks to a point. This means that the curvature is the same in all directions; there is no twist in any direction. The surface is purely spherical at that point. From the circle and Eqs. (1.4) we have (see Prob. 1.1)

$$\kappa_x + \kappa_y = \kappa_{x'} + \kappa_{y'} \tag{1.5}$$

The sum of the two curvatures in perpendicular directions at a point, called the *average curvature*, is thus *invariant* with respect to rotation of the coordinate axes. This assertion is valid at any location on the midsurface.

1.4 STRESSES AND STRESS RESULTANTS

In the case of a three-dimensional state of stress, stress and strain are related by the generalized Hooke's law, valid for an isotropic homogeneous material

$$\begin{aligned}
\varepsilon_x &= \frac{1}{E}[\sigma_x - \nu(\sigma_y + \sigma_z)] & \gamma_{xy} &= \frac{\tau_{xy}}{G} \\
\varepsilon_y &= \frac{1}{E}[\sigma_y - \nu(\sigma_x + \sigma_z)] & \gamma_{xz} &= \frac{\tau_{xz}}{G} \\
\varepsilon_z &= \frac{1}{E}[\sigma_z - \nu(\sigma_x + \sigma_y)] & \gamma_{yz} &= \frac{\tau_{yz}}{G}
\end{aligned} \tag{a}$$

where $\tau_{ij} = \tau_{ji}$ $(i, j = x, y, z)$. The constants E, ν, and G represent the modulus of elasticity, Poisson's ratio, and the shear modulus of elasticity, respectively. The connecting expression is

$$G = \frac{E}{2(1 + \nu)} \tag{1.6}$$

The double *subscript notation* for stress is interpreted as follows: the first subscript indicates the direction of a normal to the plane or face on which the stress component acts; the second subscript relates to the direction of the stress itself. Repeated subscripts will be omitted in this text. That is, the normal stresses σ_{xx}, σ_{yy}, and σ_{zz} are designated by σ_x, σ_y, and σ_z (Fig. 1.4*a*). *A face, plane, or surface is usually identified by the axis normal to it*, e.g., the x faces are perpendicular to the x axis.

The *sign convention* for stresses relies upon the relationship between the direction of an *outward normal* drawn to a particular surface, and the directions of the stress components on the same surface. If *both* the outer normal and the

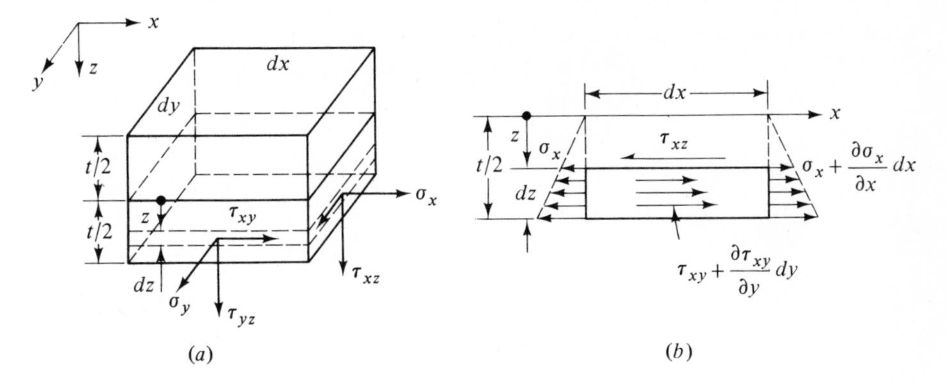

Figure 1.4

stress component are in a positive (or negative) direction relative to the coordinate axes, the stress is positive. If the outer normal points in a positive direction while the stress points in a negative direction (or vice versa), the stress is negative. On this basis, all stress components shown in Fig. 1.4*a* are positive.

Substitution of $\varepsilon_z = \gamma_{yz} = \gamma_{xz} = 0$ into Eqs. (*a*) yields the following stress-strain relations for a thin plate

$$\begin{aligned} \sigma_x &= \frac{E}{1-\nu^2}(\varepsilon_x + \nu\varepsilon_y) \\ \sigma_y &= \frac{E}{1-\nu^2}(\varepsilon_y + \nu\varepsilon_x) \\ \tau_{xy} &= G\gamma_{xy} \end{aligned} \tag{1.7}$$

Introducing the plate curvatures, Eqs. (1.3*b*) and (1.4), the above appear in the form

$$\begin{aligned} \sigma_x &= -\frac{Ez}{1-\nu^2}(\kappa_x + \nu\kappa_y) = -\frac{Ez}{1-\nu^2}\left(\frac{\partial^2 w}{\partial x^2} + \nu\frac{\partial^2 w}{\partial y^2}\right) \\ \sigma_y &= -\frac{Ez}{1-\nu^2}(\kappa_y + \nu\kappa_x) = -\frac{Ez}{1-\nu^2}\left(\frac{\partial^2 w}{\partial y^2} + \nu\frac{\partial^2 w}{\partial x^2}\right) \\ \tau_{xy} &= -\frac{Ez}{1+\nu}\kappa_{xy} = -\frac{Ez}{1+\nu}\frac{\partial^2 w}{\partial x\,\partial y} \end{aligned} \tag{1.8}$$

We observe from these formulas that the stresses vanish at the midsurface and vary *linearly* over the thickness of the plate.

The stresses distributed over the thickness of the plate produce bending moments, twisting moments, and vertical shear forces. These moments and forces *per unit length* are also called *stress resultants*. Referring to Fig. 1.4*a*, we have

$$\int_{-t/2}^{t/2} z\sigma_x\,dy\,dz = dy\int_{-t/2}^{t/2} z\sigma_x\,dz = M_x\,dy$$

Similarly, expressions for the other resultants are derived

$$\begin{Bmatrix} M_x \\ M_y \\ M_{xy} \end{Bmatrix} = \int_{-t/2}^{t/2} \begin{Bmatrix} \sigma_x \\ \sigma_y \\ \tau_{xy} \end{Bmatrix} z\, dz \tag{1.9a}$$

where $M_{xy} = M_{yx}$, and

$$\begin{Bmatrix} Q_x \\ Q_y \end{Bmatrix} = \int_{-t/2}^{t/2} \begin{Bmatrix} \tau_{xz} \\ \tau_{yz} \end{Bmatrix} dz \tag{1.9b}$$

The sign convention for shear force is the same as that for shear stress. A positive moment is one which results in positive stresses in the bottom half of the plate. Accordingly, all moments and shear forces acting on the element in Fig. 1.5 of Sec. 1.5 are positive.

It is important to mention that while the theory of thin plates omits the effect of the strain components $\gamma_{xz} = \tau_{xz}/G$ and $\gamma_{yz} = \tau_{yz}/G$ on bending, vertical forces Q_x and Q_y are not negligible. In fact, they are of the same order of magnitude as the surface loading and moments, and are included in the derivation of the equilibrium equations (Sec. 1.5).

Substituting Eqs. (1.8) into Eq. (1.9*a*) we derive the following formulas for the bending and twisting moments in terms of the curvatures and the deflection

$$\begin{aligned} M_x &= -D(\kappa_x + \nu\kappa_y) = -D\left(\frac{\partial^2 w}{\partial x^2} + \nu\frac{\partial^2 w}{\partial y^2}\right) \\ M_y &= -D(\kappa_y + \nu\kappa_x) = -D\left(\frac{\partial^2 w}{\partial y^2} + \nu\frac{\partial^2 w}{\partial x^2}\right) \\ M_{xy} &= -D(1-\nu)\kappa_{xy} = -D(1-\nu)\frac{\partial^2 w}{\partial x\, \partial y} \end{aligned} \tag{1.10}$$

where

$$D = \frac{Et^3}{12(1-\nu^2)} \tag{1.11}$$

is the *flexural rigidity* of the plate. The vertical shear forces Q_x and Q_y are related to w, upon derivation of the equilibrium equations.

It is noted that if a plate element of *unit width* and parallel to the x axis were free to move sidewise under transverse loading, the top and bottom surfaces would be deformed into saddle-shaped or *anticlastic* surfaces of curvature κ_y. The flexural rigidity would then be $Et^3/12$, as in the case of a beam. The remainder of the plate prevents the anticlastic curvature however. Owing to this action, a *plate manifests greater stiffness* than a beam by a factor $1/(1-\nu^2)$, approximately 10 percent.

The stresses are found from Eqs. (1.8) by substituting Eqs. (1.10) and by employing Eq. (1.11). In this way we obtain

$$\sigma_x = \frac{12M_x z}{t^3} \qquad \sigma_y = \frac{12M_y z}{t^3} \qquad \tau_{xy} = \frac{12M_{xy} z}{t^3} \tag{1.12}$$

The *maximum* stresses occur on the *bottom* and *top* surfaces (at $z = \pm t/2$) of the plate. It is observed from formulas (1.9*a*) and (1.12) that there is a direct correspondence between the moments and stresses. Hence, transformation equations for stress and moment are analogous. The Mohr's circle analysis and all conclusions drawn for stress therefore apply to the moments (see Prob. 1.2). Note that quantities such as stress, strain, curvature, moment (and moment of inertia), which their components transform according to a certain law, are called *tensors* of second order.[1] The order refers to the number of subscripts needed to describe a component. Mohr's circle is thus a graphical representation of a tensor transformation.

Determination of the stress components σ_z, τ_{xz}, and τ_{yz} through the use of Hooke's law is not possible, since according to Eq. (1.1) they are not related to strains. The differential equations of equilibrium for a plate element under a general state of stress serves well for this purpose, however. These equations are[1]

$$\begin{aligned} \frac{\partial \sigma_x}{\partial x} + \frac{\partial \tau_{xy}}{\partial y} + \frac{\partial \tau_{xz}}{\partial z} &= 0 \\ \frac{\partial \sigma_y}{\partial y} + \frac{\partial \tau_{xy}}{\partial x} + \frac{\partial \tau_{yz}}{\partial z} &= 0 \\ \frac{\partial \sigma_z}{\partial z} + \frac{\partial \tau_{xz}}{\partial x} + \frac{\partial \tau_{yz}}{\partial y} &= 0 \end{aligned} \tag{b}$$

From the first two of the above expressions and Eqs. (1.8), the shearing stresses τ_{xz} (Fig. 1.4*b*) and τ_{yz} are, after integration

$$\begin{aligned} \tau_{xz} &= \int_z^{t/2} \left(\frac{\partial \sigma_x}{\partial x} + \frac{\partial \tau_{xy}}{\partial y} \right) dz = -\frac{E}{2(1-\nu^2)} \left(\frac{t^2}{4} - z^2 \right) \left[\frac{\partial}{\partial x} \left(\frac{\partial^2 w}{\partial x^2} + \frac{\partial^2 w}{\partial y^2} \right) \right] \\ \tau_{yz} &= \int_z^{t/2} \left(\frac{\partial \sigma_y}{\partial y} + \frac{\partial \tau_{xy}}{\partial x} \right) dz = -\frac{E}{2(1-\nu^2)} \left(\frac{t^2}{4} - z^2 \right) \left[\frac{\partial}{\partial y} \left(\frac{\partial^2 w}{\partial x^2} + \frac{\partial^2 w}{\partial y^2} \right) \right] \end{aligned} \tag{1.13}$$

It is observed that the distribution of components τ_{xz} and τ_{yz} over the plate thickness varies according to a *parabolic* law. The component σ_z is readily determined by using the third of Eqs. (*b*), upon substitution of τ_{xz} and τ_{yz} from Eqs. (1.13) and integration

$$\sigma_z = -\frac{E}{2(1-\nu^2)} \left(\frac{t^3}{12} - \frac{t^2 z}{4} + \frac{z^3}{3} \right) \left[\left(\frac{\partial^2}{\partial x^2} + \frac{\partial^2}{\partial y^2} \right) \left(\frac{\partial^2 w}{\partial x^2} + \frac{\partial^2 w}{\partial y^2} \right) \right] \tag{1.14}$$

The normal stress σ_z thus varies as a cubic parabola over the thickness of the plate. This stress, according to assumption (4) of Sec. 1.2 is negligible. The z directed shear stress components are also regarded very small when compared with the remaining plane stresses. Equations (1.13) and (1.14) are written in concise form in Sec. 1.6, following the development of equilibrium relationships.

In the *International System of Units* (SI), the stress is measured in *newtons* per square meter (N/m^2) or so-called pascal (Pa). The moments and forces per unit length have units N·m/m or simply N and N/m, respectively.

1.5 VARIATION OF STRESS WITHIN A PLATE

The components of stress (and thus the stress resultants) generally vary from point to point in a loaded plate. These variations are governed by the conditions of equilibrium of *statics*. Fulfillment of these conditions establishes certain relationships known as the equations of equilibrium. We shall eventually reduce the latter system of equations to a single relationship expressed in terms of moments.

Consider an element $dx\,dy$ of the plate subject to a uniformly distributed load per unit area p (Fig. 1.5). We assume that inclusion of the plate weight, a small quantity, in the load p cannot affect the accuracy of the result. Note also that as the element is very small, for the sake of simplicity the force and moment components may be considered to be distributed uniformly over each face. In the figure they are shown by a single vector, representing the *mean values* applied at the center of each face.

With change of location, as for example, from upper left corner to the lower right corner, one of the moment components, say M_x, acting on the negative x face, varies in value relative to the positive x face. This variation with position may be expressed by a truncated Taylor's expansion

$$M_x + \frac{\partial M_x}{\partial x}\,dx$$

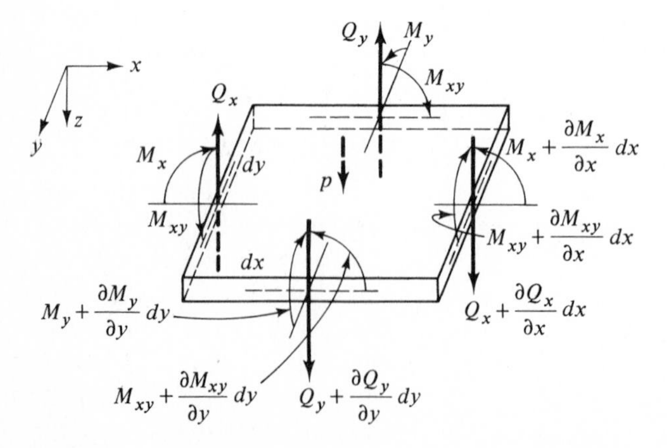

Figure 1.5

The partial derivative is used because M_x is a function of x and y. Treating all the components similarly, the state of stress resultants shown in the figure is obtained.

The condition that the sum of the z directed forces equal zero leads to

$$\frac{\partial Q_x}{\partial x} dx\,dy + \frac{\partial Q_y}{\partial y} dx\,dy + p\,dx\,dy = 0$$

from which

$$\frac{\partial Q_x}{\partial x} + \frac{\partial Q_y}{\partial y} + p = 0 \tag{a}$$

The equilibrium of moments about x axis is governed by

$$\frac{\partial M_{xy}}{\partial x} dx\,dy + \frac{\partial M_y}{\partial y} dx\,dy - Q_y\,dx\,dy = 0$$

or

$$\frac{\partial M_{xy}}{\partial x} + \frac{\partial M_y}{\partial y} - Q_y = 0 \tag{b}$$

Products of infinitesimal terms, such as the moment of p and the moment due to the change in Q_y have been omitted.

Similarly, from the equilibrium of moments about y axis, we have

$$\frac{\partial M_{xy}}{\partial y} + \frac{\partial M_x}{\partial x} - Q_x = 0 \tag{c}$$

Finally, introduction of the expression for Q_x and Q_y from Eqs. (b) and (c) into (a) yields

$$\frac{\partial^2 M_x}{\partial x^2} + 2\frac{\partial^2 M_{xy}}{\partial x\,\partial y} + \frac{\partial^2 M_y}{\partial y^2} = -p \tag{1.15}$$

This is the *differential equation* of *equilibrium* for bending of thin plates.

Expressions for vertical shear forces Q_x and Q_y may now be written in terms of deflection w, from Eqs. (b) and (c) together with Eqs. (1.10)

$$\begin{aligned} Q_x &= -D\frac{\partial}{\partial x}\left(\frac{\partial^2 w}{\partial x^2} + \frac{\partial^2 w}{\partial y^2}\right) = -D\frac{\partial}{\partial x}(\nabla^2 w) \\ Q_y &= -D\frac{\partial}{\partial y}\left(\frac{\partial^2 w}{\partial x^2} + \frac{\partial^2 w}{\partial y^2}\right) = -D\frac{\partial}{\partial y}(\nabla^2 w) \end{aligned} \tag{1.16}$$

where

$$\nabla^2 = \frac{\partial^2}{\partial x^2} + \frac{\partial^2}{\partial y^2} \tag{d}$$

is the *Laplace operator*.

Since one equation (1.15) for three unknown moments M_x, M_y, and M_{xy} is not sufficient to obtain a solution, plate problems are *internally statically indeterminate.* Reduction of unknowns to one, which follows, is made upon utilization of the moment-displacement relations.

1.6 THE GOVERNING EQUATION FOR DEFLECTION OF PLATES

The basic differential equation for the deflection of plates may readily be derived on the basis of the results obtained in the preceding sections. Introducing into Eq. (1.15) the first expressions for M_x, M_y, M_{xy} given by Eqs. (1.10), we have

$$\frac{\partial^2 \kappa_x}{\partial x^2} + 2\frac{\partial^2 \kappa_{xy}}{\partial x\,\partial y} + \frac{\partial^2 \kappa_y}{\partial y^2} = \frac{p}{D} \tag{a}$$

The above expresses the plate equilibrium in terms of the curvatures. An alternate form of Eq. (*a*) is determined by inserting the definition of curvatures from Eqs. (1.4)

$$\frac{\partial^4 w}{\partial x^4} + 2\frac{\partial^4 w}{\partial x^2\,\partial y^2} + \frac{\partial^4 w}{\partial y^4} = \frac{p}{D} \tag{1.17a}$$

This equation, first derived by Lagrange in 1811, can also be written in concise form

$$\nabla^4 w = \frac{p}{D} \tag{1.17b}$$

in which $\nabla^4 = \nabla^2\nabla^2 = (\nabla^2)^2$. When there is no lateral load acting on the plate

$$\frac{\partial^4 w}{\partial x^4} + 2\frac{\partial^4 w}{\partial x^2\,\partial y^2} + \frac{\partial^4 w}{\partial y^4} = 0 \tag{1.18}$$

Expression (1.17) is the *governing differential equation for deflection* of thin plates. To determine w, it is required to integrate this equation with the constants of integration dependent upon the appropriate boundary conditions (discussed in the next section).

Substituting Eqs. (1.16) and (1.17) into Eqs. (1.13) and (1.14) the stress components τ_{xz}, τ_{yz}, and σ_z are as follows

$$\tau_{xz} = \frac{3Q_x}{2t}\left[1 - \left(\frac{2z}{t}\right)^2\right] \qquad \tau_{yz} = \frac{3Q_y}{2t}\left[1 - \left(\frac{2z}{t}\right)^2\right]$$
$$\sigma_z = -\frac{3p}{4}\left[\frac{2}{3} - \frac{2z}{t} + \frac{1}{3}\left(\frac{2z}{t}\right)^3\right] \tag{1.19}$$

The maximum shear stress, as in the case of a beam of rectangular section, occurs at $z = 0$, and is represented by the formulas

$$\tau_{xz,\,\max} = \frac{3}{2}\frac{Q_x}{t} \qquad \tau_{yz,\,\max} = \frac{3}{2}\frac{Q_y}{t} \tag{1.20}$$

Thus, the key to determining the stress components, using the formulas derived, is the solution of Eq. (1.17) for w.

We mention that an alternative derivation of Eq. (1.17) results from equating the stress normal to the plate to the surface loading per unit area at the upper face of the plate. In this way, from Eqs. (1.14), letting $z = t/2$ and $\sigma_z = -p$, we obtain

$$\frac{Et^3}{12(1-\nu^2)}\nabla^4 w = p$$

which with Eq. (1.11) yields Eq. (1.17).

In conclusion, it is significant that the sum of the bending-moment components defined by Eqs. (1.10) is invariant. That is

$$M_x + M_y = -D(1+\nu)\left(\frac{\partial^2 w}{\partial x^2} + \frac{\partial^2 w}{\partial y^2}\right) = -D(1+\nu)\,\nabla^2 w$$

Letting M denote the *moment function* or so called *moment sum*

$$M = \frac{M_x + M_y}{1+\nu} = -D\,\nabla^2 w \tag{1.21}$$

the expressions for shear forces can be written

$$Q_x = \frac{\partial M}{\partial x} \qquad Q_y = \frac{\partial M}{\partial y} \tag{b}$$

and we may represent Eq. (1.17) as follows

$$\begin{aligned} \frac{\partial^2 M}{\partial x^2} + \frac{\partial^2 M}{\partial y^2} &= -p \\ \frac{\partial^2 w}{\partial x^2} + \frac{\partial^2 w}{\partial y^2} &= -\frac{M}{D} \end{aligned} \tag{1.22a,b}$$

The plate equation, $\nabla^4 w = p/D$, is thus reduced to two second-order partial differential equations which are sometimes preferred, depending upon the method of solution to be employed. This reduction was first introduced by Marcus.[2] Given the loading and the boundary conditions, one can solve M from Eq. (1.22*a*), then Eq. (1.22*b*) leads to w.

It can be demonstrated that Eqs. (1.22) are of the same form as the equation describing the deflection of a uniformly stretched and laterally loaded membrane.[1] Hence, an *analogy* exists between the bending of a plate and membrane problems, serving as the basis of a number of experimental and approximate numerical techniques. The latter are discussed in Chap. 5.

1.7 BOUNDARY CONDITIONS

The differential equation of equilibrium which must be satisfied within the plate is derived in Sec. 1.6. The distribution of stress in a plate must also be such as to accommodate the conditions of equilibrium with respect to prescribed forces or displacements at the boundary.

For a plate, solution of Eq. (1.17) requires that two boundary conditions be satisfied at each edge. These may be a given deflection and slope, or force and moment, or some combination. The basic difference between the boundary conditions applied to plates and those of beams is the existence along the plate edges of twisting moments. It is demonstrated below that these moments may be replaced by equivalent forces. Such a substitution causes an alteration of the distribution of stress and strain only in the immediate region of the boundary, in accordance with St. Venant's principle.[1]

We now treat the boundary conditions for a rectangular plate with edges a and b parallel to the x and y axes, as shown in Fig. 1.6. Consider two successive elemental lengths dy on edge $x = a$ (Fig. 1.6). It is seen that, on the right-hand element, a twisting moment $M_{xy}\,dy$ acts, while the left-hand element is subjected to $[M_{xy} + (\partial M_{xy}/\partial y)\,dy]\,dy$. In the figure, the moments are indicated as replaced by statically equivalent force couples. Thus in an infinitesimal region of the edge shown within the dashed line, we see that an upward directed force M_{xy} and a downward directed force $M_{xy} + (\partial M_{xy}/\partial y)\,dy$ act. The algebraic sum of these forces may be added to the shearing force Q_x to produce an effective transverse force *per unit length* for an edge parallel to the y axis, V_x. That is

$$V_x = Q_x + \frac{\partial M_{xy}}{\partial y} = -D\left[\frac{\partial^3 w}{\partial x^3} + (2-\nu)\frac{\partial^3 w}{\partial x\,\partial y^2}\right] \tag{1.23a}$$

Similarly, it can be shown that, for an edge parallel to the x axis, one has

$$V_y = Q_y + \frac{\partial M_{xy}}{\partial x} = -D\left[\frac{\partial^3 w}{\partial y^3} + (2-\nu)\frac{\partial^3 w}{\partial x^2\,\partial y}\right] \tag{1.23b}$$

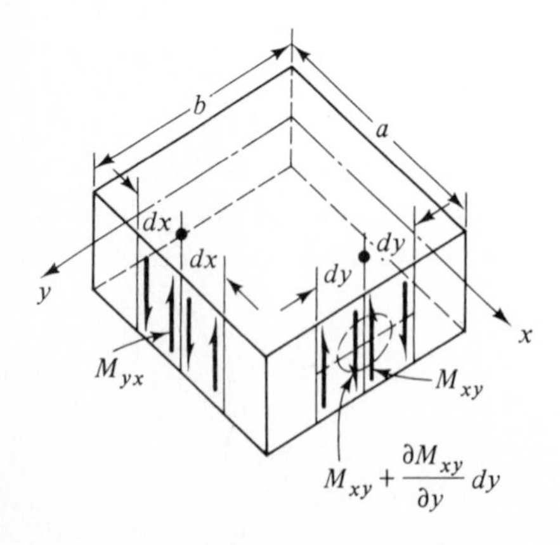

Figure 1.6

Expression (1.23) is due to *Kirchhoff*: a distribution of M_{xy} along an edge is statically equivalent to a distribution of vertical shear forces.

In addition to the edge forces described above, there may be *concentrated forces* F_c produced *at the corners*. Consider, as an example, the case of a *uniformly loaded* rectangular plate with *simply supported* edges (Fig. 1.6). At the corner (a, b) above-discussed action of twisting moments (because $M_{xy} = M_{yx}$) results in

$$F_c = 2M_{xy} = -2D(1-\nu)\frac{\partial^2 w}{\partial x\,\partial y} \qquad (x = a,\ y = b) \tag{1.24}$$

The negative sign indicates an upward direction. Owing to the symmetry of the uniform loading, this force must have the same magnitude and direction at all corners of the plate. Thus, if no anchorage is provided, the corners of the plate described tend to rise (Example 3.2).

The additional corner force for plates having various edge conditions may be determined similarly; for instance, when two adjacent plate edges are *fixed* or *free*, we have $F_c = 0$, since along these edges no twisting moment exists.

We can now formulate a variety of commonly encountered situations. The *boundary conditions* which apply along the edge $x = a$ of the rectangular plate with edges parallel to the x and y axes (Fig. 1.7) are as follows.

Clamped or Built-in Edge (Fig. 1.7*a*) In this case both the deflection and slope must vanish. That is

$$w = 0 \qquad \frac{\partial w}{\partial x} = 0 \qquad (x = a) \tag{1.25}$$

Simply Supported Edge (Fig. 1.7*b*) At the edge considered, the deflection and bending moment are both zero. Hence

$$w = 0 \qquad M_x = -D\left(\frac{\partial^2 w}{\partial x^2} + \nu\frac{\partial^2 w}{\partial y^2}\right) = 0 \qquad (x = a) \tag{1.26a}$$

The first of these equations implies that along edge $x = a$, $\partial w/\partial y = 0$,

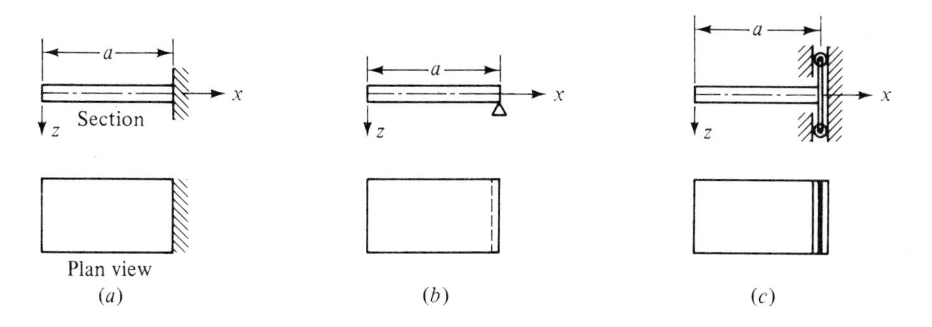

Figure 1.7

$\partial w^2/\partial y^2 = 0$. It follows that conditions expressed by Eqs. (1.26*a*) may appear in the following equivalent form

$$w = 0 \qquad \frac{\partial^2 w}{\partial x^2} = 0 \qquad (x = a) \tag{1.26b}$$

Free Edge Such an edge at $x = a$ is free of moment and vertical shear force. That is

$$\frac{\partial^2 w}{\partial x^2} + \nu \frac{\partial^2 w}{\partial y^2} = 0 \qquad \frac{\partial^3 w}{\partial x^3} + (2 - \nu)\frac{\partial^3 w}{\partial x\, \partial y^2} = 0 \qquad (x = a) \tag{1.27}$$

Sliding Edge (Fig. 1.7*c*) In this case the edge is free to move vertically, but the rotation is prevented. The support is not capable of resisting any shear force. Thus

$$\frac{\partial w}{\partial x} = 0 \qquad \frac{\partial^3 w}{\partial x^3} + (2 - \nu)\frac{\partial^3 w}{\partial x\, \partial y^2} = 0 \qquad (x = a) \tag{1.28}$$

Some other types of boundary conditions may be treated similarly. It is observed that the boundary conditions are of two basic kinds: a *geometric* or *kinematic* boundary condition describes end constraint pertaining to deflection or slope; a *static* boundary condition equates the internal forces (and moments) at the edges of the plate to the given external forces (and moments). Accordingly, in Eqs. (1.25) both conditions are kinematic; in (1.27) both are static; in (1.26) and (1.28) the conditions are *mixed*.

In addition to the *homogeneous* boundary conditions described above, it is, of course, possible to have prescribed shear, moment, rotation, or displacement at the boundary. The latter cases, *nonhomogeneous* boundary conditions, are expressed by replacing the zeros in Eqs. (1.25) to (1.28) with the specified quantity (Sec. 9.4).

1.8 METHODS FOR SOLUTION OF PLATE DEFLECTIONS

Except for simple types of loadings and shapes, such as axisymmetrically loaded circular plates (Sec. 2.3), the governing plate equation $\nabla^4 w = p/D$ yields plate deflections only with considerable difficulty. It is common to attempt a solution by the *inverse method*. The inverse method relies upon assumed solutions for w which satisfy the governing equation and the boundary conditions. Some cases may be treated by using polynomial expressions for w in x and y and undetermined coefficients. Usually, choosing the acceptable series form is laborious and requires a systematic approach. The most powerful such method is the *Fourier series*, where, once a solution has been found for sinusoidal loading, any other loading can be handled by infinite series (Secs. 3.2 and 3.4). This approach offers as an important advantage the fact that a *single expression* may apply to *the entire surface* of the plate.

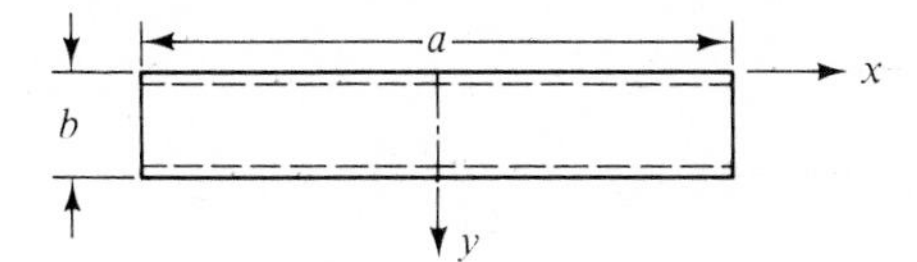

Figure 1.8

Energy methods (Sec. 1.9) should be included in a treatment of general approach. These may be employed to develop solutions, often in the form of infinite series.

The role of the foregoing methods is twofold. They can provide "exact" answers where configurations of loading and shape are simple, and they can be used as the basis of approximate techniques through numerical analysis for more practical problems.

Another approach to overcoming the difficulty involved in the solution of the governing equation is to use finite differences (Chap. 5). In this case, Eq. (1.17) or Eq. (1.22) is replaced by finite difference expressions which relate the w (and M) at nodes that are removed from one another by finite distances. The resulting equations, however, serve only for numerical treatment.

Calculation of plate deflection and stress is illustrated in the solution of the following problem.

Example 1.1 Determine the deflection and stress in a very long and narrow rectangular plate or so-called *infinite strip* (i.e., $a \gg b$), if it is simply supported at edges $y = 0$ and $y = b$ (Fig. 1.8). (*a*) The plate carries a nonuniform loading expressed by

$$p(y) = p_0 \sin \frac{\pi y}{b} \tag{a}$$

where the constant p_0 represents the load intensity along the line passing through $y = b/2$, parallel to the x axis. (*b*) The plate is under a uniform load p_0.

SOLUTION Clearly, the loading described deforms the plate into a *cylindrical surface* possessing its generating line parallel to the x axis. We thus have $\partial w/\partial x = 0$ and $\partial^2 w/\partial x \, \partial y = 0$, and Eqs. (1.10) yield

$$M_x = -\nu D \frac{d^2 w}{dy^2} \qquad M_y = -D \frac{d^2 w}{dy^2} \tag{b}$$

Expression (1.17) reduces to

$$\frac{d^4 w}{dy^4} = \frac{p}{D} \tag{c}$$

This expression is of the same form as the beam equation. Hence, the solution proceeds as in the case of a beam.

Note that since the bent plate surface is of *developable* type and the edges are free to move horizontally, the formulas derived in this example also hold for large deflections ($w \geq t$ but $w < b$).

(*a*) Substituting Eq. (*a*) into Eq. (*c*), integrating, and satisfying the boundary conditions (1.26*b*) at $y = 0$ and $y = b$, we have

$$w = \left(\frac{b}{\pi}\right)^4 \frac{p_0}{D} \sin\frac{\pi y}{b} \tag{d}$$

The maximum stresses in the plate are obtained by substituting the above with $\nu = 1/3$ into Eqs. (1.12), (1.19), and (1.20)

$$\sigma_{x,\,\max} = 0.2p_0\left(\frac{b}{t}\right)^2 \qquad \sigma_{y,\,\max} = 0.6p_0\left(\frac{b}{t}\right)^2 \qquad \left(z = \frac{t}{2},\ y = \frac{b}{2}\right)$$

$$\sigma_{z,\,\max} = -p_0 \qquad \left(z = -\frac{t}{2}\right)$$

$$\tau_{xy} = 0 \qquad \tau_{xz} = 0$$

$$\tau_{yz,\,\max} = 0.5p_0\left(\frac{b}{t}\right) \qquad (z = 0,\ y = 0)$$

To gauge the magnitude of the deviation between the stress components, consider the ratios:

$$\frac{\sigma_{z,\,\max}}{\sigma_{x,\,\max}} = 5\left(\frac{t}{b}\right)^2 \qquad \frac{\tau_{yz,\,\max}}{\sigma_{x,\,\max}} = 2.5\left(\frac{t}{b}\right)$$

If, for example, $b = 20t$, the above quotients are only $\frac{1}{80}$ and $\frac{1}{8}$ respectively. For a thin plate, $t/b < \frac{1}{20}$, and it is clear that stresses σ_z and τ_{yz} are very small compared with the normal stress components in the xy plane.

(*b*) Now Eq. (*c*) with $p = p_0$, upon integrating and satisfying $w = 0$ and $\partial w/\partial x = 0$ at $y = 0$ and $y = b$, yields

$$w = \frac{p_0 b^4}{24D}\left(\frac{y^4}{b^4} - 2\frac{y^3}{b^3} + \frac{y}{b}\right) \tag{1.29}$$

This represents the deflection of a uniformly loaded and simply supported plate strip parallel to the y axis. The maximum deflection of the plate is found by substituting $y = b/2$ in Eq. (1.29), yielding $w_{\max} = 5p_0 b^4/384D$. The largest moment and stress also occur at $y = b/2$, in the direction of the shorter span b. These are readily calculated by means of Eqs. (1.29), (1.10), and (1.12) as $p_0 b^2/8$ and $3p_0 b^2/4t^2$ respectively.

It is observed that for very long and narrow plates the supports along the short sides have little effect on the action in the plate, and hence the *plate behaves as would a simple beam* of span b.

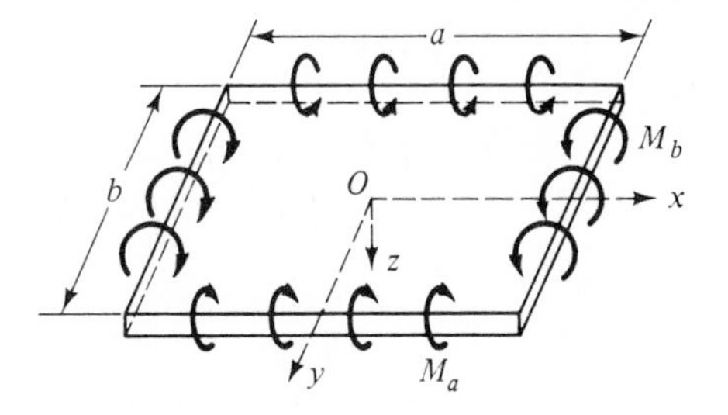

Figure 1.9

A plate which transmits a constant bending moment is said to be in *pure bending*. We consider this simple and practical case in Example 1.2. As will become evident, in this situation the deformation pattern can be established from symmetry considerations alone.

Example 1.2 A rectangular bulkhead of an elevator shaft is subjected to uniformly distributed bending moments $M_x = M_b$ and $M_y = M_a$, applied along its edges (Fig. 1.9). Derive the equation governing the surface deflection for two cases: (*a*) $M_a \neq M_b$, (*b*) $M_a = -M_b$.

SOLUTION (*a*) Substituting $M_b = M_x$ and $M_a = M_y$ into Eqs. (1.10), we obtain

$$\frac{\partial^2 w}{\partial x^2} = -\frac{M_b - \nu M_a}{D(1-\nu^2)} \qquad \frac{\partial^2 w}{\partial y^2} = -\frac{M_a - \nu M_b}{D(1-\nu^2)} \qquad \frac{\partial^2 w}{\partial x\, \partial y} = 0 \tag{e}$$

Integrating the above leads to

$$w = -\frac{M_b - \nu M_a}{2D(1-\nu^2)} x^2 - \frac{M_a - \nu M_b}{2D(1-\nu^2)} y^2 + c_1 x + c_2 y + c_3$$

If the origin of xyz is located at the center and midsurface of the *deformed plate*, the constants of integration vanish and we have

$$w = -\frac{M_b - \nu M_a}{2D(1-\nu^2)} x^2 - \frac{M_a - \nu M_b}{2D(1-\nu^2)} y^2 \tag{1.30}$$

(*b*) By letting $M_a = -M_b$ in Eqs. (*e*), the result is

$$\kappa_x = -\kappa_y = -\frac{\partial^2 w}{\partial x^2} = \frac{M_b}{D(1-\nu)} \tag{f}$$

This reveals that there is a saddle point at the center of the plate. Integrating and locating the origin xyz as before, Eq. (*f*) leads to

$$w = -\frac{M_b}{2D(1-\nu)}(x^2 - y^2) \tag{1.31}$$

It is clear that the above expression represents an *anticlastic* surface. We note that in the particular case where $M_a = M_b$, Eq. (1.30) yields a *paraboloid of revolution* (see Prob. 1.8).

1.9 STRAIN ENERGY METHODS

As an alternative to the *equilibrium methods*, the analysis of deformation and stress in an elastic body can be accomplished by employing *energy methods*. These two techniques are, respectively, the *newtonian* and *lagrangian* approaches to mechanics. The latter is predicted upon the fact that the governing equation of a deformed elastic body is derivable by minimizing the energy associated with deformation and loading. Applications of energy methods are effective in situations involving irregular shapes, nonuniform loads, variable cross sections, and anisotropic materials. We shall begin our discussion of energy techniques by treating the case of loaded thin plates.

The strain energy stored in an elastic body, for a general state of stress, is given by[1]

$$U = \frac{1}{2} \iiint_V (\sigma_x \varepsilon_x + \sigma_y \varepsilon_y + \sigma_z \varepsilon_z + \tau_{xy}\gamma_{xy} + \tau_{xz}\gamma_{xz} + \tau_{yz}\gamma_{yz})\, dx\, dy\, dz \tag{1.32}$$

Integration extends over the entire body volume. Based upon the assumptions of Sec. 1.2, for thin plates σ_z, γ_{xz}, γ_{yz} can be omitted. Thus, introducing Hooke's law, Eqs. (1.7), the above expression reduces to the following form involving only stresses and elastic constants:

$$U = \iiint_V \left[\frac{1}{2E}(\sigma_x^2 + \sigma_y^2 - 2\nu\sigma_x\sigma_y) + \frac{1}{2G}\tau_{xy}^2\right] dx\, dy\, dz \tag{1.33}$$

For a plate of *uniform thickness*, Eq. (1.33) may be written in terms of deflection w by use of Eqs. (1.8) and (1.11) as follows

$$U = \frac{1}{2} \iint_A D\left[\left(\frac{\partial^2 w}{\partial x^2}\right)^2 + \left(\frac{\partial^2 w}{\partial y^2}\right)^2 + 2\nu \frac{\partial^2 w}{\partial x^2}\frac{\partial^2 w}{\partial y^2} + 2(1-\nu)\left(\frac{\partial^2 w}{\partial x\, \partial y}\right)^2\right] dx\, dy$$

or, alternately

$$U = \frac{1}{2} \iint_A D\left\{\left(\frac{\partial^2 w}{\partial x^2} + \frac{\partial^2 w}{\partial y^2}\right)^2 - 2(1-\nu)\left[\frac{\partial^2 w}{\partial x^2}\frac{\partial^2 w}{\partial y^2} - \left(\frac{\partial^2 w}{\partial x\, \partial y}\right)^2\right]\right\} dx\, dy \tag{1.34}$$

where A is the area of the plate surface.

The second term in Eq. (1.34) is known as the *gaussian curvature*. We observe that the strain energy is a *nonlinear* (quadratic) function of deformation or stress. The principle of superposition is therefore *not* valid for the strain energy.

In the case of a plate experiencing a temperature change, the strain energy that results from heating or cooling must also be included in the above expressions.[3]

Formulas (1.32) to (1.34) are useful in the formulation of various energy techniques and the numerical finite element approaches. Next we review some commonly employed strain-energy methods based upon the potential energy and a variation in deformation of an elastic body.

The principle of virtual work Suppose that an elastic body undergoes an *arbitrary* incremental displacement or so-called *virtual displacement*. This displacement need not actually occur and need not be infinitesimal. When the displacement is taken to be infinitesimal, as is often done, it is reasonable to consider the system of forces acting on the body *as constant*. The virtual work done by surface forces T per unit area on the body in the process of bringing the body from the initial state to the equilibrium state is expressed

$$\delta W = \int_A (T_x\, \delta u + T_y\, \delta v + T_z\, \delta w)\, dA \tag{1.35}$$

Here A is the boundary surface and δu, δv, δw are the x, y, z directed virtual displacements. The notation δ denotes the *variation* of a quantity. The strain energy δU acquired by a body of volume V as a result of virtual straining is

$$\delta U = \int_V (\sigma_x\, \delta\varepsilon_x + \sigma_y\, \delta\varepsilon_y + \sigma_z\, \delta\varepsilon_z + \tau_{xy}\, \delta\gamma_{xy} + \tau_{xz}\, \delta\gamma_{xz} + \tau_{yz}\, \delta\gamma_{yz})\, dV \tag{1.36}$$

The total work done during the virtual displacement is zero: $\delta W - \delta U = 0$. The *principle of virtual work* for an elastic body is thus represented

$$\delta W = \delta U \tag{1.37}$$

The principle of minimum potential energy Inasmuch as the virtual displacements do not alter the shape of the body and the surface forces are regarded as constants, Eq. (1.37) can be written as follows:

$$\delta\Pi = \delta(U - W) = 0 \tag{1.38}$$

In this expression

$$\Pi = U - W \tag{1.39}$$

denotes the *potential energy* of the body. Equation (1.38) represents the condition of stationary potential energy of the system. It can be shown that, for *stable equilibrium*, the potential energy is a minimum. For all displacements satisfying given boundary conditions and the equilibrium conditions, the potential energy will assume a minimum value. This is referred to as the *principle of minimum potential energy*.

The potential energy stored in a plate under a *distributed lateral load* $p(x, y)$ is

$$\Pi = \iiint_V (\sigma_x \varepsilon_x + \sigma_y \varepsilon_y + \tau_{xy} \gamma_{xy})\, dx\, dy\, dz - \iint_A (pw)\, dx\, dy \tag{1.40}$$

For the case of *constant* plate *thickness*, the above may be written

$$\Pi = -\frac{1}{2} \iint_A (M_x \kappa_x + M_y \kappa_y + M_{xy} \kappa_{xy})\, dx\, dy - \iint_A (pw)\, dx\, dy \tag{1.41}$$

A physical explanation of the terms of U in this expression is as follows. As $\partial^2 w/\partial x^2 = \kappa_x$ represents the curvature of the plate in the xz plane, the angle corresponding to the moment $M_x\,dy$ equals $-(\partial^2 w/\partial x^2)\,dx$. The strain energy or work done by the moments $M_x\,dy$ is thus $-\frac{1}{2}M_x k_x\,dx\,dy$. The strain energy owing to $M_y\,dx$ and $M_{xy}\,dy$ are interpreted similarly. The principle of potential energy, referring to Eq. (1.41), is expressed in the form:

$$\delta\Pi = -\iint_A (M_x\,\delta\kappa_x + M_y\,\delta\kappa_y + M_{xy}\,\delta\kappa_{xy})\,dx\,dy - \iint_A (p\,\delta w)\,dx\,dy = 0 \qquad (1.42)$$

The Ritz method The so-called Ritz method is a convenient procedure for determining solutions by the principle of minimum potential energy. The essense of this approach is described for the case of elastic bending of plates as follows.

First choose a solution for the deflection w in the form of a series containing undetermined parameters a_{mn} $(m, n = 1, 2, \ldots)$. The deflection so selected must satisfy the geometric boundary conditions. The static boundary conditions need not be fulfilled. Clearly, a proper choice of the deflection expression is important to ensure good accuracy for the final solution. Thus, it is desirable to assume an expression for w which is nearly identical with the true bent surface of the plate. Next, employing the selected solution, determine the potential energy Π in terms of a_{mn}. (This demonstrates that the a_{mn}'s govern the variation of the potential energy.)

In order that the potential energy be a minimum at equilibrium:

$$\frac{\partial\Pi}{\partial a_{11}} = 0, \ldots, \qquad \frac{\partial\Pi}{\partial a_{mn}} = 0 \qquad (1.43)$$

The foregoing represents a system of algebraic equations which are solved to yield the parameters a_{mn}. Introducing these values into the assumed expression for deflection, one obtains the solution for a given problem. In general, a_{mn} includes only a finite number of parameters, and the final results are therefore only approximate. Of course, if the assumed w should happen to be the "exact" one, the solution will then be "exact."

Advantages of the Ritz approach lie in the relative ease with which mixed edge conditions can be handled. This method is among the simplest for solving plate and shell deflections by means of a hand calculation.

The applications of the strain-energy techniques in the treatment of bending, stretching, as well as buckling problems of plates and shells, will be discussed throughout the text.

1.10 MECHANICAL PROPERTIES AND BEHAVIOR OF MATERIALS

The ordinary mechanical properties of materials, such as the static yield, ultimate strength, and modulus of elasticity, are adequate for use in most conventional analyses. Table 1.1 furnishes average properties for some common materials. Exact values may vary with changes in composition, cold working, and heat treatment. From experiments it is known that the Poisson's ratio ν (not

Table 1.1 Average mechanical properties of materials

Material	Specific weight (kN/m^3)	Modulus of elasticity (GPa)		Yield stress (MPa)		Ultimate stress (MPa)		Coefficient of thermal expansion (10^{-6} per °C)
		Tension	Shear	Tension	Shear	Tension [Comp.]	Shear	
Aluminum alloy								
6061–T6	26.6	70	25.9	241	138	290	186	23.6
2024–T4	26.2	73	27.6	290	172	441	276	23.2
Brass	82.5	103	41	103	....	276	193	18.9
Bronze	87	103	45	138	....	345	241	18
Concrete								
Medium strength	22.8	21.4	....	....	....	[20.7]	....	10.8
Copper	80.6	117	41	245	....	345	345	16.7
Cast iron	72.3	103	41	....	....	138 [552]	207	10.8
Magnesium alloy	17.6	45	16.5	138	....	262	131	25.2
Steel								
Mild	77	200	79	248	165	410–550	331	11.7
High strength	77	200	79	345	172	483	....	11.7
Wood								
Structural	2.75–8.24	7–14	....	....	....	[28–70]	....	5.4

listed in the table) varies slightly for different metals over a relatively narrow range. We shall refer to the ordinary characteristics of materials and take $\nu = 0.3$ for metals. It is noted that the analysis and design of plate or shell elements, for instance, as components of a missile or space vehicle, embodies an unusual integration of materials, having characteristics dependent upon environmental conditions.[4,5]

A proper design includes a prediction of the circumstances under which failure is likely to take place. *Failure*, generally, refers to any action leading to an inability of a structural member to function satisfatorily. Some important variables associated with failure are the type of material, configuration and rate of loading, the shape and surface peculiarities, and the operational environment.

Yielding of plate (shell) elements, or initiation of *fracture*, are the basis of *elastic design* criteria for ductile and brittle materials, respectively. When a metal undergoes an appreciable amount of yielding or permanent deformation, it is termed *ductile*. When, prior to fracture, the material can experience only a small amount of yielding (less than 5 percent), the material is considered as brittle. Ductile materials are weak in *shearing strength*, while brittle materials are weak in *tensile strength*.

Unless we are content to overdesign plate- and shell-like structural elements, it is necessary to predict the most probable modes of failure. We observe from the formulas derived that the breakdown of elastic action is likely to originate at certain locations of these members. The formulas alone cannot predict at what loading failure will take place, however. The material strength, and an appropriate theory of failure consistent with the type of material, whether brittle or ductile, must also be considered.

A brief discussion of some *yielding theories of failure*, converting uniaxial to combined stress data, is given in the following paragraphs.[1] The discussions are limited to the case of plane stress. The *algebraically* largest and the smallest principal stresses are designated σ_1 and σ_2, respectively. The yield stress (or strain) determined in a simple tensile test is denoted by σ'_{yp} (or ε'_{yp}) and in a simple compression test, by σ''_{yp} (or ε''_{yp}). In the case of materials possessing the same yield stress in tension and compression, $\sigma'_{yp} = \sigma''_{yp} = \sigma_{yp}$ and $\varepsilon'_{yp} = \varepsilon''_{yp} = \varepsilon_{yp}$.

Maximum principal stress theory This theory predicts that a material fails by yielding when the maximum (or minimum) principal stress exceeds the tensile (or compressive) yield strength. That is, at the beginning of yielding or inelastic action

$$|\sigma_1| = \sigma'_{yp} \qquad \text{or} \qquad |\sigma_2| = \sigma''_{yp} \tag{1.44}$$

Maximum shear stress theory This theory states that yielding will begin when the maximum shear stress in the material is equal to the maximum shear stress at yielding in a simple tension test. Thus

$$|\sigma_1 - \sigma_2| = \sigma_{yp} \tag{1.45a}$$

It is observed that if σ_1 and σ_2 are of *opposite sign*, i.e., one tensile and the other compressive, the maximum shear stress is $(\sigma_1 - \sigma_2)/2$. Thus, the condition of the inelastic action is given by Eq. (1.45*a*). Interestingly, if σ_1 and σ_2 carry the *same sign* and σ_3 is *taken as zero*, one then obtains the yield conditions:

$$|\sigma_1| = \sigma_{yp} \qquad \text{and} \qquad |\sigma_2| = \sigma_{yp} \tag{1.45b}$$

for $|\sigma_1| > |\sigma_2|$ and $|\sigma_2| > |\sigma_1|$, respectively.

Maximum principal strain theory This theory predicts that a material fails by yielding when the maximum (or minimum) principal strain exceeds the tensile (or compressive) yield strain. Accordingly, upon application of Hooke's law, we have

$$\begin{aligned} |\sigma_1 - \nu\sigma_2| &= \sigma'_{yp} \\ |\sigma_2 - \nu\sigma_1| &= \sigma''_{yp} \end{aligned} \tag{1.46}$$

Maximum distortion energy theory This theory states that failure by yielding occurs when, at any location in the body, the distortion energy per unit volume in a state of combined stress will be equal to that associated with yielding in a simple tension test, that is,

$$(\sigma_1^2 - \sigma_1\sigma_2 - \sigma_2^2)^{1/2} = \sigma_{yp} \tag{1.47}$$

It is noted that the maximum energy of distortion theory agrees best with experimental data for ductile materials, and its employment in design practice is increasing. However, the maximum shear stress theory is in widespread use in some design codes because it is simple to apply and offers a conservative result. The maximum principal strain theory is not supported by experimental results and is unsafe for ductile materials. Good agreement between the maximum principal stress theory and the experiment has been realized for brittle materials. This theory is generally accepted in design practice for brittle materials.

The prediction of *failure by fracture* under combined stress may also be accomplished through the use of the maximum principal stress and maximum shear stress theories, provided the yield strengths $(\sigma'_{yp}, \sigma''_{yp})$ are replaced by the *ultimate strengths* (σ'_u, σ''_u) in Eqs. (1.44) and (1.45), respectively. The *critical values* of the ultimate yield point stresses are usually determined in a tensile experiment, where the failure of a specimen occurs by fracture.

The various failure theories are applied in a number of chapters to follow.

PROBLEMS

Secs. 1.1 to 1.5

1.1 Verify the result given by Eq. (1.5) by employing Mohr's circle.

1.2 A sheet of metal is transversely loaded so that the moment components at a critical location related to the set of axes xy are M_x, M_y, and M_{xy}. Show that the moment components $M_{x'}$, $M_{y'}$, and $M_{x'y'}$ associated with a new set of axes $x'y'$ inclined at an angle of $\theta°$ clockwise to the xy set

(Fig. 1.3) are expressed

$$\begin{Bmatrix} M_{x'} \\ M_{y'} \\ M_{x'y'} \end{Bmatrix} = \begin{bmatrix} \cos^2\theta & \sin^2\theta & 2\sin\theta\cos\theta \\ \sin^2\theta & \cos^2\theta & -2\sin\theta\cos\theta \\ -\sin\theta\cos\theta & \sin\theta\cos\theta & \cos^2\theta - \sin^2\theta \end{bmatrix} \begin{Bmatrix} M_x \\ M_y \\ M_{xy} \end{Bmatrix} \tag{P1.2a}$$

or, alternatively

$$\begin{aligned} M_{x'} &= \tfrac{1}{2}(M_x + M_y) + \tfrac{1}{2}(M_x - M_y)\cos 2\theta + M_{xy}\sin 2\theta \\ M_{y'} &= \tfrac{1}{2}(M_x + M_y) - \tfrac{1}{2}(M_x - M_y)\cos 2\theta - M_{xy}\sin 2\theta \\ M_{x'y'} &= -\tfrac{1}{2}(M_x - M_y)\sin 2\theta + M_{xy}\cos 2\theta \end{aligned} \tag{P1.2b}$$

Represent Eqs. (P1.2) by a Mohr's circle. Then ascertain that the orientation of $x'y'$ corresponding to the maximum and minimum or *principal moments* is defined by

$$\tan 2\theta = \frac{2M_{xy}}{M_x - M_y}$$

1.3 By means of Mohr's circle show that the gaussian curvature is equal to

$$\kappa_x\kappa_y - \kappa_{xy}^2 = \kappa_1\kappa_2 \tag{P1.3}$$

1.4 Demonstrate that at a corner of a polygonal, simply supported plate, $M_{xy} = 0$ unless the corner is 90°.

Secs. 1.6 to 1.10

1.5 The lateral deflection of a rectangular plate (Fig. 1.6), with built-in edges of lengths a and b and subjected to a uniform load p_0, is given by

$$w = c_0\left(\frac{x^4}{a^4} - 2\frac{x^3}{a^3} + \frac{x^2}{a^2}\right)\left(\frac{y^4}{b^4} - 2\frac{y^3}{b^3} + \frac{y^2}{b^2}\right)$$

where c_0 is a constant. Determine: (*a*) whether this deflection satisfies the boundary conditions of the plate; (*b*) the maximum plane stress components σ_x and τ_{xy} at the center, for $a = b$.

1.6 A square spacecraft panel (Fig. 1.3*a*) is subjected to uniformly distributed twisting moment $M_{xy} = M_0$ along all four edges. Determine an expression for the deflection surface w.

1.7 Determine whether the following expression satisfies the boundary conditions of a simply supported very long and narrow plate (Fig. 1.8) carrying a concentrated load P at a point $(0, b/2)$:

$$w = \frac{Pb^2}{2\pi^3 D}\sum_{n=1}^{\infty}\frac{1}{n^3}\sin\frac{n\pi}{2}\sin\alpha y(1 + \alpha x)e^{-\alpha x} \tag{P1.7}$$

Here $\alpha = n\pi/b$ and $x \geq 0$. Using $\nu = \frac{1}{3}$, obtain the corresponding expressions for: (*a*) the moment M_x in terms of M defined by Eq. (1.21); (*b*) the maximum stress σ_x for $n = 1$.

1.8 Consider the bending of the rectangular plate (Fig. 1.9) for the particular case in which $M_x = M_y = M_0$.

(*a*) Verify that, in this case, even for a plate of *any shape*, the bending moments M_0 are uniform along the boundary and the twisting moments vanish.

(*b*) Show that Eqs. (1.30) and (1.10) yield, respectively:

$$w = -\frac{M_0(x^2 + y^2)}{2D(1 + \nu)} \qquad \kappa_x = \kappa_y = -\frac{M_0}{D(1 + \nu)} \tag{P1.8}$$

It is seen that the first of the above expressions represents a paraboloid of revolution while the second implies that the surface is spherical. Explain why results are inconsistent.

1.9 Determine an expression of the strain energy for the plate described in Prob. 1.5.

CHAPTER

TWO

CIRCULAR PLATES

2.1 INTRODUCTION

In practice, members that carry transverse loads, such as end plates and closures of pressure vessels, pump diaphragms, clutches, and turbine disks, etc., are usually circular in shape. Thus, many of the significant applications fall within the scope of the formulas derived for circular plates. A number of circular-plate problems have stress distributions that are symmetrical about the center. We shall discuss some of these in Secs. 2.4 to 2.9 and 2.12. Sections 2.10 and 2.11 deal with the situations involving asymmetrical loading. In all cases the basic relationships in polar coordinates, developed in Secs. 2.2 and 2.3, are employed.

2.2 BASIC RELATIONS IN POLAR COORDINATES

In general, polar coordinates are preferred over cartesian coordinates (used exclusively thus far) where a degree of axial symmetry exists either in geometry or loading. Examples include a circular plate and a large thin plate containing holes.

The polar coordinate set (r, θ) and the cartesian set (x, y) are related by the equations (Fig. 2.1*a*)

$$x = r\cos\theta \qquad r^2 = x^2 + y^2$$

$$y = r\sin\theta \qquad \theta = \tan^{-1}\frac{y}{x}$$

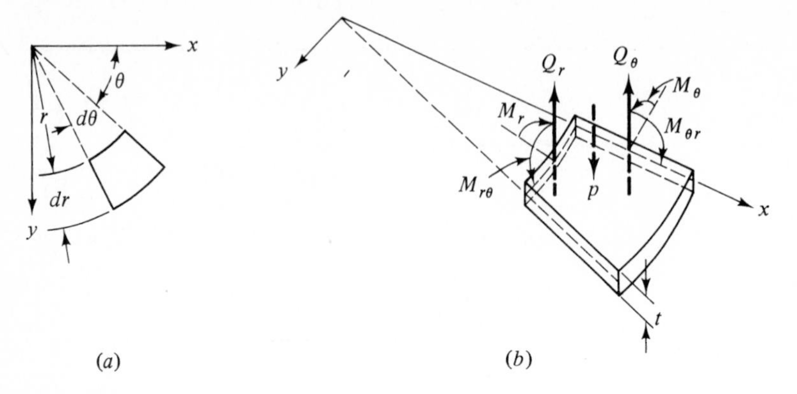

Figure 2.1

Referring to the above,

$$\frac{\partial r}{\partial x} = \frac{x}{r} = \cos\theta \qquad \frac{\partial r}{\partial y} = \frac{y}{r} = \sin\theta$$

$$\frac{\partial \theta}{\partial x} = -\frac{y}{r^2} = -\frac{\sin\theta}{r} \qquad \frac{\partial \theta}{\partial y} = \frac{x}{r^2} = \frac{\cos\theta}{r}$$

Inasmuch as the deflection is a function of r and θ, the chain rule together with the above relationships lead to

$$\frac{\partial w}{\partial x} = \frac{\partial w}{\partial r}\cos\theta - \frac{1}{r}\frac{\partial w}{\partial \theta}\sin\theta \tag{a}$$

To evaluate the expression $\partial^2 w/\partial x^2$, one uses Eq. ($a$), this time applied to $\partial w/\partial x$ rather than to w:

$$\begin{aligned}\frac{\partial^2 w}{\partial x^2} &= \cos\theta\frac{\partial}{\partial r}\left(\frac{\partial w}{\partial x}\right) - \frac{1}{r}\sin\theta\frac{\partial}{\partial \theta}\left(\frac{\partial w}{\partial x}\right) \\ &= \frac{\partial^2 w}{\partial r^2}\cos^2\theta - 2\frac{\partial^2 w}{\partial \theta\,\partial r}\frac{\sin\theta\cos\theta}{r} + \frac{\partial w}{\partial r}\frac{\sin^2\theta}{r} \\ &\qquad + 2\frac{\partial w}{\partial \theta}\frac{\sin\theta\cos\theta}{r^2} + \frac{\partial^2 w}{\partial \theta^2}\frac{\sin^2\theta}{r^2}\end{aligned} \tag{b}$$

Similarly,

$$\begin{aligned}\frac{\partial^2 w}{\partial y^2} &= \frac{\partial^2 w}{\partial r^2}\sin^2\theta + 2\frac{\partial^2 w}{\partial \theta\,\partial r}\frac{\sin\theta\cos\theta}{r} + \frac{\partial w}{\partial r}\frac{\cos^2\theta}{r} \\ &\qquad - 2\frac{\partial w}{\partial \theta}\frac{\sin\theta\cos\theta}{r^2} + \frac{\partial^2 w}{\partial \theta^2}\frac{\cos^2\theta}{r^2}\end{aligned} \tag{c}$$

$$\begin{aligned}\frac{\partial^2 w}{\partial x\,\partial y} &= \frac{\partial^2 w}{\partial r^2}\sin\theta\cos\theta + \frac{\partial^2 w}{\partial r\,\partial \theta}\frac{\cos 2\theta}{r} - \frac{\partial w}{\partial \theta}\frac{\cos 2\theta}{r^2} \\ &\qquad - \frac{\partial w}{\partial r}\frac{\sin\theta\cos\theta}{r} - \frac{\partial^2 w}{\partial \theta^2}\frac{\sin\theta\cos\theta}{r^2}\end{aligned} \tag{d}$$

Upon substitution of Eqs. (*b*) and (*c*) into Eq. (*d*) of Sec. 1.5, the laplacian operator becomes:

$$\nabla^2 w = \frac{\partial^2 w}{\partial r^2} + \frac{1}{r}\frac{\partial w}{\partial r} + \frac{1}{r^2}\frac{\partial^2 w}{\partial \theta^2} \tag{2.1}$$

Determination of the fundamental equations of a laterally loaded plate in polar coordinates requires only that the appropriate relationships of Chap. 1 be transformed from cartesian to polar coordinates. Consider now the state of moment and shear force on an infinitesimal element of thickness t, described in polar coordinates (Fig. 2.1*b*). Note that, to simplify the derivations, the x axis is taken in the direction of radius r, i.e., $\theta = 0$. The radial, tangential, twisting moments, M_θ, M_r, $M_{r\theta} = M_{\theta r}$, and the vertical shearing forces Q_r, Q_θ then have the same values as the moments M_x, M_y, M_{xy} and the shears Q_x, Q_y at the same point in the plate. Thus, letting $\theta = 0$ in Eqs. (*b*), (*c*), and (*d*), and substituting the resulting expressions into Eqs. (1.10) and (1.16), we have

$$\begin{aligned}
M_r &= -D\left[\frac{\partial^2 w}{\partial r^2} + \nu\left(\frac{1}{r}\frac{\partial w}{\partial r} + \frac{1}{r^2}\frac{\partial^2 w}{\partial \theta^2}\right)\right] \\
M_\theta &= -D\left[\frac{1}{r}\frac{\partial w}{\partial r} + \frac{1}{r^2}\frac{\partial^2 w}{\partial \theta^2} + \nu\frac{\partial^2 w}{\partial r^2}\right] \\
M_{r\theta} &= -(1-\nu)D\left(\frac{1}{r}\frac{\partial^2 w}{\partial r\,\partial \theta} - \frac{1}{r^2}\frac{\partial w}{\partial \theta}\right) \\
Q_r &= -D\frac{\partial}{\partial r}(\nabla^2 w) \\
Q_\theta &= -D\frac{1}{r}\frac{\partial}{\partial \theta}(\nabla^2 w)
\end{aligned} \tag{2.2}$$

Similarly, formulas for the plane stress components, from Eqs. (1.12), are written in the following form

$$\sigma_r = \frac{12M_r z}{t^3} \qquad \sigma_\theta = \frac{12M_\theta z}{t^3} \qquad \tau_{r\theta} = \frac{12M_{r\theta} z}{t^3} \tag{2.3}$$

where M_r, M_θ and $M_{r\theta}$ are defined by Eqs. (2.2). Clearly the *maximum stresses* take place on the surfaces (at $z = \pm t/2$) of the plate.

The effective transverse force per unit length, Eq. (1.23*a*), for an edge at $r = a$ and any θ becomes

$$V_r = Q_r + \frac{1}{r}\frac{\partial M_{r\theta}}{\partial \theta} = -D\left[\frac{\partial}{\partial r}(\nabla^2 w) + \frac{1-\nu}{r}\frac{\partial}{\partial \theta}\left(\frac{1}{r}\frac{\partial^2 w}{\partial r\,\partial \theta} - \frac{1}{r^2}\frac{\partial w}{\partial \theta}\right)\right] \tag{2.4a}$$

Expression (1.23*b*) appears as

$$V_\theta = Q_\theta + \frac{\partial M_{r\theta}}{\partial r} = -D\left[\frac{1}{r}\frac{\partial}{\partial \theta}(\nabla^2 w) + (1-\nu)\frac{\partial}{\partial r}\left(\frac{1}{r}\frac{\partial^2 w}{\partial r\,\partial \theta} - \frac{1}{r^2}\frac{\partial w}{\partial \theta}\right)\right] \tag{2.4b}$$

describing the effective transverse force on the edge at $\theta = \theta_0$ (constant) and any r.

Upon introduction of Eqs. (*b*) through (*d*) into Eq. (1.17), the governing differential equation for plate deflection in polar coordinates is derived:

$$\nabla^4 w = \left(\frac{\partial^2}{\partial r^2} + \frac{1}{r}\frac{\partial}{\partial r} + \frac{1}{r^2}\frac{\partial^2}{\partial \theta^2}\right)\left(\frac{\partial^2 w}{\partial r^2} + \frac{1}{r}\frac{\partial w}{\partial r} + \frac{1}{r^2}\frac{\partial^2 w}{\partial \theta^2}\right) = \frac{p}{D} \tag{2.5}$$

Letting w_h denote the solution of the *homogeneous* equation

$$\left(\frac{\partial^2}{\partial r^2} + \frac{1}{r}\frac{\partial}{\partial r} + \frac{1}{r^2}\frac{\partial^2}{\partial \theta^2}\right)\left(\frac{\partial^2 w_h}{\partial r^2} + \frac{1}{r}\frac{\partial w_h}{\partial r} + \frac{1}{r^2}\frac{\partial^2 w_h}{\partial \theta^2}\right) = 0 \tag{2.6}$$

and w_p the *particular solution* of Eq. (2.5), the complete solution is expressed

$$w = w_h + w_p \tag{e}$$

We assume the homogeneous or complementary solution to be expressed by the following series[6]

$$w_h = \sum_{n=0}^{\infty} f_n \cos n\theta + \sum_{n=1}^{\infty} f_n^* \sin n\theta \tag{2.7}$$

where f_n and f_n^* are functions of r only. Substituting Eq. (2.7) into Eq. (2.6) and noting the validity of the resulting expression for all r and θ, leads to two ordinary differential equations with the following solutions (Prob. 2.9)

$$\begin{aligned}
f_0 &= A_0 + B_0 r^3 + C_0 \ln r + D_0 r^2 \ln r \\
f_1 &= A_1 r + B_1 r^3 + C_1 r^{-1} + D_1 r \ln r \\
f_n &= A_n r^n + B_n r^{-n} + C_n r^{n+2} + D_n r^{-n+2} \\
f_1^* &= A_1^* r + B_1^* r^3 + C_1^* r^{-1} + D_1^* r \ln r \\
f_n^* &= A_n^* r^n + B_n^* r^{-n} + C_n^* r^{n+2} + D_n^* r^{-n+2}
\end{aligned} \tag{2.8}$$

Here $A_n, \ldots, D_n^*$ are constants, determined by satisfying the boundary conditions of the plate. Upon introducing these expressions for f_n and f_n^* into Eq. (2.7), one obtains the solution of Eq. (2.6) in a general form.

The *boundary conditions* at the edges of an annular circular plate of outer radius a and inner radius b may readily be written by referring to Eqs. (1.25) to (1.28), (2.2), and (2.4). They are listed in Table 2.1. It is noted that the inner (or outer) radius is represented by r_0. Clearly, nomenclature employed in the table parallels that defined in connection with rectangular plates in Sec. 1.7.

Table 2.1 Boundary conditions for circular plates

Edge	Clamped	Simply supported	Free	Sliding
At $r = r_0$ and	$w = 0$	$w = 0$	$M_r = 0$	$\frac{\partial w}{\partial r} = 0$
any θ	$\frac{\partial w}{\partial r} = 0$	$\frac{\partial^2 w}{\partial r^2} + \frac{\nu}{r}\frac{\partial w}{\partial r} = 0$	$V_r = 0$	$V_r = 0$

2.3 THE AXISYMMETRICAL BENDING

The deflection w of a plate will depend upon radial position r only when the applied load and end restraints are independent of the angle θ. The situation described is the *axisymmetrical bending* of the plate. For this case, *only* M_r, M_θ, and Q_r act on the circular plate element shown in Fig. 2.1*b*. The moments and shear force, in an axisymmetrically loaded circular plate, are found from Eqs. (2.2) as follows

$$M_r = -D\left(\frac{d^2w}{dr^2} + \frac{\nu}{r}\frac{dw}{dr}\right)$$

$$M_\theta = -D\left(\frac{1}{r}\frac{dw}{dr} + \nu\frac{d^2w}{dr^2}\right) \tag{2.9a–c}$$

$$Q_r = -D\frac{d}{dr}\left(\frac{d^2w}{dr^2} + \frac{1}{r}\frac{dw}{dr}\right) = -D\frac{d}{dr}\left[\frac{1}{r}\frac{d}{dr}\left(r\frac{dw}{dr}\right)\right]$$

The differential equation of the surface deflection [Eq. (2.5)] now reduces to

$$\nabla^4 w = \left(\frac{d^2}{dr^2} + \frac{1}{r}\frac{d}{dr}\right)\left(\frac{d^2w}{dr^2} + \frac{1}{r}\frac{dw}{dr}\right) = \frac{p}{D} \tag{2.10a}$$

The formulas for stress are readily obtained by substituting Eqs. (*b*), (*c*), and (*d*) of Sec. 2.2 into Eqs. (1.8) and setting θ equal to zero:

$$\sigma_r = -\frac{Ez}{1-\nu^2}\left(\frac{d^2w}{dr^2} + \frac{\nu}{r}\frac{dw}{dr}\right)$$

$$\sigma_\theta = -\frac{Ez}{1-\nu^2}\left(\frac{1}{r}\frac{dw}{dr} + \nu\frac{d^2w}{dr^2}\right) \tag{2.11}$$

To write Hooke's law in polar coordinates, it is necessary to replace subscripts x by r and y by θ in Eqs. (1.7) with the result that

$$\varepsilon_r = \frac{1}{E}(\sigma_r - \nu\sigma_\theta)$$

$$\varepsilon_\theta = \frac{1}{E}(\sigma_\theta - \nu\sigma_r) \tag{a}$$

$$\gamma_{r\theta} = \frac{\tau_{r\theta}}{G}$$

Introducing the identity

$$\nabla^2 w = \frac{d^2w}{dr^2} + \frac{1}{r}\frac{dw}{dr} = \frac{1}{r}\frac{d}{dr}\left(r\frac{dw}{dr}\right)$$

Eq. (2.10*a*) appears in the form

$$\frac{1}{r}\frac{d}{dr}\left\{r\frac{d}{dr}\left[\frac{1}{r}\frac{d}{dr}\left(r\frac{dw}{dr}\right)\right]\right\}=\frac{p}{D} \tag{2.10b}$$

The deflection w is obtained by successive integrations when $p(r)$ is given:

$$w=\int\frac{1}{r}\int r\int\frac{1}{r}\int\frac{rp}{D}\,dr\,dr\,dr\,dr \tag{2.12}$$

If the plate is under a *uniform loading* $p = p_0$, the general solution of Eq. (2.10) is (Prob. 2.10)

$$w = w_h + w_p = c_1 \ln r + c_2 r^2 \ln r + c_3 r^2 + c_4 + \frac{p_0 r^4}{64D} \tag{2.13}$$

where the c's are constants of integration. It is seen from a comparison of Eq. (2.13) and the first of Eqs. (2.8), that the homogeneous solution f_0 represents the case of axisymmetrical bending of circular plates.

2.4 UNIFORMLY LOADED CIRCULAR PLATES

Consider the case of a circular plate of radius a under a uniformly distributed load p_0. The lateral displacement w is expressed by Eq. (2.13). The constants of integration (the c's) in this equation are determined for various particular cases described below.

Plate with clamped edge (Fig. 2.2*a*) For this case, the boundary conditions are

$$w = 0 \qquad \frac{dw}{dr} = 0 \qquad (r = a) \tag{a}$$

The terms involving logarithms in Eq. (2.13) yield an infinite displacement at $r = 0$ for all values of c_1 and c_2 except zero; therefore $c_1 = c_2 = 0$. Satisfying Eqs. (*a*), we obtain

$$c_3 = -\frac{p_0 a^2}{32D} \qquad c_4 = \frac{p_0 a^4}{64D}$$

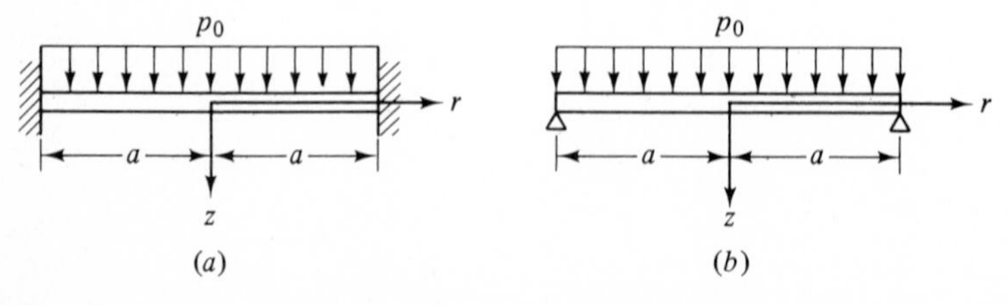

Figure 2.2

The deflection is then

$$w = \frac{p_0}{64D}(a^2 - r^2)^2 \tag{2.14}$$

The maximum displacement occurs at the center of the plate:

$$w_{\max} = \frac{p_0 a^4}{64D} \tag{b}$$

Expressions for the bending moments are calculated by means of Eqs. (2.14) and (2.9):

$$\begin{aligned} M_r &= \frac{p_0}{16}[(1 + \nu)a^2 - (3 + \nu)r^2] \\ M_\theta &= \frac{p_0}{16}[(1 + \nu)a^2 - (1 + 3\nu)r^2] \end{aligned} \tag{2.15}$$

The stresses from Eqs. (2.14) and (2.11) are

$$\begin{aligned} \sigma_r &= \frac{3p_0 z}{4t^3}[(1 + \nu)a^2 - (3 + \nu)r^2] \\ \sigma_\theta &= \frac{3p_0 z}{4t^3}[(1 + \nu)a^2 - (1 + 3\nu)r^2] \end{aligned} \tag{2.16}$$

Algebraically extreme values of the moments are found at the center and at the edge. At the edge ($r = a$), Eqs. (2.15) lead to

$$M_r = -\frac{p_0 a^2}{8} \qquad M_\theta = -\frac{\nu p_0 a^2}{8}$$

while at $r = 0$, $M_r = M_\theta = (1 + \nu)P_0 a^2/16$. It is observed that the maximum moment occurs at the edge. Thus, we have

$$\sigma_{r,\max} = \frac{6M_r}{t^2} = -\frac{3p_0}{4}\left(\frac{a}{t}\right)^2 \tag{c}$$

In Fig. 2.3 is shown the variation of stress with the ratio r/a in uniformly loaded clamped circular plates (upper base line). The curves are parabolas expressed by Eqs. (2.16) with $\nu = 1/3$.

Plate with simply supported edge (Fig. 2.2*b*) As in the previous case, displacement must be finite at $r = 0$. The values of c_1 and c_2 in Eq. (2.13) are therefore zero. The boundary conditions

$$w = 0 \qquad M_r = 0 \qquad (r = a)$$

yield the following respective expressions

$$c_3 a^2 + c_4 + \frac{p_0 a^4}{64D} = 0 \qquad c_3 = -\frac{p_0 a^2}{32D}\frac{3 + \nu}{1 + \nu}$$

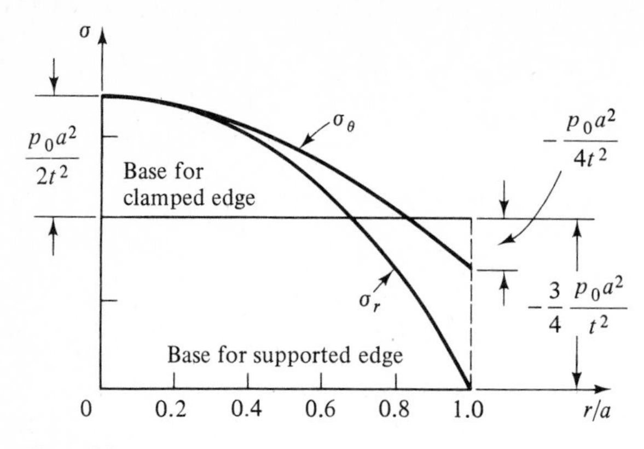

Figure 2.3

from which $c_4 = p_0 a^4(5 + \nu)/64(1 + \nu)D$. The plate deflection is then

$$w = \frac{p_0 a^4}{64D}\left(\frac{r^4}{a^4} - 2\frac{3 + \nu}{1 + \nu}\frac{r^2}{a^2} + \frac{5 + \nu}{1 + \nu}\right) \tag{2.17}$$

The maximum deflection, which occurs at $r = 0$, is thus

$$w_{\max} = \frac{p_0 a^4}{64D}\frac{5 + \nu}{1 + \nu} \tag{d}$$

If $\nu = 1/3$, comparing Eqs. (*b*) and (*d*), we see that the maximum deflection for a simply supported plate is about four times as great as that for the plate with a clamped edge.

When the load resistance of a plate is limited by the large deflections, the *results related to the simply supported case will be conservative*. In addition, the analysis of a simply supported plate is, in general, much *simpler* than that of a plate with restrained edges, recommending its use in practice. Actually, many support members tolerate some degree of flexibility, and a condition of true edge fixity is especially difficult to obtain. As a result, a partially restrained plate exhibits deformations nearly identical with those of a hinged plate. Based upon these considerations, a designer sometimes simplifies the model of the original clamped plate, using instead a hinged plate. To attain more accurate results, however, the effect of a definite amount of edge yielding or relaxation of the fixing moment can often be accommodated by the formulas for edge slope and the method of superposition.[7]

Given the deflection curve w, the distribution of moments can readily be obtained in the form

$$\begin{aligned} M_r &= \frac{p_0}{16}(3 + \nu)(a^2 - r^2) \\ M_\theta &= \frac{p_0}{16}[(3 + \nu)a^2 - (1 + 3\nu)r^2] \end{aligned} \tag{2.18}$$

The stresses are then

$$\sigma_r = \frac{3p_0 z}{4t^3}(3+\nu)(a^2 - r^2)$$
$$\sigma_\theta = \frac{3p_0 z}{4t^3}[(3+\nu)a^2 - (1+3\nu)r^2] \tag{2.19}$$

The maximum stress takes place at the center of the plate ($r = 0$) and is given by

$$\sigma_{r,\max} = \sigma_{\theta,\max} = \frac{3(3+\nu)p_0}{8}\left(\frac{a}{t}\right)^2 \tag{e}$$

The stresses, Eqs. (2.19), in a simply supported circular plate are displayed (lower base line) in Fig. 2.3 as functions of radial location (for $\nu = 1/3$).

We observe from Fig. 2.3 that the ratio of the maximum stresses for the simply supported and the clamped plates is equal to 5/3. Interestingly, the corresponding ratio for a simply supported beam and a fixed beam is 1.5. It is noted however that owing to a small degree of yielding or loosening at a nominally clamped edge, the stresses will be considerably lessened there, while the deflection and stress will increase at the center. Thus, for uniformly loaded ordinary plate structures of clamped edge, the maximum stress will be somewhat *higher* than obtained above. This is also valid for the plates with clamped edges of *any* other shape.

Example 2.1 A circular clamped edge window of a submarine is subjected to uniform pressure differential p_0 between the cabin and the outside. The plate is made of an isotropic material of tensile yield strength σ_{yp}, thickness t, and radius a. Use the maximum energy of distortion theory to predict the load-carrying capacity of the plate.

SOLUTION The principal stress components occur at the built-in edge and, referring to Fig. 2.3, are given by

$$\sigma_1 = \sigma_{r,\max} = -\frac{3}{4}\frac{p_0 a^2}{t^2} \qquad \sigma_2 = \sigma_{\theta,\max} = -\frac{p_0 a^2}{4t^2}$$

According to Eq. (1.47),

$$\left(\frac{3}{4}\frac{p_0 a^2}{t^2}\right)^2 - \left(-\frac{3}{4}\frac{p_0 a^2}{t^2}\right)\left(-\frac{p_0 a^2}{4t^2}\right) + \left(-\frac{p_0 a^2}{4t^2}\right) = \sigma_{yp}^2$$

The above yields

$$p_0 = 1.5\left(\frac{t}{a}\right)^2 \sigma_{yp} \tag{f}$$

as the value of pressure differential governing the onset of the yielding action.

2.5 EFFECT OF SHEAR ON THE PLATE DEFLECTION

In Sec. 1.2 and Example 1.1 we have observed that in the bending of plates, the influences of the shear stress τ_{rz} and the normal stress σ_z are neglected. We now demonstrate that the solutions for deflection of thin plates based upon the bending strains only, yield results of acceptable accuracy.

Denoting the deflection of the midsurface due to the shear alone by w_s, the shear strain, $\gamma_{rz} = dw_s/dr$ referring to Eqs. (1.20), is expressed:

$$\frac{dw_s}{dr} = \frac{3}{2}\frac{Q_r}{Gt}$$

Consider, for example, a simply supported circular plate under uniform load p_0 (Fig. 2.2*b*). The *deflection owing to the shear* is

$$w_s = -\frac{3}{2Gt}\int_a^r Q_r\,dr \tag{a}$$

Here the vertical shear force per unit circumferential length, at a distance r from the center, $Q_r = \pi r^2 p_0/2\pi r = rp_0/2$. Introducing this value of Q_r into Eq. (*a*), we have, upon integration,

$$w_s = \frac{3p_0}{8Gt}(a^2 - r^2) = \frac{p_0 t^2}{16(1-\nu)D}(a^2 - r^2) \tag{b}$$

The *total deflection* w_t is obtained by addition of the deflections associated with bending and shear [Eqs. (2.17) and (*b*)]:

$$w_t = \frac{p_0 a^4}{64D}\left[\frac{r^4}{a^4} - 2\frac{3+\nu}{1+\nu}\frac{r^2}{a^2} + \frac{5+\nu}{1+\nu} + \frac{4t^2}{a^2(1-\nu)}\left(1 - \frac{r^2}{a^2}\right)\right] \tag{2.20}$$

The maximum deflection occurs at the center of the plate,

$$w_{max} = \frac{p_0 a^4}{64D}\left[\frac{5+\nu}{1+\nu} + \frac{4t^2}{(1-\nu)a^2}\right]$$

The ratio of the shear deflection to the bending deflection at $r = 0$ provides a measure of plate slenderness. This ratio is, for $\nu = 1/3$:

$$\frac{p_0 t^2 a^2/16(1-\nu)D}{p_0 a^4(5+\nu)/64(1+\nu)D} = \frac{3}{2}\left(\frac{t}{a}\right)^2$$

In the case of a plate of radius 10 times its thickness, the above is small, $\frac{1}{67}$. For thin plates ($t/r < 0.1$), we thus conclude that the bending theory yields a result of sufficient accuracy.

2.6 CIRCULAR PLATES UNDER A CONCENTRATED LOAD AT ITS CENTER

When a concentrated or a (nominal) point load P acts at the plate, one can set $p_0 = 0$ in Eq. (2.13). The value of c_1 must be taken as zero, as in Sec. 2.4, in order that the deflection be finite at $r = 0$. The term involving c_2 must now be retained because of the very high shear forces present in the vicinity of the center. The deflection surface of the plate is then represented

$$w = c_2 r^2 \ln r + c_3 r^2 + c_4 \tag{2.21}$$

The constants c_2, c_3, and c_4 are calculated for two particular cases, described below.

Plate with clamped edge (Fig. 2.4*a*) The boundary condition, $w = 0$ and $\partial w/\partial r = 0$ at $r = a$, when introduced into Eq. (2.21) lead to

$$\begin{aligned} c_2 a^2 \ln a + c_3 a^2 + c_4 &= 0 \\ c_2 a(2 \ln a + 1) + 2c_3 &= 0 \end{aligned} \tag{a}$$

The additional condition is that the vertical shear force Q_r must equal $-P/2\pi r$. Thus, from Eqs. (2.9*c*) and (2.21), we obtain

$$\frac{4D}{r} c_2 = \frac{P}{2\pi r} \tag{b}$$

Solving Eqs. (*a*) and (*b*), the constants of integration are

$$c_2 = \frac{P}{8\pi D} \qquad c_3 = -\frac{P}{16\pi D}(2 \ln a + 1) \qquad c_4 = \frac{Pa^2}{16\pi D}$$

The foregoing, upon substitution into Eq. (2.21), provides the following expression for the deflection

$$w = \frac{P}{16\pi D}\left(2r^2 \ln \frac{r}{a} + a^2 - r^2\right) \tag{2.22}$$

The maximum deflection occurs at $r = 0$ and is given by

$$w_{\max} = \frac{Pa^2}{16\pi D} \tag{c}$$

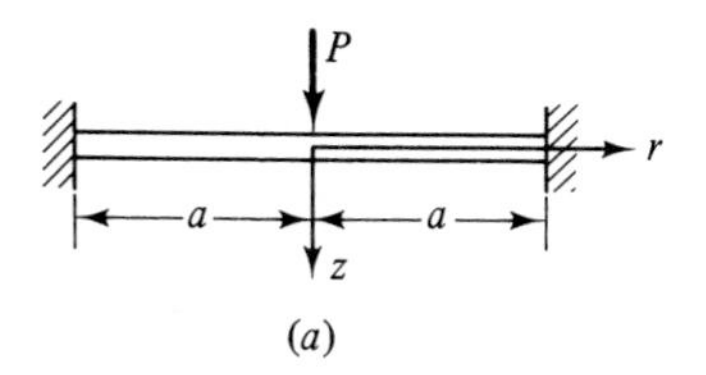

(*a*)

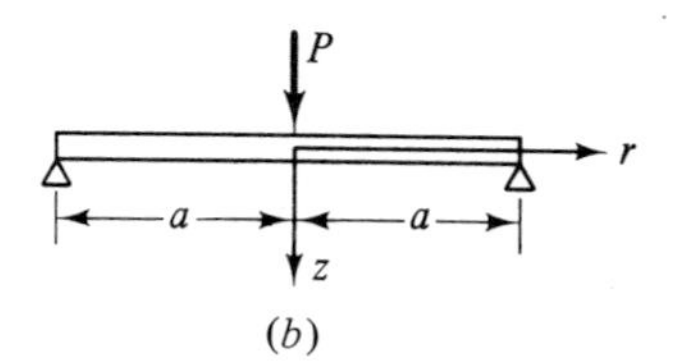

(*b*)

Figure 2.4

The stresses corresponding to Eq. (2.22), derived from Eq. (2.11), are

$$\sigma_r = \frac{3Pz}{\pi t^3}\left[(1+\nu)\ln\frac{a}{r} - 1\right]$$
$$\sigma_\theta = \frac{3Pz}{\pi t^3}\left[(1+\nu)\ln\frac{a}{r} - \nu\right] \tag{2.23}$$

It is seen from Eqs. (2.23) that σ_r and σ_θ grow without limit as $r \to 0$. Analysis by a more elaborate method indicates, however, that the actual stress caused by P on a very small area of radius r_c can be obtained by replacing the original r_c by a so-called *equivalent radius* r_e. The latter is given by the approximate formula:[7]

$$r_e = \sqrt{1.6r_c^2 + t^2} - 0.675t \tag{2.24}$$

The above applies to a plate of *any shape* for the cases in which $r_c < 1.7t$. If $r_c \geq 1.7t$ the actual r_c may be used. Upon introducing the equivalent radius $r_e = r$ in Eqs. (2.23), the maximum finite stresses produced by the concentrated load can be calculated.

Plate with simply supported edge (Fig. 2.4*b*) In this case, the deflection and the radial moment of the plate vanishes at the edge and the applied load $P = -2\pi r Q_r$. That is,

$$(w)_{r=a} = 0 \qquad (M_r)_{r=a} = 0 \qquad Q_r = -\frac{P}{2\pi r} \tag{d}$$

Substitution of Eqs. (2.21) and (2.9) into the above gives three linear equations for c_2, c_3, and c_4. Solving for the constants leads to the following relationships for deflection and stress:

$$w = \frac{P}{16\pi D}\left[2r^2\ln\frac{r}{a} + \frac{3+\nu}{1+\nu}(a^2 - r^2)\right] \tag{2.25}$$

$$\sigma_r = \frac{3Pz}{\pi t^3}(1+\nu)\ln\frac{a}{r}$$
$$\sigma_\theta = \frac{3Pz}{\pi t^3}\left[(1+\nu)\ln\frac{a}{r} + 1 - \nu\right] \tag{2.26}$$

At $r = 0$, we have the maximum deflection:

$$w_{\max} = \frac{Pa^2}{16\pi D}\frac{3+\nu}{1+\nu} \tag{e}$$

To find the largest finite stress, r_e is substituted for r in Eqs. (2.26).

Several other cases of practical importance can also be treated on the basis of the mathematical analyses described in the foregoing sections.[8] For reference purposes Table 2.2 lists the equations of deflection surfaces for three variously loaded circular plates (Probs. 2.5 to 2.7).

Table 2.2 Variously loaded circular plates

Loading	Equation of elastic surface	
A. Uniform load p_0 on outer portion 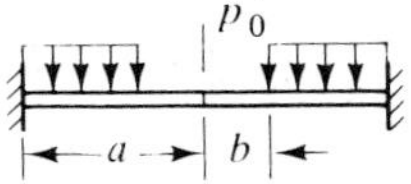	$w = \dfrac{p_0 a^4}{8D}\left\{\left[\dfrac{1}{8} + \dfrac{3}{8}\left(\dfrac{b}{a}\right)^4 - \dfrac{1}{2}\left(\dfrac{b}{a}\right)^2 - \dfrac{1}{2}\left(\dfrac{b}{a}\right)^4 \ln\dfrac{b}{a}\right] - \dfrac{r^2}{a^2}\left[\dfrac{1}{4} + \left(\dfrac{b}{a}\right)^2 \ln\dfrac{b}{a} - \dfrac{1}{4}\left(\dfrac{b}{a}\right)^4\right]\right\}$ $w_{max} = w(0) = \dfrac{p_0 a^4}{8D}\left[\dfrac{1}{8} + \dfrac{3}{8}\left(\dfrac{b}{a}\right)^4 - \dfrac{1}{2}\left(\dfrac{b}{a}\right)^2 - \dfrac{1}{2}\left(\dfrac{b}{a}\right)^4 \ln\dfrac{b}{a}\right]$	$(r \leq b)$
B. Uniform load p_0 on inner portion 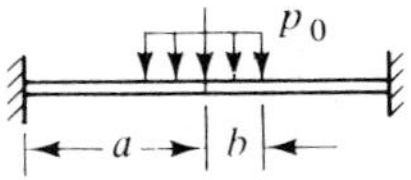	$w = \dfrac{p_0 a^4}{64D}\left(1 - \dfrac{r^2}{a^2}\right)^2 + \dfrac{p_0 a^4}{8D}\left\{-\dfrac{1}{8} - \dfrac{3}{8}\left(\dfrac{b}{a}\right)^4 + \left(\dfrac{b}{a}\right)^2 + \dfrac{1}{2}\left(\dfrac{b}{a}\right)^4 \ln\dfrac{b}{a} + \dfrac{r^2}{a^2}\left[\dfrac{1}{4} + \left(\dfrac{b}{a}\right)^2 \ln\dfrac{b}{a} - \dfrac{1}{4}\left(\dfrac{b}{a}\right)^4\right]\right\}$ $w_{max} = w(0) = \dfrac{p_0 a^4}{64D} + \dfrac{p_0 a^4}{8D}\left[-\dfrac{1}{8} - \dfrac{3}{8}\left(\dfrac{b}{a}\right)^4 + \dfrac{1}{2}\left(\dfrac{b}{a}\right)^2 + \dfrac{1}{2}\left(\dfrac{b}{a}\right)^4 \ln\dfrac{b}{a}\right]$	$(r \leq b)$
C. Line load P_1 per unit length along $2\pi b$ 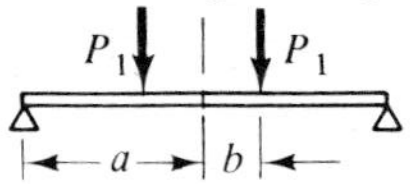	$w = \dfrac{P_1 b}{4D}\left[(a^2 - r^2)\left(1 + \dfrac{1}{2}\dfrac{1 - \nu}{1 + \nu}\dfrac{a^2 - b^2}{a^2}\right) + (b^2 + r^2)\ln\dfrac{r}{a}\right]$ $w(b) = \dfrac{P_1 b}{4D}\left[(a^2 - b^2)\left(1 + \dfrac{1}{2}\dfrac{1 - \nu}{1 + \nu}\dfrac{a^2 - b^2}{a^2}\right) + 2b^2 \ln\dfrac{b}{a}\right]$	$(r \geq b)$

2.7 ANNULAR PLATES WITH SIMPLY SUPPORTED OUTER EDGES

In this section, we discuss the bending of circular plates with a concentric circular hole, so-called *annular plates*. The inner and outer radii of a plate are denoted by b and a, respectively. Two typical cases of loading are illustrated below.

Plate loaded by edge moments (Fig. 2.5*a*) To arrive at a solution for the deflection we utilize Eq. (2.9*c*). Inasmuch as $Q_r = 0$,

$$\frac{d}{dr}\left[\frac{1}{r}\frac{d}{dr}\left(r\frac{dw}{dr}\right)\right] = 0 \tag{a}$$

Three successive integrations of the above give

$$w = \tfrac{1}{4}c_1 r^2 + c_2 \ln r + c_3 \tag{b}$$

Referring to Fig. 2.5*a* the boundary conditions are represented by

$$\begin{aligned} w &= 0 \qquad M_r = M_2 \qquad (r = a) \\ M_r &= M_1 \qquad\qquad\qquad (r = b) \end{aligned} \tag{c}$$

where the radial bending moment M_r is given by Eq. (2.9*a*). The constants are evaluated by substitution of Eq. (*b*) into Eqs. (*c*). Expressions for deflection and moments are then

$$\begin{aligned} w &= \frac{1}{2}\frac{r^2 - a^2}{a^2 - b^2}\frac{M_1 b^2 - M_2 a^2}{(1+\nu)D} + \frac{a^2 b^2}{a^2 - b^2}\frac{M_1 - M_2}{(1-\nu)D}\ln\frac{r}{a} \\ M_r &= -\frac{M_1 b^2 - M_2 a^2}{a^2 - b^2} + \frac{a^2 b^2 (M_1 - M_2)}{r^2(a^2 - b^2)} \\ M_\theta &= -\frac{M_1 b^2 - M_2 a^2}{a^2 - b^2} - \frac{a^2 b^2 (M_1 - M_2)}{r^2(a^2 - b^2)} \end{aligned} \tag{2.27}$$

The stress components are calculated by means of Eqs. (2.27) and (2.3).

Plate loaded by shear force Q_1 at inner edge (Fig. 2.5*b*) We now consider a plate under shear force Q_1 per unit circumferential length, uniformly distributed over the inner edge, resulting in a total load $2\pi b Q_1$. This must be equal to $2\pi r Q_r$,

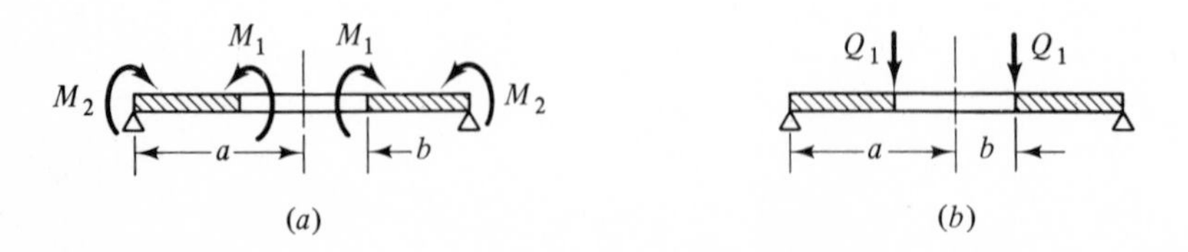

Figure 2.5

the total shear force at a distance r from the center. We thus have $Q_r = -Q_1 b/r$ and Eq. (2.9c) becomes

$$\frac{d}{dr}\left[\frac{1}{r}\frac{d}{dr}\left(r\frac{dw}{dr}\right)\right] = \frac{Q_1 b}{Dr} \tag{2.28}$$

The displacement w is obtained by successive integrations of Eq. (2.28)

$$w = \frac{Q_1 br^2}{4D}(\ln r - 1) + \frac{c_1 r^2}{4} + c_2 \ln r + c_3 \tag{d}$$

The constants c_1, c_2, and c_3 are determined from the following conditions at the outer and inner edges

$$\begin{aligned} w = 0 \qquad M_r &= 0 \qquad (r = a) \\ M_r &= 0 \qquad (r = b) \end{aligned} \tag{e}$$

Upon substitution of the constants into Eqs. (d), the following expression for the plate deflection is obtained

$$\begin{aligned} w = \frac{Q_1 a^2 b}{4D}\Bigg\{\left(1 - \frac{r^2}{a^2}\right)\left[\frac{3+\nu}{2(1+\nu)} - \frac{b^2}{a^2 - b^2}\ln\frac{b}{a}\right] \\ + \frac{r^2}{a^2}\ln\frac{r}{a} + \frac{2b^2}{a^2 - b^2}\frac{1+\nu}{1-\nu}\ln\frac{b}{a}\ln\frac{r}{a}\Bigg\} \end{aligned} \tag{2.29}$$

It is observed that if the radius of the hole becomes infinitesimally small, $b^2 \ln (b/a)$ vanishes, and Eq. (2.29), by letting $Q_1 = P/2\pi b$, reduces to Eq. (2.25), as expected.

2.8 DEFLECTION AND STRESS BY SUPERPOSITION

The integration procedures discussed in the foregoing sections for determining the elastic deflection and stress of loaded plates are generally applicable to other cases of plates. It is noted, however, that the solutions to numerous problems with simple loadings are already available. For complex configurations of loads therefore, the method of superposition may be used to good advantage to simplify the analysis.

The method is illustrated for the case of the annular plate shown in Fig. 2.6a. The plate is simply supported along its outer edge and is subjected to a

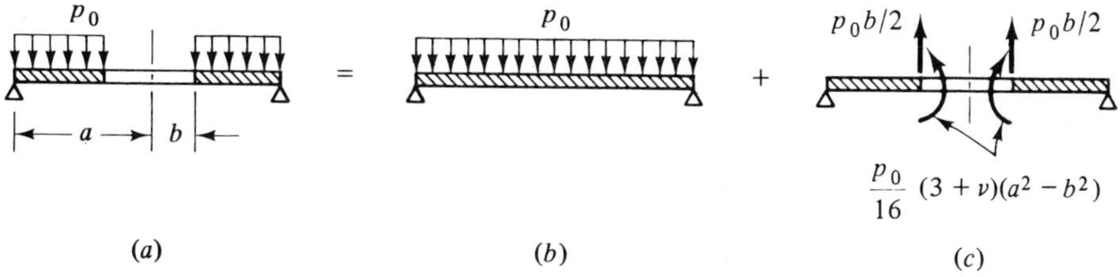

Figure 2.6

Table 2.3 Variously loaded annular plates

Uniform load: $w_{max} = k_1(p_0 a^4/Et^3)$, $\sigma_{max} = k_2(p_0 a^2/t^2)$
Concentrated load: $w_{max} = k_3(Pa^2/Et^3)$, $\sigma_{max} = k_4(P/t^2)$

A. Outer edge supported	a/b	k_1	k_2
	1.5	0.414	0.976
	2.0	0.664	1.44
	3.0	0.824	1.88
	4.0	0.830	2.08
	5.0	0.813	2.19
B. Inner and outer edges clamped	1.5	0.0062	0.273
	2.0	0.0329	0.71
	3.0	0.110	1.54
	4.0	0.179	2.23
	5.0	0.234	2.80
C. Inner and outer edges clamped	a/b	k_3	k_4
	1.5	0.0064	0.22
	2.0	0.0237	0.405
	3.0	0.062	0.703
	4.0	0.092	0.933
	5.0	0.114	1.13

uniformly distributed load p_0 at its surface. Shown in Fig. 2.6*b* is a circular plate under a uniform load p_0. Figure 2.6*c* is an annular plate carrying along its inner edge a shear force per unit circumferential length $p_0 b/2$ and a radial bending moment, defined by Eq. (2.18), $p_0(3 + \nu)(a^2 - b^2)/16$. The solutions for each of the latter two cases are known from Secs. 2.4 and 2.7. Hence, the deflection and stress *at any point* of the plate in Fig. 2.6*a* can be found by the superposition of the results at that point for the cases indicated in Fig. 2.6*b* and 2.6*c* (Prob. 2.14).

Employing similar procedures, annular plates with various load and edge conditions may be treated. Table 2.3 provides only the final results for several examples.[7,8] In all cases ν is taken as $\nu = 0.3$. Design calculations are often facilitated by this type of compilation.

2.9 THE RITZ METHOD APPLIED TO CIRCULAR PLATES ON ELASTIC FOUNDATION

In the problems discussed thus far, support was provided at the plate edges, and the plate was assumed to undergo no deflection at these supports. We now consider the case of a plate supported continuously along its bottom surface by a foundation, itself assumed to experience elastic deformation. The foundation reaction forces will be taken to be *linearly proportional* to the plate deflection at any point, i.e., wk. Here w is the plate deflection and k is a constant, termed as the *modulus of the foundation* or *bedding constant* of the foundation material, having the dimensions of force per unit surface area of plate per unit of deflection (e.g., Pa/m). The above assumption with respect to the nature of the support not only leads to equations amenable to solution, but approximates closely many real situations.[5] Examples of this type of plate include concrete slabs, bridge decks, floor structures, and airport runways.

We shall apply the Ritz method (Sec. 1.9) to treat the bending of a circular plate of radius a resting freely on an elastic foundation and subjected to a center load P. In this case of axisymmetrical bending, the expression for strain energy given by Eq. (P2.11) reduces to

$$U_1 = \pi D \int_0^a \left[\left(\frac{d^2 w}{dr^2} + \frac{1}{r}\frac{dw}{dr} \right)^2 - \frac{2(1-\nu)}{r}\frac{dw}{dr}\frac{d^2 w}{dr^2} \right] r\, dr \tag{2.30}$$

A solution can be assumed in the form of a series

$$w = c_0 + c_2 r^2 + \cdots + c_n r^n \tag{a}$$

in which c_n are to be determined from the condition that the potential energy Π of the system in stable equilibrium is minimum.

If we retain, for example, only the first two terms of Eq. (a)

$$w = c_0 + c_2 r^2 \tag{b}$$

and the strain energy, from Eq. (2.30), is then

$$U_1 = 4c_2^2 D\pi a^2 (1 + \nu) \tag{c}$$

The strain energy owing to the deformation of the elastic foundation is determined as follows

$$U_2 = \int_0^{2\pi} \int_0^a \tfrac{1}{2} k w^2 r\, dr\, d\theta = \tfrac{1}{2}\pi k (c_0 a^2 + c_0 c_2 a^4 + \tfrac{1}{3} c_2^2 a^6) \tag{d}$$

The work done by the load is given by

$$W = P \cdot (w)_{r=0} = P c_0 \tag{e}$$

The potential energy, $\Pi = U_1 + U_2 - W$, is thus

$$\Pi = 4c_2^2 D\pi a^2 (1 + \nu) + \frac{\pi k}{2} (c_0^2 a^2 + c_0 c_2 a^4 + \tfrac{1}{3} c_2^2 a^6) - P c_0$$

Applying the minimizing condition, $\partial \Pi / \partial c_n = 0$, we find that

$$c_0 = \frac{P}{\pi k a^2} \left[1 + \frac{1}{(1/3) + 32D(1+\nu)/ka^4} \right]$$

$$c_2 = -\frac{P}{\pi k a^4 [(1/6) + 16D(1+\nu)/ka^4]} \tag{f}$$

Then by substituting Eqs. (f) into (b), we obtain the maximum deflection at the center ($r = 0$):

$$w_{\max} = \frac{P}{\pi k a^2}\left[1 + \frac{1}{(1/3) + 32D(1 + \nu)/ka^4}\right] \tag{2.31}$$

An improved approximation results from retention of more terms of the series given by Eq. (a)

2.10 ASYMMETRICAL BENDING OF CIRCULAR PLATES

In the foregoing sections, our concern was with the circular plates loaded axisymmetrically. We now turn to *asymmetrical* bending. For analysis of deflection and stress we must obtain appropriate solutions of the governing differential equation (2.5).

Consider the case of a *clamped* circular plate of radius a and subjected to a linearly varying or hydrostatic loading represented by

$$p = p_0 + p_1 \frac{r}{a} \cos\theta \tag{a}$$

as shown in Fig. 2.7. The boundary conditions are

$$w = 0 \qquad \frac{\partial w}{\partial r} = 0 \qquad (r = a) \tag{b}$$

where $w = w_p + w_h$.

The particular solution corresponding to p_0, referring to Sec. 2.4, is $w'_p = p_0 r^4/64D$. For the linear portion of the loading,

$$w''_p = A\frac{p_1 r^5 \cos\theta}{a}$$

Introduction of the above into Eq. (2.5) yields $A = 1/192D$. We thus have

$$w_p = \frac{p_0 r^4}{64D} + \frac{p_1 r^5 \cos\theta}{192aD} \tag{c}$$

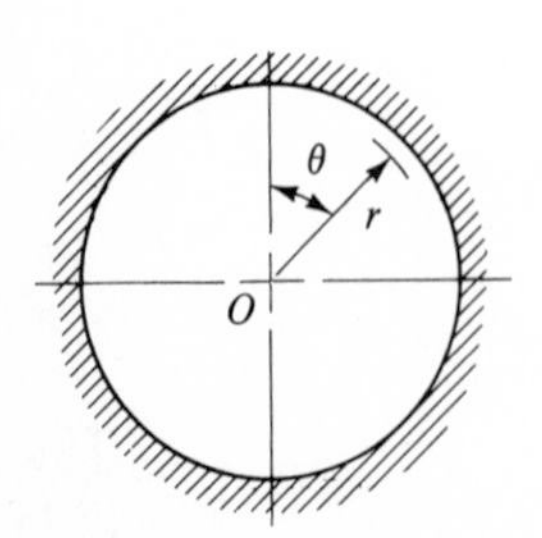

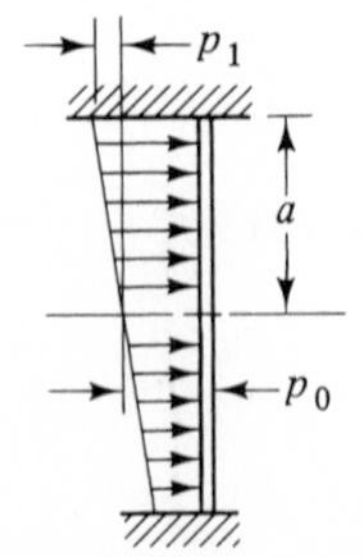

Figure 2.7

It is noted that the general method of obtaining the particular solution of Eq. (2.5), given in Prob. 2.20, follows a procedure identical with that described in Sec. 3.4 (Lévy's solution for rectangular plates).

The homogeneous solution w_h will be symmetrical in θ; thus f_n^* in Eq. (2.7) vanishes. Owing to the nature of p and w_p, we take only the terms of series (2.8) containing the function f_0 and f_1. The deflection w_h and its derivatives (or slope, moment, and shear) must be finite at the center ($r = 0$). It follows that $B_0 = D_0 = B_1 = D_1 = 0$ in expressions for f_0 and f_1. Hence,

$$w_h = A_0 + C_0 r^2 + (A_1 r + C_1 r^3) \cos\theta \tag{d}$$

The conditions (*b*) combined with Eqs. (*c*) and (*d*) yield two equations in the four unknown constants

$$\frac{p_0 a^4}{64D} + A_0 + C_0 a^2 + \left(\frac{p_1 a^4}{192D} + A_1 a + C_1 a^3\right)\cos\theta = 0$$

$$\frac{4p_0 a^3}{64D} + 2C_0 a + \left(\frac{5p_1 a^3}{192D} + A_1 + 3C_1 a^2\right)\cos\theta = 0$$

Since the term in each pair of parentheses is independent of $\cos\theta$, a solution exists for all θ provided that

$$\frac{p_0 a^4}{64D} + A_0 + C_0 a^2 = 0 \qquad \frac{p_1 a^4}{192D} + A_1 a + C_1 a^3 = 0$$

$$\frac{4p_0 a^3}{64D} + 2C_0 a = 0 \qquad \frac{5p_1 a^3}{192D} + A_1 + 3C_1 a^2 = 0$$

which upon solution, leads to

$$A_0 = -\frac{p_0 a^4}{64D} \qquad C_0 = -\frac{p_0 a^2}{32D} \qquad C_1 = -\frac{2p_1 a}{192D} \qquad A_1 = \frac{p_1 a^3}{192D} \tag{e}$$

The deflection is therefore

$$w = \frac{p_0}{64D}(a^2 - r^2)^2 + \frac{p_1}{192D}\frac{r}{a}(a^2 - r^2)^2 \cos\theta \tag{2.32}$$

The center displacement is

$$w_c = \frac{p_0 a^4}{64D} \tag{f}$$

We observe that, when the loading is uniform, $p_1 = 0$ and Eq. (2.32) reduces to Eq. (2.14) as expected. The expressions for the bending and twisting moments are, from Eqs. (2.32) and (2.2)

$$
\begin{aligned}
M_r &= \frac{p_0}{16}[a^2(1+\nu) - r^2(3+\nu)] - \frac{p_1}{48}\left[\frac{r^3}{a}(5+\nu) - ar(3+\nu)\right]\cos\theta \\
M_\theta &= \frac{p_0}{16}[a^2(1+\nu) - r^2(1+3\nu)] - \frac{p_1}{48}\left[\frac{r^3}{a}(5\nu+1) - ar(3\nu+1)\right]\cos\theta \qquad (2.33) \\
M_{r\theta} &= -\frac{1-\nu}{48}p_1 ra\left(1 - \frac{r^2}{a^2}\right)\sin\theta
\end{aligned}
$$

The case of the simply supported plate under hydrostatic loading can be treated in a similar way.

2.11 DEFLECTION BY THE RECIPROCITY THEOREM

Presented in this section is a practical approach for computation of the *center* deflection of a circular plate with *symmetrical* edge conditions under asymmetrical or nonuniform loading. The method utilizes the *reciprocity* theorem together with expressions for deflection of axisymmetrically bent plates.

Consider, for example, the forces P_1 and P_2 acting at the center and at r (any θ) of a circular plate with simply supported edge (Fig. 2.8). According to the reciprocity theorem,[1] due to E. Betti and Lord Rayleigh, we may write:

$$P_1 w_{21} = P_2 w_{12} \qquad (a)$$

That is, the work done by P_1 owing to displacement w_{21} due to P_2, is equal to the work done by P_2 owing to displacement w_{12} due to P_1.

For the sake of simplicity let $P_1 = 1$, $w_{21} = w_c$, and $w_{12} = \bar{w}(r)$. The deflection at the *center* w_c of a circular plate with a nonuniform loading $p(r, \theta)$ but symmetric boundary conditions may therefore be determined through application of Eq. (a) as follows

$$w_c = \int_0^{2\pi}\int_0^a p(r, \theta)\bar{w}(r) r\, dr\, d\theta \qquad (2.34)$$

Clearly, $\bar{w}(r)$ is the deflection at r due to a *unit force* at the center. In the cases of fixed and simply supported plates, $\bar{w}(r)$ is given by the expressions obtained by setting $P = 1$ in Eqs. (2.22) and (2.25), respectively.

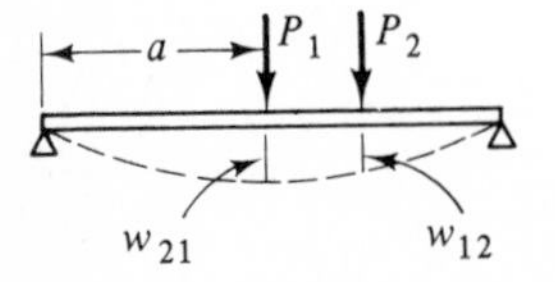

Figure 2.8

To illustrate the application of the approach, reconsider the bending of the plate described in Sec. 2.10. Upon substituting $p(r, \theta) = p_0 + p_1(r/a) \cos \theta$ and Eq. (2.22) into Eq. (2.34), setting $P = 1$ and integrating, we have

$$w_c = \frac{1}{16\pi D} \int_0^{2\pi} \int_0^a \left(p_0 + p_1 \frac{r}{a} \cos \theta\right)\left(2r^2 \ln \frac{r}{a} + a^2 - r^2\right) r \, dr \, d\theta = \frac{p_0 a^4}{64D}$$

The above is identical with the value given by Eq. (f) of the preceding section.

2.12 CIRCULAR PLATES OF VARIABLE THICKNESS UNDER NONUNIFORM LOAD

In this section we discuss an approximate method for computing stresses and deflections in *solid* or *annual* circular plates of variable thickness, subjected to arbitrary lateral loading.[9] Except for the requirement of axisymmetry, no special restrictions are placed on the manner in which either the thickness or the lateral loading vary with the radial coordinate. Several applications immediately come to mind: turbine disks, clutches, and pistons of reciprocating machinery.

Consider a circular plate (Fig. 2.9*a*), and the division of the plate into small (finite) ring segments, as in Fig. 2.9*b*. Note that radial lengths of the segments need not be equal, but that the thickness is taken as constant for each. For each element defined as in Fig. 2.9*b*, the development of Sec. 2.3 applies. In order to accommodate the substitution of a series of constant-thickness elements for the original structure of varying thickness, it is necessary to match slopes and moments at the boundary between adjacent segments. The boundary conditions are handled in the usual manner. As the method treats the plate as a collection of constant-thickness disks, it is unnecessary to determine an analytical expression for thickness as a function of radius. Prior to illustrating the technique by means of a numerical example, the general calculation procedure is outlined. A knowledge of the derivation of the basic relationships [Sec. 2.12(b)] is not essential in applying the method.

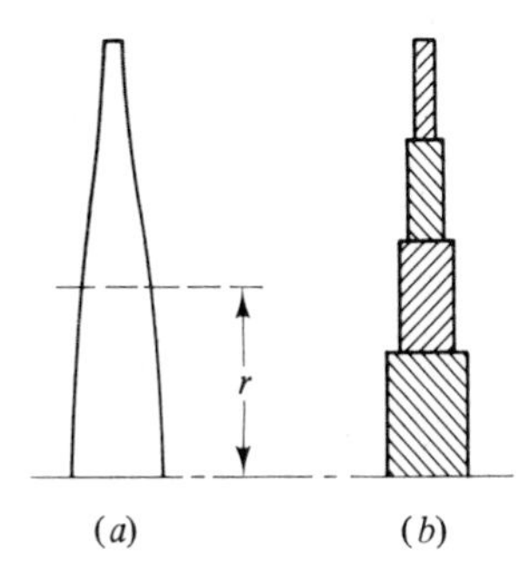

Figure 2.9

(a) Calculation Procedure

The expressions developed in Sec. 2.12(*b*) may be so arranged as to facilitate the calculation process. Consider, with this end in mind, a plate subdivided into a number of annular elements with the applied lateral loading on each element denoted $\bar{Q}$, the average load on that element (Fig. 2.9). We shall apply the notation

$$
\begin{aligned}
\rho &= \frac{r_i}{r_o} & \alpha &= \tfrac{1}{2}(1-\rho^2) \\
\beta_r &= \frac{1+\nu}{4\pi}\ln\rho - \frac{1-\nu}{8\pi}(1-\rho^2) & D &= \frac{Et^3}{12(1-\nu^2)} \\
\lambda_r &= \frac{1-\rho^2}{4(1+\nu)} + \frac{\rho^2}{2(1-\nu)}\ln\rho & \beta_\theta &= \frac{1+\nu}{4\pi}\ln\rho + \frac{1-\nu}{8\pi}(1-\rho^2) \\
\lambda_Q &= \frac{1-\rho^2}{8\pi} + \frac{1+\rho^2}{8\pi}\ln\rho & \lambda_\theta &= \frac{1-\rho^2}{4(1+\nu)} - \frac{\rho^2}{2(1-\nu)}\ln\rho
\end{aligned}
\tag{2.35}
$$

In addition, for each joint between adjacent elements,

$$\eta = \frac{t_{n+1}^3}{t_n^3} - 1 \tag{2.36}$$

The notation thus introduced is next applied in the determination of the following quantities.

Bending moments The change in bending moment in proceeding from the inner edge to the outer edge of any element may be ascertained by rearranging Eqs. (2.44) as follows:

$$
\begin{aligned}
\Delta M_r &= \alpha(M_{\theta i} - M_{ri}) + \beta_r \bar{Q} \\
\Delta M_\theta &= -\alpha(M_{\theta i} - M_{ri}) + \beta_\theta \bar{Q}
\end{aligned}
\tag{2.37}
$$

At the inner edge, the moments are either given or assumed. The outer moments acting on an element are then

$$
\begin{aligned}
M_{ro} &= M_{ri} + \Delta M_r \\
M_{\theta o} &= M_{\theta i} + \Delta M_\theta
\end{aligned}
\tag{2.38}
$$

Similarly, the moment increments corresponding to the interface between adjacent elements are

$$\Delta M_r = 0 \qquad \Delta M_\theta = \eta(M_{\theta o} - \nu M_{ro}) \tag{2.39}$$

It now follows that the moments at the inner dege of the next element are found from

$$
\begin{aligned}
(M_{ri})_{n+1} &= (M_{ro})_n \\
(M_{\theta i})_{n+1} &= (M_{\theta o})_n + \Delta M_\theta
\end{aligned}
\tag{2.40}
$$

Table 2.4

Given inner boundary conditions	Assumed values at inner boundary of plate (A and B are any arbitrary values $B \neq 0$)	
	Particular solution	Homogenous solution
M_{ra}	$M'_{ra} = M_{ra}$	$M''_{ra} = 0$
	$M'_{\theta a} = A$	$M''_{\theta a} = B$
Clamped	$M'_{ra} = A$	$M''_{ra} = B$
	$M'_{\theta a} = \nu M'_{ra}$	$M''_{\theta a} = \nu M''_{ra}$
Solid plate	$M'_{ra} = A$	$M''_{ra} = B$
	$M'_{\theta a} = M'_{ra}$	$M''_{\theta a} = M''_{ra}$

When Eqs. (2.37) to (2.40) are applied successively, beginning with the innermost element, the moments at any intermediate edge may be found in terms of the moment at the inner boundary of the disk.

Boundary conditions The following steps are taken to satisfy the boundary conditions at the inner and outer edges of the disk:

Step 1. Apply Eqs. (2.37) to (2.40) to obtain a particular solution (denoted by a single prime). Begin the calculations with the appropriate inner boundary values specified in Table 2.4.

Step 2. Repeat step 1 with $\bar{Q} = 0$ to obtain a homogeneous solution (denoted by a double prime).

Step 3. Superimpose the values found in steps 1 and 2 to obtain the general solution:

$$
\begin{aligned}
M_r &= M'_r + kM''_r \\
M_\theta &= M'_\theta + kM''_\theta
\end{aligned}
\tag{2.41}
$$

The constant k is calculated as indicated in Table 2.5 from the given boundary conditions at the outer edge of the disk.

Table 2.5

Given outer boundary conditions	Formula for k in Eq. (2.41)
M_{rb}	$k = (M_{rb} - M'_{rb})/M''_{rb}$
Clamped	$k = (M'_{\theta b} - \nu M'_{rb})/(M''_{\theta b} - \nu M''_{rb})$

To verify the correctness of the final results, use the values found in the final step to perform the calculations indicated in step 1. The results thus obtained should be the same as those already found at the conclusion of step 3.

Stresses On the basis of the moments now known, the bending stresses are calculated from:

$$\sigma_r = \frac{6M_r}{t^2} \qquad \sigma_\theta = \frac{6M_\theta}{t^2} \tag{a}$$

The stresses thus found will show a stepped distribution throughout the disk owing to the nature of the analysis. The actual distribution of stress may be approximated adequately by drawing a smooth curve through the calculated points.

Deflections The change in deflection, Δw, is for each element found by rearranging Eq. (2.45):

$$\Delta w = \frac{r_o^2}{D}(\lambda_r + M_{ri} + \lambda_\theta M_{\theta i} + \lambda_Q \bar{Q}) \tag{2.42}$$

where M_r and M_θ are given by Eqs. (2.41). The total deflection at any point in the plate is thus computed by adding the increments given by Eq. (2.42).

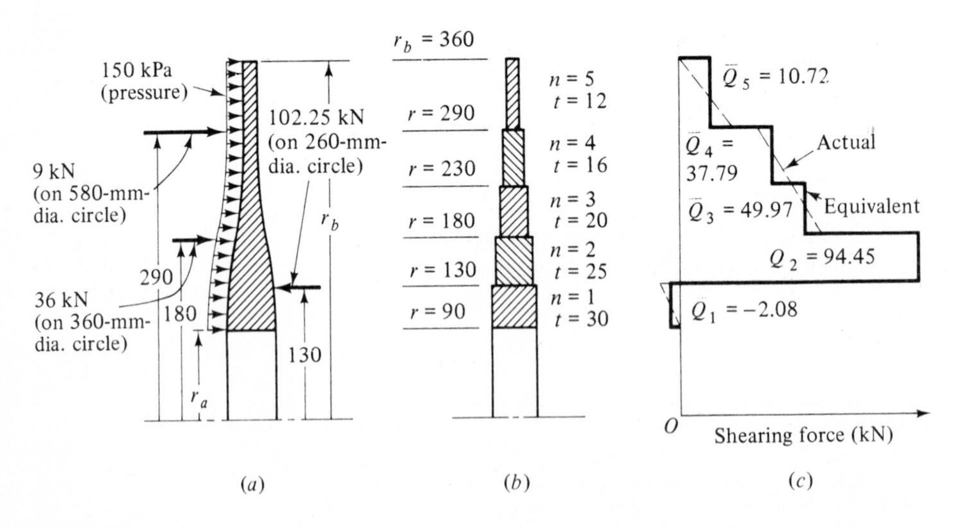

Figure 2.10

The foregoing procedures apply to statically determinate problems. When indeterminate situations are encountered, the above procedure remains applicable but requires the superposition of a number of determinate cases.

Example 2.2 Calculate the stress distribution and the deflections of the steel disk shown in Fig. 2.10*a*. The member is free at the inner and outer edges. Divide the plate into five segments as in Fig. 2.10*b*. Let $E = 207$ GPa and $\nu = 0.3$. All dimensions are in millimeters.

SOLUTION The uniform lateral loading applied to each segment is shown in Fig. 2.10*c*. The results of the complete calculation are presented in Tables 2.6 through 2.9. Table 2.6 lists the various coefficients calculated on the basis of Eqs. (2.35). By assuming for the free inner edge, $M'_{ra} = M_{ra} = 0$, $M'_{\theta a} = 100$ (arbitrary), Eqs. (2.37) through (2.40) provide the particular solution (Table 2.7). The foregoing step is repeated for $\bar{Q} = 0$ to find the complementary solution (Table 2.8). The constant, k, is next obtained by applying the expression for the outer edge (Table 2.5):

$$k = \frac{M_{rb} - M'_{rb}}{M''_{rb}} = \frac{0 - (-5954.571)}{25.125} = 236.995$$

The final bending moments M_r and M_θ are then calculated using Eqs. (2.41). Following this, Eqs. (*a*) and (2.42) provide the final stresses and deflections (Table 2.9).

Table 2.6 Plate coefficients

		Element number, n				
Symbol	Units	1	2	3	4	5
r_i	m	0.090	0.130	0.180	0.230	0.290
r_o	m	0.130	0.180	0.230	0.290	0.360
t	m	0.030	0.025	0.020	0.016	0.012
$\bar{Q}$	N	−2080	94450	49970	37790	10720
α	.....	0.26035	0.23920	0.19376	0.18550	0.17554
β_r	.....	−0.05254	−0.04699	−0.03615	−0.03431	−0.03215
β_θ	.....	−0.02354	−0.02024	−0.01456	−0.01365	−0.01259
η	.....	−0.42130	−0.48800	−0.48800	−0.57813	
λ_r	.....	−0.02575	−0.02924	−0.03271	−0.03280	−0.03271
λ_θ	.....	0.22603	0.21324	0.18176	0.17549	0.16774
λ_Q	.....	−0.00093	−0.00067	−0.00031	−0.00026	−0.00022
$(r_o^2/D)10^8$	m/N	3.305	10.948	34.912	108.40	395.97

Table 2.7 Particular solution

Symbol	Element number, n				
	1	2	3	4	5
M'_{ri}	0	135.318	−4314.148	−5587.744	−6267.813
M'_{ro}	135.318	−4313.148	−5587.744	−6267.813	−5954.571
$M'_{\theta i}$	100	88.241	−1564.256	−2264.261	−2520.007
$M'_{\theta o}$	122.928	−1821.611	−2824.638	−3396.600	−3312.862

Table 2.8 Homogeneous solution

Symbol	Element number, n				
	1	2	3	4	5
M''_{ri}	0	26.035	30.833	29.830	27.639
M''_{ro}	26.035	30.833	29.830	27.639	25.125
$M''_{\theta i}$	100	46.094	25.657	18.017	13.319
$M''_{\theta o}$	73.965	41.296	26.660	20.208	15.833

Table 2.9 Final moments, stresses, and deflections

Symbol	Element number, n				
	1	2	3	4	5
M_{ri}	0	6305.48	2993.14	1481.89	282.49
M_{ro}	6305.48	2993.14	1481.89	282.49	0
$M_{\theta i}$	23799.5	11012.31	4516.44	2005.73	636.53
$M_{\theta o}$	17652.26	7965.33	3493.72	1392.71	439.41
$(\sigma_{ri}) \times 10^{-6}$	0	60.533	44.897	34.732	11.770
$(\sigma_{ro}) \times 10^{-6}$	42.036	28.734	22.228	6.621	0
$(\sigma_{\theta i}) \times 10^{-6}$	158.663	105.718	67.747	47.009	26.522
$(\sigma_{\theta o}) \times 10^{-6}$	117.682	76.467	52.406	32.642	18.309
$(\Delta w) \times 10^{4}$	1.778	2.999	2.470	3.182	3.769
$(w_o) \times 10^{4}$	1.778	4.078	6.548	9.730	13.499

(b) Basic Derivation

The differential equation governing a circular plate of constant thickness, subjected to a constant shearing force $\bar{Q} = 2\pi r Q_r$, from Eq. (2.9*c*) is

$$\frac{d}{dr}\left[\frac{1}{r}\frac{d}{dr}\left(r\frac{dw}{dr}\right)\right] = -\frac{\bar{Q}}{2\pi r D} \tag{2.43}$$

Integrating this expression twice, we obtain

$$\frac{dw}{dr} = \frac{1}{D}\left[-\frac{\bar{Q}r}{8\pi}(2\ln r - 1) + \frac{c_1 r}{2} + \frac{c_2}{r}\right] \tag{b}$$

where c_1 and c_2 are constants of integration which may vary from element to element. The derivative of Eq. (b) is

$$\frac{d^2w}{dr^2} = \frac{1}{D}\left[-\frac{\bar{Q}}{8\pi}(2\ln r + 1) + \frac{c_1}{r} - \frac{c_2}{r^2}\right] \tag{c}$$

Substituting Eqs. (b) and (c) into Eqs. (2.9a) and (2.9b), we have

$$\begin{aligned} M_r &= -\frac{\bar{Q}}{8\pi}[2(1+\nu)\ln r + (1-\nu)] + \frac{(1+\nu)c_1}{2} - (1-\nu)\frac{c_2}{r^2} \\ M_\theta &= -\frac{\bar{Q}}{8\pi}[2(1+\nu)\ln r - (1-\nu)] + \frac{(1+\nu)c_1}{2} + (1-\nu)\frac{c_2}{r^2} \end{aligned} \tag{d}$$

Solving for c_1 and c_2 at the inner edge r_i of the element in terms of bending moments M_{ri}, $M_{\theta i}$, yields

$$\begin{aligned} c_1 &= \frac{M_{\theta i} + M_{ri}}{1+\nu} + \frac{\bar{Q}}{2\pi}\ln r_i \\ c_2 &= \left[\frac{M_{\theta i} - M_{ri}}{1-\nu} - \frac{\bar{Q}}{4\pi}\right]\frac{r_i^2}{2} \end{aligned} \tag{e}$$

Next, Eqs. (e) are substituted into Eqs. (d) and the bending moments M_{ro} and $M_{\theta o}$ at the outer edge of the same element are found:

$$\begin{aligned} M_{ro} &= \frac{1}{2}(M_{\theta i} + M_{ri}) - \frac{1}{2}\frac{r_i^2}{r_o^2}(M_{\theta i} - M_{ri}) \\ &\quad + \frac{\bar{Q}}{4\pi}(1+\nu)\ln\frac{r_i}{r_o} - \frac{\bar{Q}}{8\pi}(1-\nu)\left(1 - \frac{r_i^2}{r_o^2}\right) \\ M_{\theta o} &= \frac{1}{2}(M_{\theta i} + M_{ri}) + \frac{1}{2}\frac{r_i^2}{r_o^2}(M_{\theta i} - M_{ri}) \\ &\quad + \frac{\bar{Q}}{4\pi}(1+\nu)\ln\frac{r_i}{r_o} + \frac{\bar{Q}}{8\pi}(1-\nu)\left(1 - \frac{r_i^2}{r_o^2}\right) \end{aligned} \tag{f}$$

Continuity across the joint between adjacent elements is satisfied by equating the radial moments M_r and slopes dw/dr at each side of the interface. To accomplish this, d^2w/dr^2 is eliminated between Eqs. (2.9a) and (2.9b) with the result for adjacent elements

$$
\begin{aligned}
(M_{\theta o} - \nu M_{ro})_n &= (1 - \nu^2) D_n \left[\frac{1}{r_o} \left(\frac{dw}{dr} \right)_o \right]_n \\
(M_{\theta i} - \nu M_{ri})_n &= (1 - \nu^2) D_{n+1} \left[\frac{1}{r_i} \left(\frac{dw}{dr} \right)_i \right]_{n+1}
\end{aligned}
\tag{2.44}
$$

Because the quantities r and dw/dr must be equal in both of the above expressions,

$$
\left[\frac{1}{r_o} \left(\frac{dw}{dr} \right)_o \right]_n = \left[\frac{1}{r_i} \left(\frac{dw}{dr} \right)_i \right]_{n+1}
$$

It then follows that the second of Eqs. (2.44) divided by the first leads to

$$
\frac{(M_{\theta i} - \nu M_{ri})_{n+1}}{(M_{\theta o} - \nu M_{ro})_n} = \frac{D_{n+1}}{D_n} = \frac{t_{n+1}^3}{t_n^3} \tag{g}
$$

Expressions (2.44) also indicate that in the event the edge of a disk is fixed,

$$
M_\theta = \nu M_r \qquad \left(\text{since} \quad \frac{dw}{dr} = 0 \right) \tag{h}
$$

This relationship is applied in the construction of Table 2.4.

The change in deflection w in proceeding from one edge to the other of the same element is found by first integrating Eq. (b) over the radial length of an element:

$$
D \int_{w_i}^{w_o} \frac{dw}{dr} dr = -\frac{\bar{Q}}{4\pi} \int_{r_i}^{r_o} r \ln r \, dr + \frac{\bar{Q}}{8\pi} \int_{r_i}^{r_o} r \, dr + \frac{c_1}{2} \int_{r_i}^{r_o} r \, dr + c_2 \int_{r_i}^{r_o} \frac{1}{r} dr \tag{i}
$$

Performing the indicated operations, we obtain

$$
D \, \Delta w = \frac{\bar{Q}}{8\pi} [r_i^2 \ln r + r_o^2 - r_o^2 \ln r_o - r_i^2] + \frac{c_1}{2} \left(\frac{r_o^2 - r_i^2}{2} \right) - c_2 \ln \frac{r_i}{r_o} \tag{j}
$$

Introducing c_1 and c_2 from Eqs. (e), the above assumes the form

$$
\begin{aligned}
D \, \Delta w = {} & \frac{M_{\theta i} + M_{ri}}{2} \frac{r_o^2 - r_i^2}{2(1 + \nu)} - \frac{M_{\theta i} - M_{ri}}{2} \frac{r_i^2}{1 - \nu} \ln \frac{r_i}{r_o} \\
& + \frac{\bar{Q}}{4\pi} \left[\left(\frac{r_o^2 + r_i^2}{2} \right) \ln \frac{r_i}{r_o} + \frac{r_o - r_i}{2} \right]
\end{aligned}
\tag{2.45}
$$

The boundary conditions at the inner and outer edges of the plate are related to the above expressions in the following manner. By direct substitution, the general solution of dw/dr in Eq. (2.43) may be shown to be expressed

$$
\frac{dw}{dr} = \left(\frac{dw}{dr} \right)' + k \left(\frac{dw}{dr} \right)'' \tag{2.46}
$$

where the single prime denotes the *particular* solution, the double prime indicates the *homogeneous* solution, and k represents a constant. The same relationship applies to the moments given by Eqs. (f). Referring to Table 2.4, note that the particular and homogeneous solutions may be selected so as to satisfy the conditions imposed at the inner edge of the disk regardless of the value of k. The constant k can thus be selected to satisfy the prescribed condition at the outer edge only, as indicated in Table 2.5.

PROBLEMS

Secs. 2.1 to 2.8

2.1 A pressure control system includes a thin steel disk which is to close an electrical circuit by deflecting 1 mm at the center when the pressure attains a value of 3 MPa. Calculate the required disk thickness if it has a radius of 0.030 m and is built-in at the edge. Let $\nu = 0.3$ and $E = 200$ GPa.

2.2 A cylindrical thick-walled vessel of 0.25 m radius and flat thin plate head is subjected to an internal pressure 7 MPa. Determine: (a) the thickness of the cylinder head if the allowable stress is limited to 90 MPa; (b) the maximum deflection of the cylinder head. Use $E = 200$ GPa and $\nu = 0.3$.

2.3 An aluminum alloy (6061-T6) flat simply supported disk valve of 0.2 m diameter and 10 mm thickness is subject to a water pressure of 0.5 MPa. What is the factor of safety, assuming failure to take place according to the maximum principal stress theory. The yield strength of the material is 241 MPa.

2.4 The flat head of a piston is considered to be a clamped circular plate of radius a. The head is under a pressure

$$p = p_0\left(\frac{r}{a}\right)^2$$

where p_0 is constant. Derive the equation

$$w = \frac{p_0 a^4}{576D}\left[\left(\frac{r}{a}\right)^6 - 3\left(\frac{r}{a}\right)^2 + 2\right] \tag{P2.4}$$

for the resulting displacement.

2.5 to 2.7 For the circular plates loaded as shown in the Figs. A, B, and C of Table 2.2, verify the results provided for the maximum deflections.

2.8 A simply supported circular plate is under a rotationally symmetric lateral load which increases from the center to the edge:

$$p = p_0\frac{r}{a}$$

Show that the expression

$$w = \frac{p_0 r^5}{225Da} + c_4 + c_3 r^2 \tag{P2.8}$$

where

$$c_4 = \frac{p_0 a^4}{45D}\left[\frac{4+\nu}{2(1+\nu)} - \frac{1}{5}\right]$$

$$c_3 = -\frac{p_0 a^2}{90D}\left(\frac{4+\nu}{1+\nu}\right)$$

represent the resulting displacement.

2.9 Verify the result given by Eq. (2.8). [*Hint:* Introduction of Eq. (2.7) into Eq. (2.6) leads to

$$\frac{d^4 f_n}{dr^4} + \frac{2}{r}\frac{d^3 f_n}{dr^3} - \frac{1+2n^2}{r^2}\frac{d^2 f_n}{dr^2} + \frac{1+2n^2}{r^3}\frac{df_n}{dr} + \frac{n^2(n^2-4)}{r^4} f_n = 0 \qquad (n = 0, 1, 2, \ldots)$$

$$\frac{d^4 f_n^*}{dr^4} + \frac{2}{r}\frac{d^3 f_n^*}{dr^3} - \frac{1+2n^2}{r^2}\frac{d^2 f_n^*}{dr^2} + \frac{1+2n^2}{r^3}\frac{df_n^*}{dr} + \frac{n^2(n^2-4)}{r^4} f_n^* = 0 \qquad (n = 1, 2, \ldots) \tag{P2.9}$$

Solution of these equidimensional equations can be taken as:[10] $f_n(r) = a_n r^\lambda$ and $f_n^*(r) = b_n r^\lambda$, wherein a_n and b_n are constants and the λ's are the roots of auxiliary equation of Eqs. (P2.9).]

2.10 Verify the result given by Eq. (2.13): (*a*) by integrating Eq. (2.12); (*b*) by expanding Eq. (2.10*a*), setting $t = \ln r$, and thereby transforming the resulting expression into an ordinary differential equation with constant coefficients.

2.11 Show that Eq. (1.34) for the strain energy results in the following form in terms of the polar coordinates

$$U = \frac{D}{2}\iint_A \left[\left(\frac{\partial^2 w}{\partial r^2} + \frac{1}{r}\frac{\partial w}{\partial r} + \frac{1}{r^2}\frac{\partial^2 w}{\partial \theta^2}\right)^2 - 2(1-\nu)\frac{\partial^2 w}{\partial r^2}\left(\frac{1}{r}\frac{\partial w}{\partial r} + \frac{1}{r^2}\frac{\partial^2 w}{\partial \theta^2}\right)\right.$$
$$\left. + 2(1-\nu)\left(\frac{1}{r}\frac{\partial^2 w}{\partial r\,\partial \theta} - \frac{1}{r^2}\frac{\partial w}{\partial \theta}\right)^2\right] r\,dr\,d\theta \tag{P2.11}$$

2.12 Calculate the maximum deflection w in the annular plate loaded as shown in Fig. 2.5*a* by setting $a = 2b$, $M_1 = 2M_2$, and $\nu = 0.3$.

2.13 Determine the maximum displacement in the annular plate loaded as shown in Fig. 2.5*b* by setting $a = 2b$ and $\nu = 0.3$.

2.14 A pump diaphragm can be approximated as an annular plate under a uniformly distributed surface load p_0 and with outer edge simply supported (case A in Table 2.3). Compute, using the method of superposition, the maximum plate deflection for $b = a/4$ and $\nu = 0.3$. Compare the result with that given in the table.

Secs. 2.9 to 2.12

2.15 An aircraft window is approximated as a simply supported circular plate of radius a. The window is subject to a uniform cabin pressure p_0. Determine its maximum deflection, assuming that a diametrical section of the bent plate is parabolic. Employ the Ritz method. Take $\nu = 0.3$.

2.16 Redo Prob. 2.15 for a simply supported circular plate that is loaded only by a concentrated center force P.

2.17 Determine the maximum deflection of a structural steel circular plate with free end resting on a gravel-sand mixture foundation and submitted to a load P at its center. Use the Ritz method, taking the first three terms in Eq. (*a*) of Sec. 2.9. Let $k = 200$ MPa/m, $a = 0.5$ m, $t = 40$ mm, $E = 200$ GPa, and $\nu = 0.3$.

2.18 A simply supported circular plate is loaded by asymmetrically distributed edge couples described by

$$M_r = \sum_{n=1,2,\ldots}^{\infty} M_n \cos n\theta \qquad (r = a)$$

In this case, it is observed that $w_p = 0$ and thus $w = w_h$ reduces to

$$w = \sum_{n=0,1,\ldots}^{\infty} (A_n r^n + c_n r^{n+2}) \cos n\theta$$

Verify that the resulting deflection is

$$w = -\frac{1}{2D} \sum_{n=0,1,\ldots}^{\infty} \frac{M_n(r^{n+2} - a^2 r^n)}{a^n(2n + 1 + \nu)} \cos n\theta \tag{P2.18}$$

2.19 Determine the expression for the radial stress in the plate described in Prob. 2.18 by taking $n = 0, 1$.

2.20 The particular solution of Eq. (2.5), for an arbitrary loading $p(r, \theta)$ expanded in a Fourier series

$$p(r, \theta) = p_0(r) + \sum_{n=1}^{\infty} [P_n(r) \cos n\theta + R_n(r) \sin n\theta] \tag{a}$$

where

$$P_n(r) = \frac{1}{\pi} \int_{-\pi}^{\pi} p(r, \theta) \cos n\theta \, d\theta \qquad (n = 0, 1, \ldots)$$

$$R_n(r) = \frac{1}{\pi} \int_{-\pi}^{\pi} p(r, \theta) \sin n\theta \, d\theta \qquad (n = 1, 2, \ldots)$$

is expressed in the general form:[6]

$$w_p = F_0(r) + \sum_{n=1}^{\infty} [F_n(r) \cos n\theta + G_n(r) \sin n\theta \tag{P2.20}$$

Here $F_0(r)$, $F_n(r)$, and $G_n(r)$ are functions of r. Demonstrate that substitution of Eqs. (a) and (P2.20) into Eq. (2.5) leads to

$$\frac{d^4F_0}{dr^4} + \frac{2}{r}\frac{d^3F_0}{dr^3} - \frac{1}{r^2}\frac{d^2F_0}{dr^2} + \frac{1}{r^3}\frac{dF_0}{dr} = \frac{p_0}{D}$$

$$\frac{d^4F_0}{dr^4} + \frac{2}{r}\frac{d^3F_n}{dr^3} - \frac{1 + 2n^2}{r^2}\frac{d^2F_n}{dr^2} + \frac{1 + 2n^2}{r^2}\frac{dF_n}{dr} + \frac{n^2(n^2 - 4)}{r^4} F_n = \frac{P_n}{D} \tag{b}$$

$$\frac{d^4G_n}{dr^4} + \frac{2}{r}\frac{d^3G_n}{dr^3} - \frac{1 + 2n^2}{r^2}\frac{d^2G_n}{dr^2} + \frac{1 + 2n^2}{r^2}\frac{dG_n}{dr} + \frac{n^2(n^2 - 4)}{r^4} G_n = \frac{R_n}{D}$$

Thus, for a prescribed loading $p(r, \theta)$, Eqs. (b) are solved for F_0, F_n, and G_n. The particular solution is then obtained from Eq. (P2.20).

2.21 A simply supported circular plate is subjected to hydrostatic loading. Employ the reciprocity theorem to find the maximum deflection.

2.22 A clamped circular plate carries a concentrated downward load P at a point located at a distance b from its center. Apply the reciprocity theorem to obtain the center deflection.

2.23 Redo Prob. 2.22 for the case of the plate with simply supported edge.

2.24 For the turbine disk shown in Fig. 2.10*a*, compute the stress distribution and the deflections, assuming the disk to be clamped at the inner edge and free at the outer edge.

CHAPTER

THREE

RECTANGULAR PLATES

3.1 INTRODUCTION

In this chapter consideration is given stresses and deflections in thin rectangular plates. As observed in Chap. 1, the rectangular-plate element is an excellent model for development of the basic relationships in cartesian coordinates. On the other hand, we shall see that rectangular plates in bending frequently lead to solutions in the form of series that are unsuited to hand computation of numerical values. That is, the deflections and moments are often described by unwieldy infinite series, and the summation of these series presents difficulties. Such computations are, of course, readily performed by digital computer.

Rectangular plates are generally classified in accordance with the type of support used. We are here concerned with the bending of simply supported plates, clamped or built-in plates, plates having *mixed* support conditions, plates on an elastic foundation, and continuous plates. The latter often refer to structures consisting of a single plate supported by intermediate beams or columns. All cases are treated by relationships derived in Chap. 1. The strip method of Sec. 3.8 is a discussion of bending of rectangular plates based upon elementary beam theory.

3.2 NAVIER'S SOLUTION FOR SIMPLY SUPPORTED RECTANGULAR PLATES

Consider the rectangular plate of sides a and b, simply supported on all edges and subjected to a distributed load $p(x, y)$. The origin of coordinates is placed at the upper left corner of the plate as shown in Fig. 3.1a. In general, solution of the

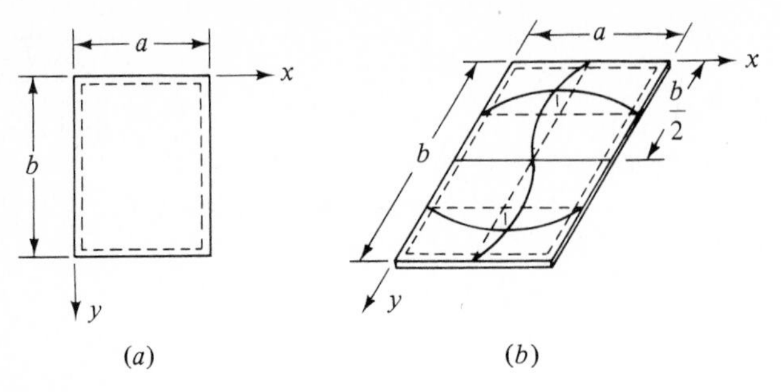

Figure 3.1

bending problem employs the following *Fourier series* (App. A) for load and deflection:

$$p(x, y) = \sum_{m=1}^{\infty} \sum_{n=1}^{\infty} p_{mn} \sin \frac{m\pi x}{a} \sin \frac{n\pi y}{b} \tag{3.1a}$$

$$w(x, y) = \sum_{m=1}^{\infty} \sum_{n=1}^{\infty} a_{mn} \sin \frac{m\pi x}{a} \sin \frac{n\pi y}{b} \tag{3.1b}$$

where p_{mn} and a_{mn} represent coefficients to be determined. This approach was introduced by Navier in 1820. The deflection must satisfy the differential equation (1.17) with the following boundary conditions (1.26*b*)

$$\begin{aligned} w = 0 \qquad \frac{\partial^2 w}{\partial x^2} = 0 \qquad (x = 0,\ x = a) \\ w = 0 \qquad \frac{\partial^2 w}{\partial y^2} = 0 \qquad (y = 0,\ y = b) \end{aligned} \tag{a}$$

Clearly, these edge restraints are fulfilled by Eq. (3.1*b*) and the coefficients a_{mn} must be such as to satisfy Eq. (1.17). The solution corresponding to loading $p(x, y)$ thus requires determination of p_{mn} and a_{mn}.

As a physical interpretation of Eq. (3.1*b*), consider the true deflection surface of the plate to be the superposition of sinusoidal curves of m and n different configurations in the x and y directions, respectively. The coefficients a_{mn} of the series are the maximum central coordinates of the sine curves, and the m's and the n's indicate the number of half-sine curves in the x and y directions, respectively. For example, the term $a_{12} \sin (\pi x/a) \sin (2\pi y/b)$ of the series is illustrated in Fig. 3.1*b*. By increasing the number of terms in the series, the accuracy can, of course, be improved.

We proceed by dealing first with a general load configuration, subsequently treating specific loadings. To determine the *Fourier coefficients* p_{mn}, each side of Eq. (3.1*a*) is multiplied by

$$\sin \frac{m'\pi x}{a} \sin \frac{n'\pi y}{b}\, dx\, dy$$

and integrated between limits 0, a and 0, b:

$$\int_0^b \int_0^a p(x, y) \sin \frac{m'\pi x}{a} \sin \frac{n'\pi y}{b} \, dx \, dy$$

$$= \sum_{m=1}^{\infty} \sum_{n=1}^{\infty} p_{mn} \int_0^b \int_0^a \sin \frac{m\pi x}{a} \sin \frac{n\pi y}{b} \sin \frac{m'\pi x}{a} \sin \frac{n'\pi y}{b} \, dx \, dy \qquad (b)$$

It can be shown by direct integration that

$$\int_0^a \sin \frac{m\pi x}{a} \sin \frac{m'\pi x}{a} \, dx = \begin{cases} 0 & (m \neq m') \\ a/2 & (m = m') \end{cases}$$
$$\int_0^b \sin \frac{n\pi y}{b} \sin \frac{n'\pi y}{b} \, dy = \begin{cases} 0 & (n \neq n') \\ b/2 & (n = n') \end{cases} \qquad (3.2)$$

The coefficients of the double Fourier expansion are therefore

$$p_{mn} = \frac{4}{ab} \int_0^b \int_0^a p(x, y) \sin \frac{m\pi x}{a} \sin \frac{n\pi y}{b} \, dx \, dy \qquad (3.3)$$

Evaluation of a_{mn} in Eq. (3.1b) requires substitution of Eqs. (3.1) into Eq. (1.17), with the result:

$$\sum_{m=1}^{\infty} \sum_{n=1}^{\infty} \left\{ a_{mn} \left[\left(\frac{m\pi}{a} \right)^4 + 2 \left(\frac{m\pi}{a} \right)^2 \left(\frac{n\pi}{b} \right)^2 + \left(\frac{n\pi}{b} \right)^4 \right] - \frac{p_{mn}}{D} \right\} \sin \frac{m\pi x}{a} \sin \frac{n\pi y}{b} = 0$$

This equation must apply for all x and y. We conclude therefore that

$$a_{mn} \pi^4 \left(\frac{m^2}{a^2} + \frac{n^2}{b^2} \right)^2 - \frac{p_{mn}}{D} = 0$$

from which

$$a_{mn} = \frac{1}{\pi^4 D} \frac{p_{mn}}{[(m/a)^2 + (n/b)^2]^2} \qquad (3.4)$$

Substituting Eq. (3.4) into Eq. (3.1b), the equation of the deflection surface of the plate becomes

$$w = \frac{1}{\pi^4 D} \sum_{m=1}^{\infty} \sum_{n=1}^{\infty} \frac{p_{mn}}{[(m/a)^2 + (n/b)^2]^2} \sin \frac{m\pi x}{a} \sin \frac{n\pi y}{b} \qquad (3.5)$$

in which p_{mn} is given by Eq. (3.3). It can be shown, by noting that $|\sin (m\pi x/a)| \leq 1$ and $|\cos (n\pi y/b)| \leq 1$ for every x and y and for m and n, that series (3.5) is convergent. Thus Eq. (3.5) is a valid solution for bending of simply supported rectangular plates under various kinds of loadings. Application of the Navier's method to several particular cases is presented in the next section.

3.3 SIMPLY SUPPORTED RECTANGULAR PLATES UNDER VARIOUS LOADINGS

When a rectangular plate is subjected to a *uniformly distributed load* $p(x, y) = p_0$, the results of the previous section are simplified considerably. Now Eq. (3.3), after integration, yields

$$p_{mn} = \frac{4p_0}{\pi^2 mn}(1 - \cos m\pi)(1 - \cos n\pi) = \frac{4p_0}{\pi^2 mn}[1 - (-1)^m][1 - (-1)^n]$$

or

$$p_{mn} = \frac{16p_0}{\pi^2 mn} \qquad (m, n = 1, 3, \ldots) \tag{a}$$

It is observed that because $p_{mn} = 0$ for even values of m and n, these integers assume only odd values. Introducing p_{mn} into Eq. (3.5), we have

$$w = \frac{16p_0}{\pi^6 D} \sum_m^\infty \sum_n^\infty \frac{\sin(m\pi x/a)\sin(n\pi y/b)}{mn[(m/a)^2 + (n/b)^2]^2} \qquad (m, n = 1, 3, \ldots) \tag{3.6}$$

Clearly, based upon physical considerations, the uniformly loaded plate must deflect into a symmetrical shape. Such a configuration results when m and n are odd. The maximum deflection occurs at the center of the plate ($x = a/2$, $y = b/2$) and its value, from Eq. (3.6), is

$$w_{\max} = \frac{16p_0}{\pi^6 D} \sum_m^\infty \sum_n^\infty \frac{(-1)^{[(m+n)/2]-1}}{mn[(m/a)^2 + (n/b)^2]^2} \tag{b}$$

Note that in Eq. (3.6), $\sin m\pi/2$ and $\sin n\pi/2$ are replaced by $(-1)^{(m-1)/2}$ and $(-1)^{(n-1)/2}$, respectively.

Introducing Eq. (3.6) into Eqs. (1.10), the components of the moment are derived:

$$\begin{aligned} M_x &= \frac{16p_0}{\pi^4} \sum_m^\infty \sum_n^\infty \frac{(m/a)^2 + \nu(n/b)^2}{mn[(m/a)^2 + (n/b)^2]^2} \sin\frac{m\pi x}{a} \sin\frac{n\pi y}{b} \\ M_y &= \frac{16p_0}{\pi^4} \sum_m^\infty \sum_n^\infty \frac{\nu(m/a)^2 + (n/b)^2}{mn[(m/a)^2 + (n/b)^2]^2} \sin\frac{m\pi x}{a} \sin\frac{n\pi y}{b} \\ M_{xy} &= -\frac{16(1-\nu)}{\pi^4 ab} \sum_m^\infty \sum_n^\infty \frac{1}{[(m/a)^2 + (n/b)^2]^2} \cos\frac{m\pi x}{a} \cos\frac{n\pi y}{b} \end{aligned} \tag{3.7}$$

We observe that the bending moments M_x and M_y are both zero at ($x = 0$, $x = a$) and ($y = 0$, $y = b$) respectively. However, the twisting moment M_{xy} does not vanish at the edges and at the corners of the plate. The presence of M_{xy} causes a modification of the distribution of the reactions on the supports (Sec. 1.7). Recall, however, that St. Venant's principle permits us to regard the stress distribution unaltered for sections away from the edges and the corners.

Situations involving simply supported rectangular plates carrying sinusoidal, partial, and concentrated loadings are discussed in the examples 3.2 and 3.3.

Example 3.1 A square wall-panel is taken to be simply supported on all edges and subjected to a uniform pressure differential p_0. Determine the maximum deflection, moment, and stress.

SOLUTION The first term ($m = 1, n = 1$) of Eq. (b) yields, for $a = b$:

$$w_{\max} = 0.00416 p_0 \frac{a^4}{D}$$

Very rapid convergence of Eq. (b) is demonstrated by noting that retaining the first four terms ($m = 1$, $n = 1$, 3; $m = 3$, $n = 1$, 3) results in what is essentially the "exact" solution, $w_{\max} = 0.00406 p_0 a^4/D$.

The maximum bending moments, found at the center of the plate, are determined by applying Eqs. (3.7). The first term of the series yields

$$M_{x,\max} = M_{y,\max} = 0.0534 p_0 a^2$$

while the first four terms result in

$$M_{x,\max} = M_{y,\max} = 0.0469 p_0 a^2 \tag{c}$$

It is thus observed from a comparison of the above that the series for the bending moments given by Eqs. (3.7) does not converge as rapidly as that of Eq. (b). The maximum bending stress produced by the moment of Eq. (c), by application of Eqs. (1.12), is determined to be $\sigma_{\max} = 0.281 p_0 a^2/t^2$.

Example 3.2 A rectangular warehouse floor slab of sides a and b is simply supported on all edges. Determine the reactions at the supports if the material is stored on the entire floor in such a way that the loading is expressed in the following approximate form

$$p(x, y) = p_0 \sin \frac{\pi x}{a} \sin \frac{\pi y}{b} \tag{d}$$

Here p_0 represents the intensity of the load at the center of the plate, as shown in Fig. 3.2.

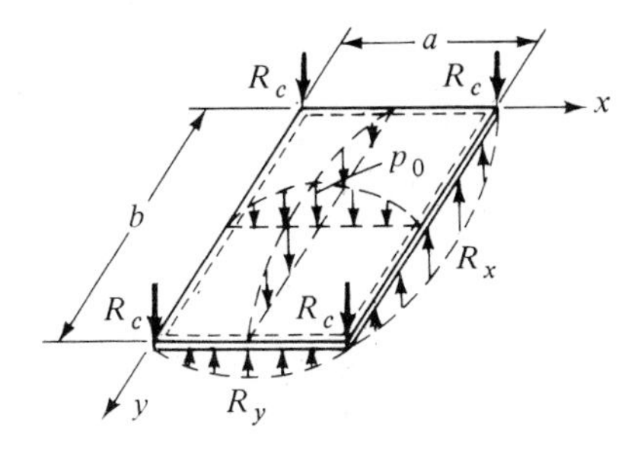

Figure 3.2

SOLUTION Substituting Eq. (d) into Eq. (3.3) and letting $m = n = 1$, we have $p_{mn} = p_0$. Expression (3.5) in this case leads to

$$w = \frac{p_0}{\pi^4 D(1/a^2 + 1/b^2)^2} \sin \frac{\pi x}{a} \sin \frac{\pi y}{b}$$

Introducing w from the above into Eqs. (1.10), we have

$$\begin{aligned} M_x &= \frac{p_0}{\pi^2(1/a^2 + 1/b^2)^2}\left(\frac{1}{a^2} + \frac{\nu}{b^2}\right) \sin \frac{\pi x}{a} \sin \frac{\pi y}{b} \\ M_y &= \frac{p_0}{\pi^2(1/a^2 + 1/b^2)^2}\left(\frac{\nu}{a^2} + \frac{1}{b^2}\right) \sin \frac{\pi x}{a} \sin \frac{\pi y}{b} \\ M_{xy} &= -\frac{p_0(1-\nu)}{\pi^2(1/a^2 + 1/b^2)^2 ab} \cos \frac{\pi x}{a} \cos \frac{\pi y}{b} \end{aligned} \tag{e}$$

The shear forces are determined from Eqs. (1.16)

$$\begin{aligned} Q_x &= \frac{p_0}{\pi a(1/a^2 + 1/b^2)} \cos \frac{\pi x}{a} \sin \frac{\pi y}{b} \\ Q_y &= \frac{p_0}{\pi b(1/a^2 + 1/b^2)} \sin \frac{\pi x}{a} \cos \frac{\pi y}{b} \end{aligned} \tag{f}$$

The total load carried by the plate, neglecting its weight, is equal to

$$\int_0^b \int_0^a p_0 \sin \frac{\pi x}{a} \sin \frac{\pi y}{b}\, dx\, dy = \frac{4p_0 ab}{\pi^2} \tag{g}$$

The reactions can act only vertically inasmuch as the edges sit on rollers. Equation (1.23a) is applied to obtain the *reaction* $R_x = -V_x$ for the edge $x = a$:

$$R_x = -\frac{p_0}{\pi a(1/a^2 + 1/b^2)^2}\left(\frac{1}{a^2} + \frac{2-\nu}{b^2}\right) \sin \frac{\pi y}{b} \tag{3.8}$$

Similarly, for the edge $y = b$, setting $R_y = -V_y$ from Eq. (1.23b), one has

$$R_y = -\frac{p_0}{\pi b(1/a^2 + 1/b^2)^2}\left(\frac{1}{b^2} + \frac{2-\nu}{a^2}\right) \sin \frac{\pi x}{a} \tag{3.9}$$

The edge support reactions thus also vary sinusoidally (Fig. 3.2). The minus sign indicates that they are directed upward as shown in the figure. Conditions of symmetry dictate that the reactions along the support at $x = 0$ and $y = 0$ are identical with those given by Eqs. (3.8) and (3.9), respectively.

The sum of distributed reactions

$$2\int_0^a R_y\, dx + 2\int_0^b R_x\, dy$$

is, after substituting Eqs. (3.8) and (3.9) and integrating,

$$-\frac{4p_0 ab}{\pi^2} - \frac{8p_0(1-\nu)}{\pi^2 ab(1/a^2 + 1/b^2)^2} \tag{h}$$

It is seen from Eqs. (*g*) and (*h*) that the distributed reactions are larger than necessary to compensate for the p loading. Thus, as already shown in Sec. 1.7, there will be a concentrated reaction $R_c = -F_c$ at each corner having a value (Eq. 1.24)

$$R_c = \frac{2p_0(1-\nu)}{\pi^2 ab(1/a^2 + 1/b^2)^2} \tag{3.10}$$

The condition that the resultant of the laterally applied and the reactional forces in the plate equal to zero is thus satisfied (Fig. 3.2).

The reason for the corner reactions can be recognized intuitively. When a rectangular plate, supported freely by a rectangular frame, is subjected to center loading, it tends to deform in such a way as to have the corners rise, with contact made only at the middle of the sides. To prevent this, the corners must be held down.

Example 3.3 Find the equations of the elastic surface of a simply supported rectangular plate (Fig. 3.3) for two particular cases: (*a*) the plate is subjected to a load P uniformly distributed over a subregion $4cd$; (*b*) the plate carries a (nominal) point load P at $x = x_1$, $y = y_1$.

SOLUTION (*a*) Since $p(x, y) = P/4cd$, we have through the use of Eq. (3.3)

$$p_{mn} = \frac{P}{abcd}\int_{y_1-d}^{y_1+d}\int_{x_1-c}^{x_1+c} \sin\frac{m\pi x}{a}\sin\frac{n\pi y}{b}\,dx\,dy$$

from which

$$p_{mn} = \frac{4P}{\pi^2 mncd}\sin\frac{m\pi x_1}{a}\sin\frac{n\pi y_1}{b}\sin\frac{m\pi c}{a}\sin\frac{n\pi d}{b} \tag{i}$$

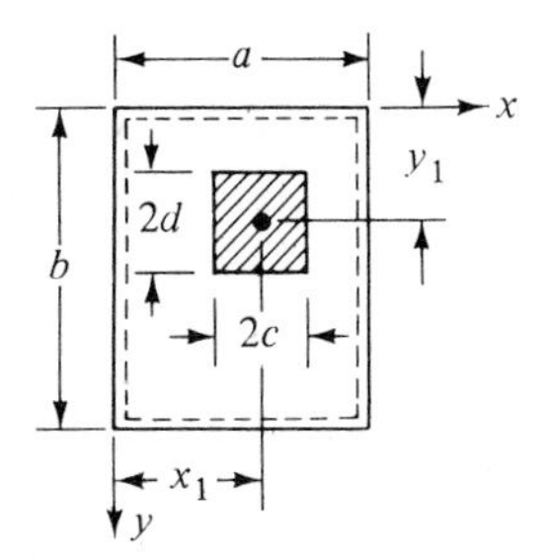

Figure 3.3

Clearly, for $x_1 = a/2$, $y_1 = b/2$, $c = a/2$, and $d = b/2$, the foregoing reduces to Eq. (*a*). Introduction of Eq. (*i*) into Eq. (3.1*b*) gives the required deflection surface.

(*b*) When c and d are made to approach zero Eq. (*i*) appears in the form (Prob. 3.5)

$$p_{mn} = \frac{4P}{ab} \sin \frac{m\pi x_1}{a} \sin \frac{n\pi y_1}{b} \tag{j}$$

Note that the load P is an idealization of the distributed load $p(x, y)$ concentrated in a very small area of size defined by Eq. (2.24). Inserting Eq. (3.4) together with Eq. (*j*) into Eq. (3.1*b*) results in the plate deflection:

$$w = \frac{4P}{\pi^4 Dab} \sum_m^\infty \sum_n^\infty \frac{\sin(m\pi x_1/a) \sin(n\pi y_1/b)}{[(m/a)^2 + (n/b)^2]^2} \sin \frac{m\pi x}{a} \sin \frac{n\pi y}{b} \tag{3.11}$$

Eq. (3.11) may be used with the method of superposition (Secs. 2.8 and 3.7) to determine the deflection of simply supported rectangular plates under various types of loadings.

When the load P is applied at the center of the plate ($x_1 = a/2$, $y_1 = b/2$), Eq. (3.11) reduces to

$$w = \frac{4P}{\pi^4 Dab} \sum_m^\infty \sum_n^\infty (-1)^{[(m+n)/2]-1} \frac{\sin(m\pi x/a) \sin(n\pi y/b)}{[(m/a)^2 + (n/b)^2]^2} \tag{3.12}$$

Furthermore, if the plate is square ($a = b$), the maximum deflection, which occurs at the center, is from Eq. (3.12):

$$w_{\max} = \frac{4Pa^2}{\pi^4 D} \sum_m^\infty \sum_n^\infty \frac{1}{(m^2 + n^2)^2} \tag{k}$$

Retaining the first nine terms of this series ($m = 1$, $n = 1, 3, 5$; $m = 3$, $n = 1, 3, 5$; $m = 5$, $n = 1, 3, 5$) we obtain

$$w_{\max} = \frac{4Pa^2}{\pi^4 D} \left[\frac{1}{2^2} + \frac{2}{10^2} + \frac{1}{18^2} + \frac{2}{25^2} + \frac{2}{34^2} + \frac{1}{50^2} \right] = 0.01142 \frac{Pa^2}{D}$$

The "exact" value is $w_{\max} = 0.01159Pa^2/D$ and the error is thus 1.56 percent.

3.4 LÉVY'S SOLUTION FOR RECTANGULAR PLATES

It is seen in the foregoing section that the calculation of bending moments using Eq. (3.7) is not very satisfactory because of the slow convergence of the series. An important approach which overcomes this difficulty was developed by M. Lévy in 1900. Another advantage of the Lévy's solution as compared with the Navier's methods is that instead of a double series one has to deal with a single

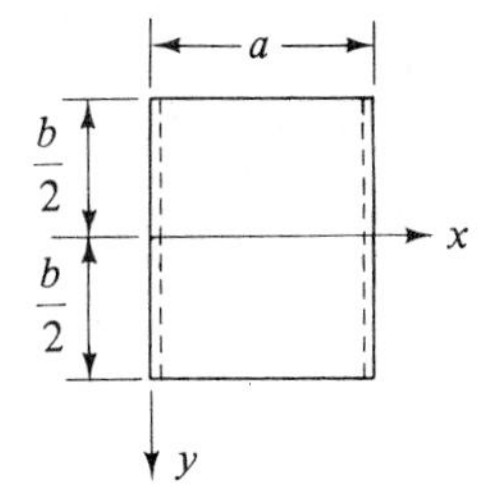

Figure 3.4

series (App. A). In general, it is easier to perform numerical calculations for single series than for double series.

The Lévy's method is applicable to the bending of rectangular plates with *particular* boundary conditions on the two opposite sides (say, at $x = 0$ and $x = a$) and *arbitrary* conditions of support on the remaining edges (at $y = \pm b/2$) (Fig. 3.4). The total solution consists of the homogeneous solution w_h of Eq. (1.18) and the particular solution w_p of Eq. (1.17):

$$w = w_h + w_p \tag{a}$$

Inasmuch as $\nabla^4 w_h = 0$ is independent of the loading, a single expression can be derived for w_h that is valid for all rectangular plates having particular boundary conditions on the two opposite sides. Clearly, *for each specific loading $p(x, y)$ a solution w_p must be obtained.*

The homogeneous solution is selected of the following general form

$$w_h = \sum_{m=1}^{\infty} f_m(y) \begin{cases} \sin (m\pi x/a) \\ \cos (m\pi x/a) \end{cases} \tag{b}$$

where the function $f_m(y)$ must be obtained such as to fulfill the conditions of the supports at $y = \pm b/2$, and to satisfy Eq. (1.18). We proceed with the description of the method by assuming that *two opposite sides* of the rectangular plate at $x = 0$ and $x = a$ are *simply supported* as shown in Fig. 3.4. In this case, Eq. (*b*) appears as

$$w_h = \sum_{m=1}^{\infty} f_m(y) \sin \frac{m\pi x}{a} \tag{3.13}$$

Eq. (3.13) satisfies the conditions (1.26*b*) along the x edges. To complete the solution, we must now apply to w the boundary conditions on the *two arbitrary sides* at $y = \pm b/2$.

Substituting Eq. (3.13) into Eq. (1.18) yields

$$\sum_{m=1,3,\ldots}^{\infty} \left[\frac{d^4 f_m}{dy^4} - 2\left(\frac{m\pi}{a}\right)^2 \frac{d^2 f_m}{dy^2} + \left(\frac{m\pi}{a}\right)^4 f_m \right] \sin \frac{m\pi x}{a} = 0$$

In order that this equation be valid for any x,

$$\frac{d^4 f_m}{dy^4} - 2\left(\frac{m\pi}{a}\right)^2 \frac{d^2 f_m}{dy^2} + \left(\frac{m\pi}{a}\right)^4 f_m = 0$$

The general solution of the above is (see Prob. 3.9)

$$f_m = A'_m e^{m\pi y/a} + B'_m e^{-m\pi y/a} + C'_m y e^{m\pi y/a} + D'_m y e^{-m\pi y/a} \tag{3.14}$$

or by employing trigonometric identities

$$f_m = A_m \sinh \frac{m\pi y}{a} + B_m \cosh \frac{m\pi y}{a} + C_m y \sinh \frac{m\pi y}{a} + D_m y \cosh \frac{m\pi y}{a} \tag{3.15}$$

The homogeneous solution is therefore

$$w_h = \sum_{m=1}^{\infty} \left(A_m \sinh \frac{m\pi y}{a} + B_m \cosh \frac{m\pi y}{a} + C_m y \sinh \frac{m\pi y}{a} + D_m y \cosh \frac{m\pi y}{a} \right) \sin \frac{m\pi x}{a} \tag{3.16}$$

where A_m, B_m, C_m, and D_m are constants, to be determined later for specified cases.

It is observed that the boundary conditions (1.26*b*) along the edges $x = 0$ and $x = a$ are satisfied if the particular solution is expressed by the following single Fourier series

$$w_p = \sum_{m=1}^{\infty} k_m(y) \sin \frac{m\pi x}{a} \tag{3.17}$$

where the $k_m(y)$'s are functions of y only. Let us also expand $p(x, y)$ in terms of a single Fourier series

$$p(x, y) = \sum_{m=1}^{\infty} p_m(y) \sin \frac{m\pi x}{a} \tag{3.18}$$

wherein

$$p_m(y) = \frac{2}{a} \int_0^a p(x, y) \sin \frac{m\pi x}{a}\, dx \tag{3.19}$$

Substituting Eqs. (3.17) and (3.18) into Eq. (1.17) and noting the validity of the resulting expression for all values of x between 0 and a, we find that

$$\frac{d^4 k_m}{dy^4} - 2\left(\frac{m\pi}{a}\right)^2 \frac{d^2 k_m}{dy^2} + \left(\frac{m\pi}{a}\right)^4 k_m = \frac{p_m}{D} \tag{c}$$

Upon determination of a particular solution, k_m, to this ordinary differential equation, we then obtain w_p from Eq. (3.17). The method is illustrated by considering the following commonly referred to example.

Simply Supported Rectangular Plate Under Uniform Loading

For this case $p(x, y) = p_0$ for which Eq. (3.19) yields, upon integration

$$p_m = \frac{4p_0}{m\pi} \qquad (m = 1, 3, \ldots) \tag{d}$$

Then Eq. (c) becomes

$$\frac{d^4k_m}{dy^4} - 2\left(\frac{m\pi}{a}\right)^2 \frac{d^2k_m}{dy^2} + \left(\frac{m\pi}{a}\right)^4 k_m = \frac{4p_0}{m\pi D}$$

The particular solution of the above is $k_m = 4p_0 a^4/m^5\pi^5 D$. Expression (3.17) is therefore

$$w_p = \frac{4p_0 a^4}{\pi^5 D} \sum_{m=1}^{\infty} \frac{1}{m^5} \sin \frac{m\pi x}{a} \tag{3.20a}$$

This represents the deflection of a uniformly loaded, *simply supported strip* parallel to the x axis and may be rewritten in the following alternate form (see Example 1.1):

$$w_p = \frac{p_0}{24D}(x^4 - 2ax^3 + a^3x) \tag{3.20b}$$

The condition that the plate deflection must be symmetrical with respect to the x axis [i.e., it must have the same values for $+y$ and $-y$ (Fig. 3.4)] is satisfied by Eq. (3.16) if we let $A_m = D_m = 0$. Then, combining Eqs. (3.16) and (3.20) we have

$$w = \sum_{m=1,3,\ldots} \left(B_m \cosh \frac{m\pi y}{a} + C_m y \sinh \frac{m\pi y}{a} + \frac{4p_0 a^4}{m^5\pi^5 D}\right) \sin \frac{m\pi x}{a} \tag{3.21}$$

This expression satisfies Eq. (1.17) and the simple support restraints at $x = 0$, $x = a$. The remaining edge conditions are

$$w = 0 \qquad \frac{\partial^2 w}{\partial y^2} = 0 \qquad \left(y = \pm\frac{b}{2}\right)$$

Application of the above to w leads to two expressions. These will be satisfied for all values of x when

$$\begin{aligned} &B_m \cosh \alpha_m + C_m \frac{b}{2} \sinh \alpha_m + \frac{4p_0 a^4}{m^5\pi^5 D} = 0 \\ &2\left(\frac{B_m \alpha_m}{b} + C_m\right) \cosh \alpha_m + C_m \alpha_m \sinh \alpha_m = 0 \end{aligned} \tag{e}$$

in which

$$\alpha_m = \frac{m\pi b}{2a}$$

Solution of Eqs. (e) gives the unknown constants

$$B_m = -\frac{4p_0 a^4 + m\pi p_0 a^3 b \tanh \alpha_m}{m^5\pi^5 D \cosh \alpha_m} \qquad C_m = \frac{2p_0 a^3}{m^4\pi^4 D \cosh \alpha_m}$$

The deflection surface of the plate [Eq. (3.21)] may thus be expressed

$$w = \frac{4p_0 a^4}{\pi^5 D} \sum_{m=1,3,\ldots}^{\infty} \frac{1}{m^5} \left[1 - \frac{\alpha_m \tanh a_m + 2}{2 \cosh \alpha_m} \cosh \frac{2\alpha_m y}{b} + \frac{1}{2 \cosh \alpha_m} \frac{m\pi y}{a} \sinh \frac{2\alpha_m y}{b} \right] \sin \frac{m\pi x}{a} \tag{3.22}$$

The maximum displacement occurs at the center of the plate ($x = a/2$, $y = 0$). That is, from Eq. (3.22),

$$w_{\max} = \frac{4p_0 a^4}{\pi^5 D} \sum_{m=1,3,\ldots}^{\infty} \frac{(-1)^{(m-1)/2}}{m^5} \left[1 - \frac{\alpha_m \tanh \alpha_m + 2}{2 \cosh \alpha_m} \right]$$

Since

$$\sum_{m=1,3,\ldots}^{\infty} \frac{(-1)^{(m-1)/2}}{m^5} = \frac{5\pi^5}{2^9(3)}$$

we can write the following expression for the maximum deflection of the plate:

$$w_{\max} = \frac{5p_0 a^4}{384D} - \frac{4p_0 a^4}{\pi^5 D} \sum_{m=1,3,\ldots}^{\infty} \frac{(-1)^{(m-1)/2}}{m^5} \frac{\alpha_m \tanh \alpha_m + 2}{2 \cosh \alpha_m} \tag{3.23}$$

The first term above represents the deflection $w_{\max}$ of the middle of a uniformly loaded, simply supported strip. The second term is a very rapidly converging series. For example, in the case of a square plate ($a = b$ and $\alpha_m = m\pi/2$), the maximum displacement is given by

$$w_{\max} = \frac{5p_0 a^4}{384D} - \frac{4p_0 a^4}{\pi^5 D}(0.68562 - 0.00025 + \cdots) = 0.00406 \frac{p_0 a^4}{D}$$

It is observed that the result obtained, even retaining only the first term of the series in the parentheses, will be accurate to the third significant figure.

Introducing the following notation into Eq. (3.23)

$$\delta_1 = \frac{5}{384} - \frac{4}{\pi^5} \sum_{m=1,3,\ldots}^{\infty} \frac{(-1)^{(m-1)/2}}{m^5} \frac{\alpha_m \tanh \alpha_m + 2}{2 \cosh \alpha_m} \tag{f}$$

the maximum deflection of the plate is found to be

$$w_{\max} = \delta_1 \frac{p_0 a^4}{D} \qquad \left(x = \frac{a}{2}, y = 0 \right) \tag{3.24a}$$

Expressions for the moments, edge forces, and stresses of the plate can be derived by following a procedure similar to that described in Sec. 3.3. The maximum moments in the plate can also be put into the form

$$M_{x,\max} = \delta_2 p_0 a^2 \qquad M_{y,\max} = \delta_3 p_0 a^2 \qquad \left(x = \frac{a}{2}, y = 0 \right) \tag{3.24b}$$

Numerical values of the coefficients δ_1, δ_2, and δ_3 are given in Table 3.1 for

Table 3.1

b/a	1.0	2.0	3.0	4.0	5.0
δ_1	0.00406	0.01013	0.01223	0.01282	0.01297
δ_2	0.0479	0.1017	0.1189	0.1235	0.1246
δ_3	0.0479	0.0464	0.0406	0.0384	0.0375

various aspect ratios b/a of the plate sides.[11] It is seen from the table that as b/a increases $w_{\max}$ and $M_{x,\max}$ increase while $M_{y,\max}$ decreases.

Example 3.4. A window of a high-rise building is approximated by a rectangular plate with three edges simply supported and one edge clamped. The plate is under uniform wind-loading of intensity p_0. Derive an expression for the deflection surface.

SOLUTION Let the uniformly loaded plate be bounded as shown in Fig. 3.5 and assume that the edge $y = 0$ is clamped and the remaining edges are simply supported. Thus, the *deflection is symmetrical* about the line $x = a/2$, and w may be summed only for odd integers of m. The general solution is obtained by combining Eqs. (3.16) and (3.20)

$$w = \sum_{m=1,3,\ldots}^{\infty} \left(A_m \sinh \frac{m\pi y}{a} + B_m \cosh \frac{m\pi y}{a} + C_m y \sinh \frac{m\pi y}{a} + D_m y \cosh \frac{m\pi y}{a} + \frac{4p_0 a^4}{m^5 \pi^5 D} \right) \sin \frac{m\pi x}{a} \tag{3.25}$$

Boundary conditions are represented by

$$\begin{aligned} w &= 0 \qquad \frac{\partial w}{\partial y} = 0 \qquad (y = 0) \\ w &= 0 \qquad \frac{\partial^2 w}{\partial y^2} = 0 \qquad (y = b) \end{aligned} \tag{g}$$

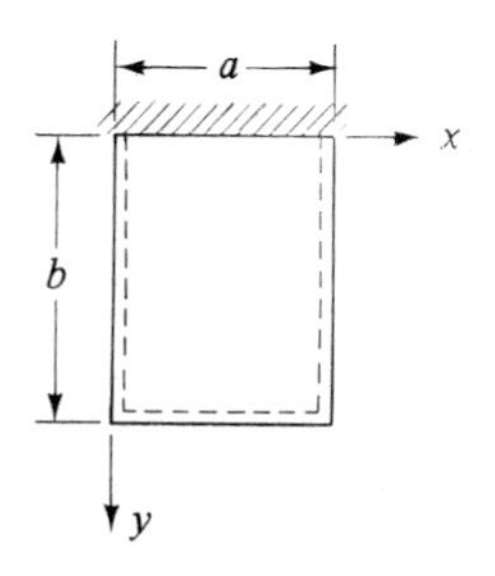

Figure 3.5

Application of Eqs. (g) to w leads to values of A_m, B_m, C_m, and D_m:

$$A_m = -\frac{a}{m\pi} D_m = \frac{2p_0 a^4}{m^5\pi^5 D} \frac{2\cosh^2\beta_m - 2\cosh\beta_m - \beta_m \sinh\beta_m}{\cosh\beta_m \sinh\beta_m - \beta_m}$$

$$B_m = -\frac{4p_0 a^4}{m^5\pi^5 D} \tag{h}$$

$$C_m = -\frac{1}{2} B_m \cdot \frac{2\dfrac{m\pi}{a}\sinh\beta_m \cosh\beta_m - \dfrac{m\pi}{a}\sinh\beta_m - \left(\dfrac{m\pi}{a}\right)^2 b\cosh\beta_m}{\cosh\beta_m \sinh\beta_m - \beta_m}$$

where $\beta_m = m\pi b/a$. When Eqs. (h) are inserted into Eq. (3.25), an expression for the plate deflection is established.

In the case of a square plate ($a = b$) the center deflection and the maximum bending moment are found to be (Prob. 3.10)

$$w = 0.0028 \frac{p_0 a^4}{D} \qquad \left(x = \frac{a}{2}, y = \frac{b}{2}\right)$$

$$M_y = 0.08 p_0 a^2 = M_{\max} \qquad \left(x = \frac{a}{2}, y = 0\right) \tag{i}$$

Situations involving other combinations of boundary conditions, on the two opposite arbitrary sides of the plate, may be treated similarly as illustrated in the following example.

Example 3.5 The uniform load p_0 acts on a rectangular balcony reinforcement plate with opposite edges $x = 0$ and $x = a$ simply supported, the third edge $y = b$ free, and the fourth edge $y = 0$ clamped (Fig. 3.6). Outline the derivation of the expression for the deflection surface w.

SOLUTION For the situation described, the boundary conditions, Eqs. (1.25) to (1.27), are

$$w = 0 \qquad \frac{\partial^2 w}{\partial x^2} = 0 \qquad (x = 0, x = a) \tag{j}$$

$$w = 0 \qquad \frac{\partial w}{\partial y} = 0 \qquad (y = 0) \tag{k}$$

$$\frac{\partial^2 w}{\partial y^2} + \nu\frac{\partial^2 w}{\partial x^2} = 0 \qquad \frac{\partial^3 w}{\partial y^3} + (2 - \nu)\frac{\partial^3 w}{\partial x^2\,\partial y} = 0 \qquad (y = b) \tag{l}$$

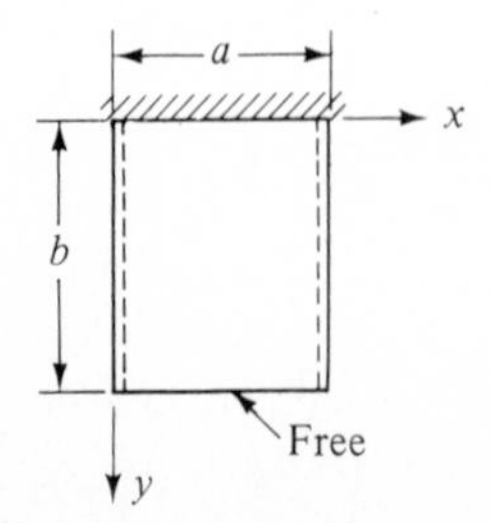

Figure 3.6

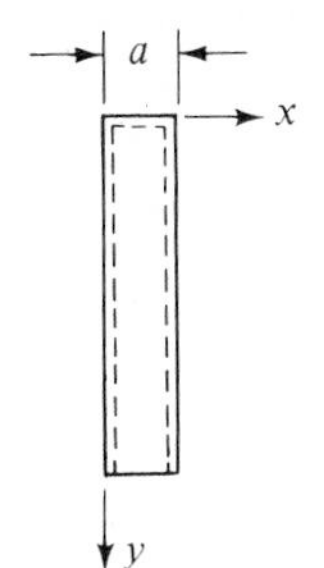

Figure 3.7

The particular and the homogeneous solutions are given by Eqs. (3.20) and (3.16), respectively, both of which satisfy the conditions (j). Applications of Eqs. (k) and (l) to $w_h + w_p$ leads to definite values of the constants A_m, B_m, C_m, D_m. The deflection is then obtained by adding Eqs. (3.16) and (3.20).

Example 3.6 Derive an expression for the deflection surface of a very long and narrow rectangular floor-panel subjected to a uniform load of intensity p_0 (Fig. 3.7). Assume the edges $x = a$, $x = a$, and $y = 0$ are simply supported.

SOLUTION The plate deflection can readily be obtained by superposing the solution for an infinite strip w_p given by Eq. (3.20) with a suitable solution w_h of Eq. (1.18). It is observed from Eq. (3.14) that in order for the coefficients of w_h, f_m, and its derivatives to vanish at $y = \infty$, A'_m and C'_m should be equated to zero. Hence, the homogeneous solution w_h can be represented by the following expression

$$w_h = \sum_{m=1,3,\ldots}^{\infty} (B'_m + D'_m y)e^{-m\pi y/a} \sin \frac{m\pi x}{a} \tag{3.26}$$

The above, of course, satisfies Eq. (1.18).

The boundary conditions of sides $x = 0$ and $x = a$ are fulfilled by Eqs. (3.20) and (3.26). It remains now to determine B'_m and D'_m in such a manner as to satisfy the boundary conditions on side $y = 0$. Substituting $w = w_h + w_p$ into $w = 0$ and $\partial^2 w/\partial y^2 = 0$ and setting $y = 0$, we obtain two equations which after solution yield $B'_m = -4p_0 a^4/\pi^5 D m^5$ and $D'_m = m\pi B'_m/2a$. It then follows that the elastic surface is given by

$$w = \frac{p_0 a^4}{24D}\left[\frac{x^4}{a^4} - 2\frac{x^2}{a^2} + \frac{x}{a} - \frac{96}{\pi^5}\sum_{m=1,3,\ldots}^{\infty}\frac{1}{m^5}\left(1 + \frac{m\pi y}{2a}\right)e^{-m\pi y/a}\sin\frac{m\pi x}{a}\right] \tag{3.27}$$

The cases involving a *clamped edge* at $y = 0$ or a *free edge* at $y = 0$ may be treated in a like manner, applying Eqs. (1.25) and (1.27), respectively.

3.5 LÉVY'S METHOD APPLIED TO NONUNIFORMLY LOADED RECTANGULAR PLATES

Lévy's approach is now applied to the treatment of bending problems of rectangular plates under nonuniform loading which is a function of x only. Bounding the plate as shown in Fig. 3.4, and assuming that the edges $x = 0$ and $x = a$ are simply supported, the loading is expressed by the Fourier series:

$$p(x) = \sum_{m=1,2,\ldots}^{\infty} p_m \sin \frac{m\pi x}{a} \tag{3.28}$$

where

$$p_m = \frac{2}{a}\int_0^a p(x) \sin \frac{m\pi x}{a}\, dx \tag{3.29}$$

Proceeding as in Sec. 3.4, we obtain

$$w_p = \frac{a^4}{\pi^4 D} \sum_{m=1,2,\ldots}^{\infty} \frac{p_m}{m^4} \sin \frac{m\pi x}{a} \tag{3.30}$$

The above represents the deflection of a strip under load $p(x)$ and satisfies Eq. (1.17) as well as the simple support conditions (1.26b) at $x = 0$ and $x = a$.

Let us assume that the two arbitrary edges $y = \pm b/2$ are also simply supported. The total deflection expression (3.21) then becomes

$$w = \sum_{m=1,2,\ldots}^{\infty} \left(B_m \cosh \frac{m\pi y}{a} + C_m y \sinh \frac{m\pi y}{a} + \frac{p_m a^4}{m^4 \pi^4 D} \right) \sin \frac{m\pi x}{a} \tag{3.31}$$

where the constants B_m and C_m are to be determined from the conditions at $y = \pm b/2$: $w = 0$, $\partial^2 w/\partial y^2 = 0$. Finally, we obtain

$$w = \frac{a^4}{\pi^4 D} \sum_{m=1,2,\ldots}^{\infty} \frac{p_m}{m^4} \left(1 - \frac{2 + \alpha_m \tanh \alpha_m}{2 \cosh \alpha_m} \cosh \frac{m\pi y}{a} + \frac{(m\pi y/a) \sinh (m\pi y/a)}{2 \cosh \alpha_m} \right) \sin \frac{m\pi x}{a} \tag{3.32}$$

in which, as before, $\alpha_m = m\pi b/2a$.

Introduction of a given load distribution $p(x)$ into Eq. (3.29) yields p_m, following which Eq. (3.32) results in the displacements. The moments and stresses are found by applying the usual procedure. Table 3.2 furnishes[8] the values of p_m for various types of load distributions (Prob. 3.12).

Consider, as an example, the bending of a hydrostatically loaded plate (Fig. A of Table 3.2):

$$p_m = \frac{2}{a}\int_0^a \frac{p_0 x}{a} \sin \frac{m\pi x}{a}\, dx = \frac{2p_0}{m\pi}(-1)^{m+1} \qquad (m = 1, 2, \ldots) \tag{a}$$

Equation (3.32) together with Eq. (a) represents the deflection. Suppose now that the plate is square ($a = b$). The deflection occurring at the center ($x = a/2$, $y = 0$) is

$$w = 0.00203 \frac{p_0 a^4}{D}$$

Table 3.2 Variously loaded simply supported plates

	Geometry	Type of loading and Expression for p_m
A	p_0; b; $\frac{b}{2}$; x; a; y	Hydrostatic loading: $p = p_0 \frac{x}{a}$ $p_m = \frac{2p_0}{m\pi}(-1)^{m+1} \qquad (m = 1, 2, \ldots)$
B	$2e$; e; p_0; x_1; x; y	Uniform load from $(x_1 - e)$ to $(x_1 + e)$ $p_m = \frac{4p_0}{m\pi} \sin \frac{m\pi x_1}{a} \sin \frac{m\pi e}{a} \qquad (m = 1, 2, \ldots)$
C	p_0; x_1; x; y	Line load p_0 at $x = x_1$ $p_m = \frac{2p_0}{a} \sin \frac{m\pi x_1}{a} \qquad (m = 1, 2, \ldots)$

The above result is one-half the displacement of a simply supported rectangular plate under uniform load (Table 3.1), a result intuitively appreciated.

3.6 RECTANGULAR PLATES UNDER DISTRIBUTED EDGE MOMENTS

Consider a simply supported rectangular plate subjected to *symmetrically* distributed edge moments at $y = \pm b/2$, described by a Fourier sine series (Fig. 3.8)

$$f(x) = \sum_{m=1}^{\infty} M_m \sin \frac{m\pi x}{a} \qquad \left(y = \pm \frac{b}{2}\right) \tag{a}$$

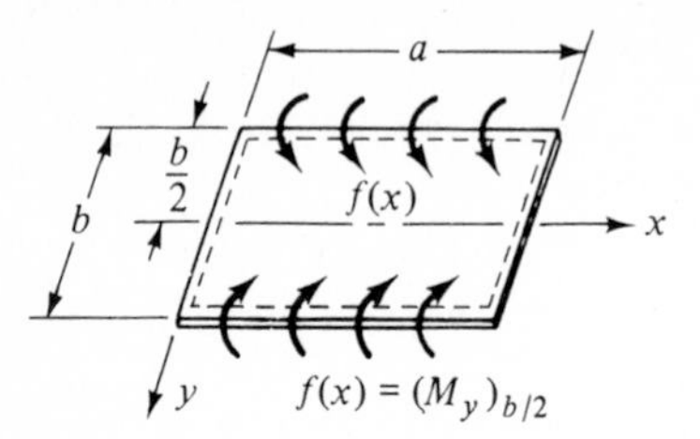

Figure 3.8

Here M_m represents the unknown set of coefficients

$$M_m = \frac{2}{a}\int_0^a f(x) \sin \frac{m\pi x}{a}\, dx \tag{b}$$

The boundary conditions are

$$w = 0 \qquad \frac{\partial^2 w}{\partial x^2} = 0 \qquad (x = 0,\ x = a) \tag{c}$$

$$w = 0 \qquad \left(y = \pm\frac{b}{2}\right) \tag{d}$$

$$-D\frac{\partial^2 w}{\partial y^2} = f(x) \qquad \left(y = \pm\frac{b}{2}\right) \tag{e}$$

The solution next proceeds with an assumption of the deflection surface in the form of Eq. (3.21) with $p_0 = 0$, except that summation goes over $m = 1, 2, 3, \ldots$:

$$w = \sum_{m=1}^{\infty}\left(B_m \cosh\frac{m\pi y}{a} + C_m y \sinh\frac{m\pi y}{a}\right)\sin\frac{m\pi x}{a} \tag{f}$$

This equation satisfies Eq. (1.17) and conditions (*c*) as already verified in Sec. 3.4. The condition (*d*) is fulfilled if the terms in parentheses in the above are set equal to zero. As before, setting $\alpha_m = m\pi b/2a$, we have

$$B_m \cosh \alpha_m + C_m \frac{b}{2} \sinh \alpha_m = 0$$

from which

$$B_m = -C_m\frac{b}{2} + \tanh \alpha_m$$

Equation (*f*) now takes the form

$$w = \sum_{m=1}^{\infty} C_m\left(y \sinh\frac{m\pi y}{a} - \frac{b}{2}\tanh\alpha_m \cosh\frac{m\pi y}{a}\right)\sin\frac{m\pi x}{a} \tag{g}$$

Substitution of Eqs. (*g*) and (*a*) into Eq. (*e*) leads to

$$-2D\sum_{m=1}^{\infty}\frac{m\pi}{a}C_m \cosh\alpha_m \sin\frac{m\pi x}{a} = \sum_{m=1}^{\infty} M_m \sin\frac{m\pi x}{a}$$

It follows that

$$C_m = -\frac{aM_m}{2m\pi D\cosh\alpha_m}$$

The deflection is therefore

$$w = \frac{a}{2\pi D}\sum_{m=1}^{\infty}\frac{\sin(m\pi x/a)}{m\cosh\alpha_m}M_m\left(\frac{b}{2}\tanh\alpha_m\cosh\frac{m\pi y}{a} - y\sinh\frac{m\pi y}{a}\right) \tag{3.33}$$

The moments and the stresses are determined from expression (3.33) for w.

In the case of *uniformly* distributed moments, we have $f(x) = M_0$, and Eq. (*b*) gives $M_m = 4M_0/m\pi$. Equation (3.33) then becomes

$$w = \frac{2M_0 a}{\pi^2 D} \sum_{m=1,3,\ldots}^{\infty} \frac{\sin(m\pi x/a)}{m^2 \cosh \alpha_m}\left(\frac{b}{2}\tanh \alpha_m \cosh \frac{m\pi y}{a} - y \sinh \frac{m\pi y}{a}\right) \tag{3.34}$$

For a square plate ($a = b$), deflection and bending moments at the center are determined by the use of Eqs. (3.34) and (1.10) in the form

$$w = 0.0368\frac{M_0 a^2}{D} \qquad M_x = 0.394M_0 \qquad M_y = 0.256M_0$$

The displacement occurring along the axis of symmetry is

$$w = \frac{M_0 ab}{\pi^2 D} \sum_{m=1,3,\ldots}^{\infty} \frac{1}{m^2}\frac{\tanh \alpha_m}{\cosh \alpha_m} \sin \frac{m\pi x}{a} \qquad (y = 0) \tag{3.35}$$

When $a \gg b$, we can set $\tanh \alpha_m \approx \alpha_m$ and $\cosh \alpha_m \approx 1$, and the above reduces to

$$w = \frac{M_0 b^2}{2\pi D} \sum_{m=1,3,\ldots}^{\infty} \frac{1}{m} \sin \frac{m\pi x}{a} = \frac{1}{8}\frac{M_0 b^2}{D}$$

Interestingly, this result is the same as that for the *center deflection* of a strip of length b, subjected to two equal and opposite moments at the ends.

The particular case in which the plate is loaded by an *antimetric* moment distribution $(M_y)_{y=b/2} = -(M_y)_{y=-b/2}$ can be treated similarly by taking the solution of Eq. (1.17) in the form of Eq. (3.16) and modifying the conditions (*e*) as follows:

$$-D\left(\frac{\partial^2 w}{\partial y^2}\right)_{y=b/2} = (M_y)_{y=b/2} \qquad -D\left(\frac{\partial^2 w}{\partial y^2}\right)_{y=b/2} = -(M_y)_{y=-b/2}$$

Furthermore, the general case can be obtained by combination of symmetric and antimetric situations. Solutions for the symmetric [Eq. (3.33)], and the antimetric moment distributions are useful in dealing with plates with various edge conditions (Sec. 3.7).

3.7 METHOD OF SUPERPOSITION APPLIED TO BENDING OF RECTANGULAR PLATES

The deflection and stress in a rectangular plate with any edge conditions and arbitrary loading can efficiently be determined by the method of superposition (Sec. 2.8). According to this procedure, first a given complex problem is replaced by several simpler situations, each of which can be treated by the Navier's or the Lévy's approaches. The deflections obtained for each replacement plate are then superposed in such a way that the governing equation $\nabla^4 w = p/D$ and the boundary conditions are fulfilled for the original case.

Consider, for example, the bending of a plate under any lateral load, with one edge clamped and the other edges simply supported (Fig. 3.5). The solution

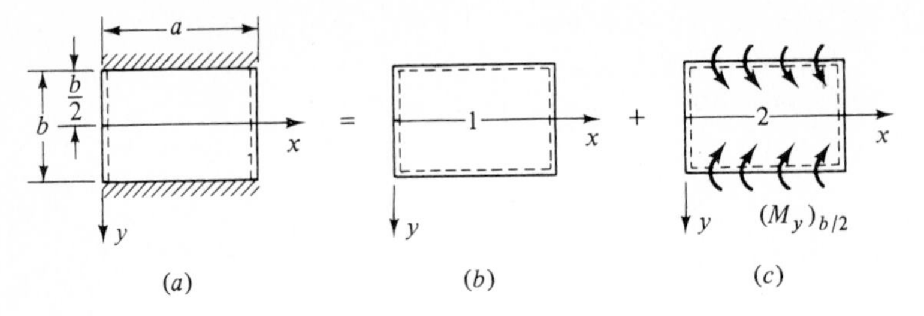

Figure 3.9

begins with the assumption that all edges are simply supported. Then, a bending moment along edge $y = 0$ is applied of such a magnitude as to eliminate the rotations due to the lateral load.

The solution procedure for the following problem serves to further illustrate the method.

Example 3.7 A rectangular plate has opposite edges at $x = 0$ and $x = a$ simply supported, and the other two edges at $y = \pm b/2$ clamped (Fig. 3.9*a*). The plate is subjected to a uniformly distributed load of intensity p_0. We wish to derive an expression for the deflection surface and the moments.

SOLUTION We shall proceed by superimposing the solutions of each of the two plates illustrated in Figs. 3.9*b* and 3.9*c*. Plate 1 has its edges simply supported and under uniform load p_0. Plate 2 also has all edges simply supported but in addition the two edges at $y = \pm b/2$ are subjected to uniformly distributed moments yet to be determined.

For plate 1, from Eq. (3.22), we have

$$w_1 = \frac{4p_0 a^4}{\pi^5 D} \sum_{m=1,3,\ldots}^{\infty} \frac{1}{m^5} \left[1 - \frac{\alpha_m \tanh \alpha_m + 2}{2 \cosh \alpha_m} \cosh \frac{2\alpha_m y}{b} + \frac{1}{2 \cosh \alpha_m} \frac{m\pi y}{a} \sinh b \frac{2\alpha_m y}{b}\right] \sin \frac{m\pi x}{a} \tag{3.36}$$

The rotation of the bent plate along side $y = b/2$ is then

$$\frac{\partial w_1}{\partial y} = \frac{2p_0 a^3}{\pi^4 D} \sum_{m=1,3,\ldots}^{\infty} \frac{1}{m^4} [\alpha_m - \tanh \alpha_m (1 + \alpha_m \tanh \alpha_m)] \sin \frac{m\pi x}{a} \tag{a}$$

In order to prevent this rotation and thus fulfill the actual conditions of the boundary of the initial plate, the following bending moments are applied along side $y = \pm b/2$ of plate 2:

$$M_y = \sum_{m=1}^{\infty} M_m \sin \frac{m\pi x}{a} \tag{b}$$

wherein the coefficients M_m are obtained so as to make the slope owing to these moments equal and opposite to that represented by Eq. (*a*). The deflection of plate 2 is given by Eq. (3.33), from which for sides $y = \pm b/2$, we have

$$\frac{\partial w_2}{\partial y} = \frac{a}{2\pi D} \sum_{m=1,3,\ldots}^{\infty} \frac{1}{m} M_m[\tanh \alpha_m(\alpha_m \tanh \alpha_m - 1) - \alpha_m] \sin \frac{m\pi x}{a} \qquad (c)$$

The requirement that the slopes for both plates at $y = \pm b/2$ have the same value but be of opposite sign is satisfied if

$$\frac{\partial w_1}{\partial y} = -\frac{\partial w_2}{\partial y} \qquad \left(y = \pm\frac{b}{2}\right)$$

Upon inserting Eqs. (a) and (c) into the above and solving, we obtain

$$M_m = \frac{4p_0 a^2}{m^3\pi^3} \frac{\alpha_m - \tanh \alpha_m(1 + \alpha_m \tanh \alpha_m)}{\alpha_m - \tanh \alpha_m(\alpha_m \tanh \alpha_m - 1)} \tag{3.37}$$

With the expression for M_m determined, the deflection of plate 2 can be obtained by introducing Eq. (3.37) into Eq. (3.33). Hence

$$w_2 = -\frac{2p_0 a^4}{\pi^5 D} \sum_{m=1,3,\ldots}^{\infty} \frac{\sin (m\pi x/a)}{m^5 \cosh \alpha_m} \frac{\alpha_m - \tanh \alpha_m(1 + \alpha_m \tanh \alpha_m)}{\alpha_m - \tanh \alpha_m(\alpha_m \tanh \alpha_m - 1)} \times \left(\frac{m\pi y}{a} \sinh \frac{m\pi y}{a} - \alpha_m \tanh \alpha_m \cosh \frac{m\pi y}{a}\right) \tag{3.38}$$

This series converges very rapidly and the first few terms will give a satisfactory result. For instance, when $a = b$, we find that the center deflection given by only the first term in the series is equal to

$$w_2 = 0.00214\frac{p_0 a^4}{D} \qquad \left(x = \frac{a}{2}, y = 0\right)$$

The center deflection for plate 1 is (from Table 3.1):

$$w_1 = 0.00406\frac{p_0 a^4}{D} \qquad \left(x = \frac{a}{2}, y = 0\right)$$

The maximum displacement, which takes place at the center of a uniformly loaded square plate with two simply supported and two fixed edges, is thus

$$w = w_1 - w_2 = 0.00192\frac{p_0 a^4}{D} = w_{\max}$$

The values of bending moments for plate 2 is found by employing the usual procedure while those for plate 1 are listed in Table 3.1. To ascertain the bending moments for the original plate, the results of the replacement cases are superimposed. It can be shown that the maximum moment occurs at the middle of the fixed edges and for $a = b$ is given by

$$M_y = -0.0697p_0 a^2 = M_{\max}$$

Deflections and moments at any other points are calculated in a like manner.

Employing similar procedures, the solutions for other cases of practical importance may be found. In the case of a *clamped square plate*, for instance, the largest deflection also occurs at the center and the largest bending moments are again found at the middle of the fixed edges. Their numerical values are:

$$w_{max} = 0.00126\frac{p_0 a^4}{D} \tag{d}$$

$$M_x = M_y = 0.0513 p_0 a^2 = M_{max} \tag{e}$$

Upon comparison of the above with the results obtained in Examples 3.4 and 3.7, we observe that *as the number of built-in plate edges increases, the deflection and moment produced by the same loading decreases* considerably. A discussion of a number of practical aspects of the edge fixity is found in Sec. 2.4.

3.8 THE STRIP METHOD

We now present a simple approximate approach, due to H. Grashof, for computing deflection and moment in a rectangular plate with arbitrary boundary conditions. In this so-called *strip method*, the plate is assumed to be divided into two systems of strips at right angles to one another, each strip regarded as functioning as a beam (Fig. 3.10*a*). The method permits *qualitative analysis* of the plate behavior with ease but is less adequate, in general, in obtaining accurate quantitative results. Note, however, that because this method always gives conservative values for both deflection and moment, it is often employed in practice. A very efficient engineering approach[12] to design of the rectangular floor slabs is also based upon the strip method.

Before proceeding to a description of the method, it will prove useful to introduce maximum deflection and moments of a beam with various end conditions. Expressions for such quantities for a beam of length L subjected to a uniform load p, derived from the mechanics of materials, are given in Table 3.3.

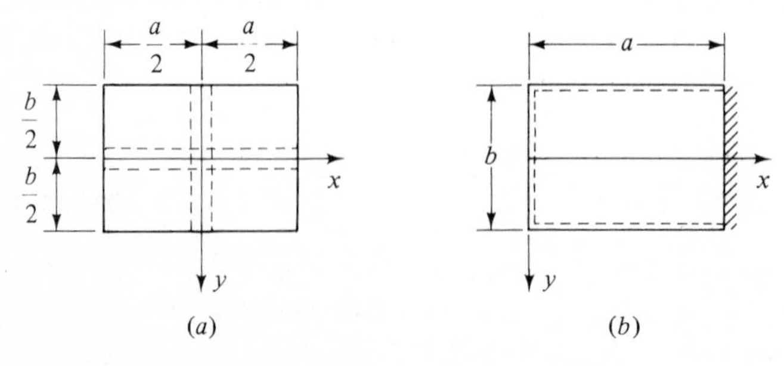

Figure 3.10

Table 3.3

(simply supported beam A–B, uniform load p, center C at $\frac{L}{2}$, span L)	$w_c = \frac{5}{384}\frac{pL^4}{EI}$	$M_c = \frac{1}{8}pL^2$	$M_a = M_b = 0$
(beam clamped at both ends, uniform load p)	$w_c = \frac{1}{384}\frac{pL^4}{EI}$	$M_c = \frac{1}{24}pL^2$	$M_a = M_b = \frac{1}{12}pL^2$
(beam clamped at one end, simply supported at other, uniform load p, D at $\frac{5}{8}L$)	$w_c = \frac{1}{192}\frac{pL^4}{EI}$	$M_d = M_{\max} = \frac{9}{128}pL^2$	$M_a = \frac{1}{8}pL^2$ $M_b = 0$

Consider a rectangular plate under a uniform load p_0, and assume that the plate is divided into strips of spans a and b, carrying the uniform loads p_a and p_b, respectively. The loaded system of beams will be impossible to arrange in such a way as to compose the plate unless the following conditions are met:

$$w_a = w_b \qquad p_0 = p_a + p_b \qquad (x = y = 0) \tag{3.39a, b}$$

In the case of a plate with simply supported edges, from Table 3.3, we have: $w_a = 5p_a a^4/384EI$, $w_b = 5p_b b^4/384EI$, and Eq. (3.39a) yields $p_a a^4 = p_b b^4$. Then Eq. (3.39b) leads to

$$p_a = p_0 \frac{b^4}{a^4 + b^4} \qquad p_b = p_0 \frac{a^4}{a^4 + b^4} \tag{3.40}$$

Expressions (3.40) are also valid in the case of a clamped plate, since from Table 3.3 we again find that $p_a a^4 = p_b b^4$ for $w_a = w_b$.

The deflection and bending moments of the plate are obtained by means of Eqs. (3.40). Referring to the Eqs. (3.24), we can represent the maximum deflection, occurring at the center of the plate, as follows

$$w_{\max} = k_1 \frac{p_a a^4}{D} = k_1 \frac{p_0 a^4 b^4}{D(a^4 + b^4)} \tag{3.41}$$

The bending moments at the center are written

$$M_x = k_2 p_a a^2 = k_2 \frac{p_0 a^2 b^4}{a^4 + b^4} \qquad M_y = k_2 \frac{p_0 a^4 b^2}{a^4 + b^4} \tag{3.42}$$

and the bending moments at the edges are

$$M_x = k_3 p_a a^2 = k_3 \frac{p_0 a^2 b^4}{a^4 + b^4} \qquad M_y = k_3 \frac{p_0 a^4 b^2}{a^4 + b^4} \tag{3.43}$$

Here k_1, k_2, and k_3 are constants dependent upon the side conditions of the plate. Values for these factors, referring to Table 3.3, are listed in Table 3.4. The deflection and moment at any other location in the plate can be expressed in a similar way.

Table 3.4

Edge condition	k_1	k_2	k_3
Simply supported	5/384	1/8	0
Clamped	1/384	1/24	1/12

For example, in the case of a uniformly loaded *clamped square plate* ($a = b$), we find from Eq. (3.41) and Table 3.4 that

$$w_{\max} = 0.0013\frac{p_0 a^4}{D} \qquad (x = 0,\ y = 0)$$

This is accurate to the fourth significant figure as compared with the value given by Eq. (*d*) in Sec. 3.7. The results for moments are crude approximations, however.

Example 3.8 A rectangular plate with three edges simply supported and one edge built in is subjected to a uniform load p_0 (Fig. 3.10*b*). Determine the deflection at the center and the maximum bending moment of the plate.

SOLUTION The plate is divided into simply supported and clamped simply supported beams. Thus, Eq. (3.39*a*) and Table 3.3 give

$$\frac{p_a a^4}{192EI} = \frac{5p_b b^4}{384EI}$$

From Eq. (3.39*b*) we then have

$$p_a = p_0\frac{5b^4}{2a^4 + 5b^4} \qquad p_b = p_0\frac{2a^4}{2a^4 + 5b^4}$$

The plate deflection at the center is therefore

$$w = \frac{5}{384}\frac{p_b b^4}{D} = \frac{5}{192}\frac{p_0 a^4 b^4}{(2a^4 + 5b^4)D} \tag{a}$$

We also obtain the following maximum bending moments

$$M_x = \frac{45}{128}\frac{p_0 a^2 b^4}{2a^4 + 5b^4} \qquad (x = \tfrac{3}{8}L)$$

$$M_y = \frac{1}{4}\frac{p_0 a^4 b^2}{2a^4 + 5b^4} \qquad (y = 0) \tag{b}$$

and the bending moment at the fixed edge

$$M_f = \frac{5}{8}\frac{p_0 a^2 b^4}{2a^4 + 5b^4} \qquad (x = a) \tag{c}$$

In the case of a square plate ($a = b$), Eqs. (a) to (c) reduce to

$$w = 0.00372 p_0 a^4/D$$

$$M_x = 0.0502 p_0 a^2$$

$$M_y = 0.0357 p_0 a^2$$

$$M_f = 0.0893 p_0 a^2$$

The deflection at the center of the plate is 33 percent greater and the bending moment M_f is 11 percent greater than the values obtained by the bending theory of plates (Example 3.4).

3.9 SIMPLY SUPPORTED CONTINUOUS RECTANGULAR PLATES

When a uniform plate extends over a support and has more than one span along its length or width, it is termed *continuous*. In a continuous plate the several spans may be of varying length. Intermediate supports are provided in the form of beams or columns. Only a continuous plate with a *rigid* intermediate beam is treated in this section, i.e., the plate has zero deflection along the axis of the supporting beam. We shall assume that the beam does not prevent rotation of the plate. Hence, the intermediate beam represents a simple support to the plate.

Consider the two-span simply supported continuous plate, half of which is subjected to a uniform load of intensity p_0 (Fig. 3.11a). A convenient way of looking at the problem is to draw a free body diagram of each rectangular panel as shown in Fig. 3.11b. The distributed moment along the common edge may be represented by a Fourier series

$$f(y) = \sum_{m=1,3,\ldots}^{\infty} M_m \sin \frac{m\pi y}{b} \tag{a}$$

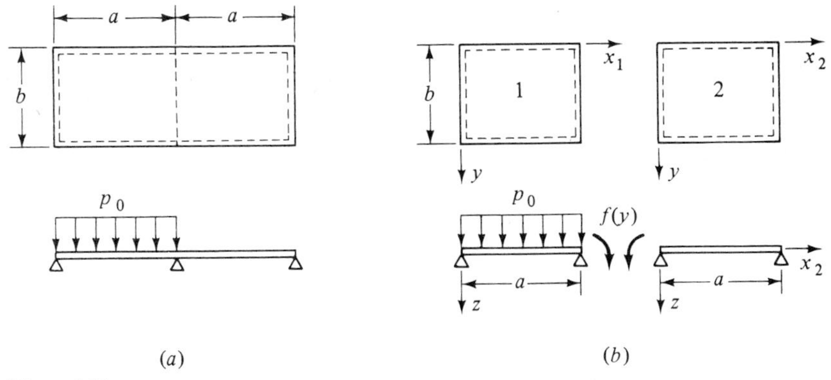

Figure 3.11

When the set of coefficients M_m is determined, the expressions for the boundary conditions can be handled with ease. The lateral deflection of each simply supported replacement plate may be obtained by application of Lévy's method.

Referring to Figs. 3.5 and 3.11*b* we conclude from the symmetry in deflections that the general solution for plate 1 is given by Eq. (3.25) if y is replaced by x_1, x by y, and a by b. That is

$$w_1 = \sum_{m=1,3,\ldots}^{\infty} \left(A_m \sinh \lambda_m x_1 + B_m \cosh \lambda_m x_1 + C_m x_1 \sinh \lambda_m x_1 + D_m x_1 \cosh \lambda_m x_1 + \frac{4p_0 b^4}{m^5 \pi^5 D} \right) \sin \lambda_m y \tag{3.44}$$

where

$$\lambda_m = \frac{m\pi}{b} \tag{b}$$

Similarly, for plate 2, setting $p_0 = 0$, the deflection expressed in terms of the coordinates x_2 and y, and for a different set of constants is

$$w_2 = \sum_{m=1,3,\ldots}^{\infty} [E_m \sinh \lambda_m x_2 + F_m \cosh \lambda_m x_2 + G_m x_2 \sinh \lambda_m x_2 + H_m x_2 \cosh \lambda_m x_2] \sin \lambda_m y \tag{3.45}$$

The boundary conditions for plate 1 and plate 2 are represented as follows, respectively:

$$\begin{aligned} &w_1 = 0 \qquad \frac{\partial^2 w_1}{\partial x_1^2} = 0 && (x_1 = 0) \\ &w_1 = 0 \qquad -D\frac{\partial^2 w_1}{\partial x_1^2} = \sum_{m=1,3,\ldots}^{\infty} M_m \sin \lambda_m y && (x_1 = a) \end{aligned} \tag{c}$$

and

$$\begin{aligned} &w_2 = 0 \qquad \frac{\partial^2 w_2}{\partial x_2^2} = 0 && (x_2 = a) \\ &w_2 = 0 \qquad -D\frac{\partial^2 w_2}{\partial x_2^2} = \sum_{m=1,3,\ldots}^{\infty} M_m \sin \lambda_m y && (x_2 = 0) \end{aligned} \tag{d}$$

We thus have eight equations (*c*) and (*d*) containing nine unknowns $A_m, \ldots, H_m, M_m$. The required additional equation is obtained by expressing the condition that the slopes must be the same for each panel at the middle support. This continuity requirement is expressed

$$\left.\frac{\partial w_1}{\partial x_1}\right|_{x_1=a} = \left.\frac{\partial w_2}{\partial x_2}\right|_{x_2=0} \tag{e}$$

Introducing Eqs. (3.44) and (3.45) into the above we have

$$A_m \lambda_m \cosh \lambda_m a + B_m \lambda_m \sin \lambda_m a + C_m(\sinh \lambda_m a + \lambda_m a \cosh \lambda_m a) + D_m(\cosh \lambda_m a + \lambda_m a \sinh \lambda_m a) = E_m \lambda_m + H_m \tag{3.46}$$

Application of the edge conditions (c) and (d) to Eqs. (3.44) and (3.45) leads to values for the constants as follows:

$$\begin{aligned} A_m &= \frac{1}{D}\left\{a \frac{\coth \lambda_m a}{\sinh \lambda_m a}\left[\frac{M_m}{2\lambda_m} + \frac{2p_0}{\lambda_m^4 b}(-1 + \cosh \lambda_m a)\right] + \frac{4p_0}{\lambda_m^5 b}\left(\coth \lambda_m a - \frac{\lambda_m a}{2} - \operatorname{csch} \lambda_m a\right)\right\} \\ B_m &= -\frac{4p_0}{\lambda_m^5 Db} \\ C_m &= \frac{2p_0}{\lambda_m^4 Db} \\ D_m &= -\frac{\operatorname{csch} \lambda_m a}{D}\left[\frac{M_m}{2\lambda_m} + \frac{2p_0}{\lambda^4 b}(-1 + \cosh \lambda_m a)\right] \end{aligned} \tag{3.47}$$

and

$$\begin{aligned} E_m &= -\frac{M_m a}{2\lambda_m D}(1 + \coth^2 \alpha_m a) \\ F_m &= 0 \\ G_m &= \frac{M_m}{2\lambda_m D} \\ H_m &= -\frac{M_m}{2\lambda_m D} \coth \lambda_m a \end{aligned} \tag{3.48}$$

Having Eqs. (3.47) and (3.48) available we obtain, from Eq. (3.46), the moment coefficients M_m. Equations (3.44) and (3.45) then give the deflection of the continuous plate from which moments and stresses can also be computed.

The foregoing approach may be extended to include the case of long rectangular plates with many supports, subjected to loading which is symmetric in y. In so doing, an equation similar to that of the three-moment equation of continuous beams is obtained.[8] It is noted that there are situations where the intermediate beams are relatively flexible compared with the flexibility of the plate. The deflections and rotations of the plate are not then taken as zero along the supports, but are functions of the bending and torsional stiffnesses of the supporting beams.[6]

The design methods used in connection with continuous plates utilize the solutions derived in the foregoing sections and a number of approximations.[11]

3.10 RECTANGULAR PLATES SUPPORTED BY INTERMEDIATE COLUMNS

In this section we consider the bending of a thin continuous plate over many columns. To attain a simplified expression for the lateral deflection it is *assumed* that: the plate is subjected to a uniform load p_0, the column cross sections are so small that their reactions on the plate are regarded as point loads, the columns are equally spaced in mutually perpendicular directions, and the dimensions of the plate are large as compared with the column spacing. The foregoing set of assumptions enables one to assume that the bending in all panels, away from the boundary of the plate, is the same. We can therefore restrict our attention to the bending of one panel alone, and consider it as a uniformly loaded rectangular plate ($a \times b$) supported at the corners by the columns. The origin of coordinates is placed at the center of a panel shown by the shaded area in Fig. 3.12a. Clearly, the maximum deflection occurs at the center of the panel.

A solution can be obtained utilizing Lévy's approach (Sec. 3.4). Accordingly, the deflection may be expressed as a combination of that associated with a strip with uniform load and fixed ends $y = \pm b/2$ and that associated with a rectangular plate. That is,

$$w = w_p + w_h = \frac{p_0 b^4}{384D}\left(1 - \frac{4y^2}{b^2}\right)^2 + B_0 + \sum_{m=2,4,\ldots}^{\infty}\left(B_m \cosh\frac{m\pi y}{a} + E_m \frac{m\pi y}{a}\sinh\frac{m\pi y}{a}\right)\cos\frac{m\pi x}{a} \tag{3.49}$$

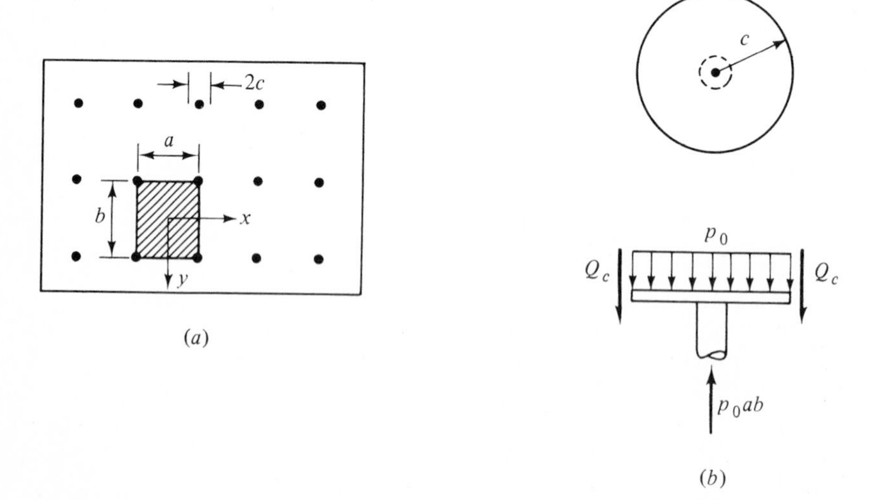

Figure 3.12

where B_0, B_m, and E_m are constants of w_h, yet to be determined. The above satisfies the boundary conditions for the rectangular panel along the x edges,

$$\frac{\partial w}{\partial x} = 0 \qquad Q_x = -D\left(\frac{\partial^3 w}{\partial x^3} + \frac{\partial^3 w}{\partial y^2\, \partial x}\right) = 0 \qquad \left(x = \pm\frac{a}{2}\right) \tag{a}$$

and Eqs. (1.17) and (1.18).

For purposes of simplifying the analysis, the support forces are regarded as acting over short (infinitesimal) line segments, between $x = a/2 - c$ and $x = a/2 + c$ or $2c$ (Fig. 3.12*a*). The plate loading is transmitted to the columns by the vertical shear forces. From the conditions of symmetry of the bent panel we are led to conclude that the slope in the direction of the normal to the boundary and the vertical shear force vanish everywhere on the edges of the panel with the exception of the corner points. Thus,

$$\begin{aligned} &Q_y = 0 \qquad \left(0 < x < \frac{a}{2} - c\right) \\ &\int_{a/2-c}^{a/2} Q_y\, dx = -\frac{p_0 ab}{4} \end{aligned} \tag{b}$$

The shear force Q_y can be represented by

$$Q_y = C_0 + \sum_{m=2,4,\ldots}^{\infty} C_m \cos\frac{m\pi x}{a} \tag{c}$$

Expression (*c*) and conditions (*b*), referring to Eq. (A.4), lead to

$$\begin{aligned} C_0 &= -\frac{p_0 b}{2} \\ C_m &= \frac{4}{a}\int_0^{a/2} Q_y \cos\frac{m\pi x}{a}\, dx = -p_0 b(-1)^{m/2} \end{aligned} \tag{d}$$

We are now in a position to represent the boundary conditions for the rectangular panel along the y edges:

$$\begin{aligned} &\frac{\partial w}{\partial y} = 0 \qquad \left(y = \pm\frac{b}{2}\right) \\ &Q_y = -D\left(\frac{\partial^3 w}{\partial y^3} + \frac{\partial^3 w}{\partial x^2\, \partial y}\right) \\ &\quad = -p_0 b\left[\frac{1}{2} - \sum_{m=2,4,\ldots}^{\infty} (-1)^{m/2}\cos\frac{m\pi x}{a}\right] \qquad \left(y = \pm\frac{b}{2}\right) \end{aligned} \tag{e}$$

Inasmuch as $\partial w/\partial y = 0$, the second term in the parentheses above is zero. Upon substituting Eq. (3.49) into Eqs. (*e*) we then determine the constants B_m and E_m.

Finally, we find B_0 such that the deflection w is zero at the corners

$$w = 0 \qquad \left(x = \frac{a}{2},\ y = \frac{b}{2} \right) \tag{f}$$

It follows that the values of the constants are

$$\begin{aligned}
B_0 &= -\frac{p_0 b^4}{2\pi^3 D}\left(\frac{a}{b}\right)^3 \sum_{m=2,4,\ldots}^{\infty} \frac{1}{m^3}\left(\alpha_m - \frac{\alpha_m + \tanh \alpha_m}{\tanh^2 \alpha_m}\right) \\
B_m &= -\frac{p_0 b}{2D}(-1)^{m/2}\left(\frac{a}{m\pi}\right)^3 \frac{\alpha_m + \tanh \alpha_m}{\sinh \alpha_m \tanh \alpha_m} \\
E_m &= \frac{p_0 b}{2D}(-1)^{m/2}\left(\frac{a}{m\pi}\right)^3 \frac{1}{\sinh \alpha_m}
\end{aligned} \tag{3.50}$$

where $\alpha_m = m\pi b/2a$. Equation (3.49) together with Eqs. (3.50) represent the deflection surface of the panel.

The maximum deflection takes place at the center of the plate ($x = y = 0$) and is

$$\begin{aligned}
w_{\max} = \frac{p_0 b^4}{384D} &- \frac{p_0 a^3 b}{2\pi^3 D} \sum_{m=2,4,\ldots}^{\infty} \frac{(-1)^{m/2}}{m^3} \frac{\alpha_m + \tanh \alpha_m}{\sinh \alpha_m \tanh \alpha_m} \\
&- \frac{p_0 a^3 b}{2\pi^3 D} \sum_{m=2,4,\ldots}^{\infty} \frac{1}{m^3}\left(\alpha_m - \frac{\alpha_m + \tanh \alpha_m}{\tanh^2 \alpha_m}\right)
\end{aligned} \tag{3.51}$$

By inserting w from Eq. (3.49) together with Eqs. (3.50) into Eqs. (1.10) and setting $x = y = 0$ one also determines the largest bending moments. The resulting expressions may be put into the following form

$$w_{\max} = \delta_1 \frac{p_0 b^4}{D} \qquad M_{x,\max} = \delta_2 p_0 b^2 \qquad M_{y,\max} = \delta_3 p_0 b^2 \tag{g}$$

Table 3.5 lists values[11] of the constants δ_1, δ_2, and δ_3 for different values of the aspect ratio b/a.

The moments near the columns as calculated by means of Eq. (3.49) are much larger than the moments some distance away. The assumption made above, pertaining to support force distribution in an infinitesimal length $2c$, may thus be dispensed with. It can now be assumed that each support reaction is distributed over a small concentric circular area of the radius c, representing the domain around the ends of the columns. An equivalent circular plate separated

Table 3.5

b/a	1.0	1.2	1.4	1.5	2.0
δ_1	0.00581	0.00428	0.00358	0.00337	0.00292
δ_2	0.0331	0.0210	0.0149	0.0131	0.0092
δ_3	0.0331	0.0363	0.0384	0.0387	0.0411

from the panel and subjected to uniform load p_0, shear forces Q_c around the periphery, and reactional force $p_0 ab$ at the inner fixed boundary, is shown in Fig. 3.12*b*. The radius of this plate and the shear force per unit length, as in cases of nearly square panels, are given by[8]

$$c = 0.22a \qquad Q_c = 0.723 p_0 b - 0.11 p_0 a \tag{h}$$

Referring to Fig. 3.12*b*, the moment and stress can thus be obtained at the supports (see Sec. 2.8).

3.11 RECTANGULAR PLATES ON ELASTIC FOUNDATION

This section is concerned with the bending of the rectangular plates on elastic foundation. As in Sec. 2.9, the intensity of the foundation reaction q at any point of the plate is assumed proportional to the deflection w at that point. Hence, $q = -kw$, where k is the modulus of the foundation.

The deflection w of a rectangular plate subjected to load p per unit surface area and reaction q must satisfy Eq. (1.17)

$$\frac{\partial^4 w}{\partial x^4} + 2\frac{\partial^4 w}{\partial x^2\,\partial y^2} + \frac{\partial^4 w}{\partial y^4} = \frac{p - kw}{D} \tag{3.52}$$

and the given boundary conditions.

Consider the bending of a rectangular plate of sides a and b subjected to arbitrary loading $p(x, y)$ (Fig. 3.1*a*). The plate is resting on an elastic subgrade and simply supported along its edges. The problem is treated by Navier's approach described in Sec. 3.2. Hence, the load and the deflection are represented by Eqs. (3.1). To evaluate a_{mn} in Eq. (3.1*b*) we substitute Eqs. (3.1) into (3.52). It follows that

$$a_{mn} = \frac{p_{mn}}{\pi^4 D[(m/a)^2 + (n/b)^2]^2 + k} \tag{3.53}$$

This value of a_{mn} and Eq. (3.1*b*) yields the equation of the plate deflection of the form

$$w = \sum_{m=1}^{\infty}\sum_{n=1}^{\infty} \frac{p_{mn}}{\pi^4 D[(m/a)^2 + (n/b)^2]^2 + k} \sin\frac{m\pi x}{a} \sin\frac{n\pi y}{b} \tag{3.54}$$

where p_{mn} is given by Eq. (3.3).

It is seen that for $k = 0$, Eq. (3.54) reduces to the solution for the rectangular simply supported plates [Eq. (3.5)]. Expressions for deflection and stress of a rectangular plate resting on an elastic foundation and at the same time simply supported at its edges, under any specific load, may thus easily be determined as illustrated in Sec. 3.3.

3.12 THE RITZ METHOD APPLIED TO BENDING OF RECTANGULAR PLATES

We now apply the Ritz method to the bending of rectangular plates. The strain energy U associated with the bending of a plate is given by Eq. (1.34), Sec. 1.9. One can represent the work done by the lateral surface loading $p(x, y)$ as follows

$$W = \iint_A wp \, dx \, dy \tag{3.55}$$

where A is the area of the plate surface. The potential energy $\Pi = U - W$ is therefore

$$\Pi = \frac{D}{2} \iint_A \left\{ \left(\frac{\partial^2 w}{\partial x^2} + \frac{\partial^2 w}{\partial y^2} \right)^2 - 2(1 - \nu) \left[\frac{\partial^2 w}{\partial x^2} \frac{\partial^2 w}{\partial y^2} - \left(\frac{\partial^2 w}{\partial x \, \partial y} \right)^2 \right] - wp \right\} dx \, dy \tag{3.56}$$

The application of the method is illustrated by considering the bending of a clamped rectangular plate of sides a and b, carrying a uniform load of intensity p_0 (Fig. 3.13). The boundary conditions are

$$\begin{aligned} w &= 0 \qquad \frac{\partial w}{\partial x} = 0 \qquad (x = 0, x = a) \\ w &= 0 \qquad \frac{\partial w}{\partial y} = 0 \qquad (y = 0, y = b) \end{aligned} \tag{a}$$

Integration by parts of the last term in Eq. (3.56) leads to

$$\begin{aligned} \iint_A \frac{\partial^2 w}{\partial x \, \partial y} \frac{\partial^2 w}{\partial x \, \partial y} \, dx \, dy &= \int_S \frac{\partial^2 w}{\partial x \, \partial y} \frac{\partial w}{\partial x} \, dx - \iint_A \frac{\partial w}{\partial x} \frac{\partial^3 w}{\partial x \, \partial y^2} \, dx \, dy \\ &= \int_S \frac{\partial^2 w}{\partial x \, \partial y} \frac{\partial w}{\partial x} \, dx - \int_S \frac{\partial w}{\partial x} \frac{\partial^2 w}{\partial y^2} \, dy + \iint_A \frac{\partial^2 w}{\partial x^2} \frac{\partial^2 w}{\partial y^2} \, dx \, dy \end{aligned} \tag{b}$$

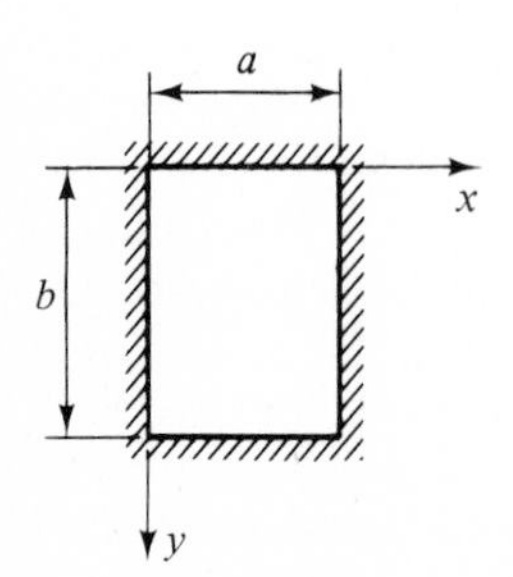

Figure 3.13

According to Eqs. (a), the first two integrals in Eq. (b) become identically zero. Hence

$$\iint_A \left[\frac{\partial^2 w}{\partial x^2}\frac{\partial^2 w}{\partial y^2} - \left(\frac{\partial^2 w}{\partial x\,\partial y}\right)^2\right] dx\,dy = 0$$

The bending strain energy therefore reduces to

$$U = \frac{D}{2}\iint_A \left(\frac{\partial^2 w}{\partial x^2} + \frac{\partial^2 w}{\partial y^2}\right)^2 dx\,dy \tag{3.57}$$

Assuming a deflection expression of the form

$$w = \sum_{m=1}^{\infty}\sum_{n=1}^{\infty} a_{mn}\left(1 - \cos\frac{2m\pi x}{a}\right)\left(1 - \cos\frac{2n\pi y}{b}\right) \tag{3.58}$$

the boundary conditions (a) are satisfied. Introduction of the above into Eq. (3.57) yields

$$U = \frac{D}{2}\int_0^b\int_0^a \left\{\sum_{m=1}^{\infty}\sum_{n=1}^{\infty} 4\pi^2 a_{mn}\left[\frac{m^2}{a^2}\cos\frac{2m\pi x}{a}\left(1 - \cos\frac{2n\pi y}{b}\right) + \frac{n^2}{b^2}\cos\frac{2n\pi y}{b}\left(1 - \cos\frac{2m\pi x}{a}\right)\right]\right\}^2 dx\,dy$$

from which

$$U = 2\pi^4 abD\left\{\sum_{m=1}^{\infty}\sum_{n=1}^{\infty}\left[3\left(\frac{m}{a}\right)^4 + 3\left(\frac{n}{b}\right)^4 + 2\left(\frac{m}{a}\right)^2\left(\frac{n}{a}\right)^2\right]a_{mn}^2 + \sum_{m=1}^{\infty}\sum_{r=1}^{\infty}\sum_{s=1}^{\infty}2\left(\frac{m}{a}\right)^4 a_{mr}a_{ms} + \sum_{r=1}^{\infty}\sum_{s=1}^{\infty}\sum_{n=1}^{\infty}2\left(\frac{n}{b}\right)^4 a_{rn}a_{sn}\right\} \tag{3.59}$$

which is valid for $r \neq s$. The work done by p_0 is, by application of Eq. (3.55)

$$W = p_0\int_0^b\int_0^a\left[\sum_{m=1}^{\infty}\sum_{n=1}^{\infty}a_{mn}\left(1 - \cos\frac{2m\pi x}{a}\right)\left(1 - \cos\frac{2n\pi y}{b}\right)\right]dx\,dy$$

$$= p_0 ab\sum_{m=1}^{\infty}\sum_{n=1}^{\infty} \tag{3.60}$$

From the minimizing conditions $\partial\Pi/\partial a_{mn} = 0$, it follows that

$$4D\pi^4 ab\left\{\left[3\left(\frac{m}{a}\right)^4 + 3\left(\frac{n}{b}\right)^4 + 2\left(\frac{m}{a}\right)^2\left(\frac{n}{b}\right)^2\right]a_{mn} + \sum_{r=1}^{\infty}2\left(\frac{m}{a}\right)^4 a_{mr} + \sum_{r=1}^{\infty}2\left(\frac{n}{b}\right)^4 a_{rn}\right\} - p_0 ab = 0 \tag{c}$$

which is valid for $r \neq n$ and $r \neq m$.

Dropping all but the first term a_{11}, Eq. (*c*) gives

$$a_{11} = \frac{p_0 a^4}{4\pi^4 D} \frac{1}{3 + 3(a/b)^4 + 2(a/b)^2}$$

In the case of a square plate ($a = b$), $a_{11} = p_0 a^4/32\pi^4 D$. The maximum deflection takes place at the center of the plate and is obtained by substituting this value of a_{11} into Eq. (3.58)

$$w_{\max} = 0.00128 \frac{p_0 a^4}{D}$$

This is approximately 1.5 percent greater than the value, given by Eq. (*d*) of Sec. 3.7, based upon a more elaborate approach. It should be noted that the result obtained, taking only one term of the series, is remarkably accurate. So few terms will not, in general, result in such accuracy when applying the Ritz method.

Let us express the deflection w by retaining seven parameters a_{11}, a_{12}, a_{21}, a_{22}, a_{13}, a_{31}, and a_{33}. Now Eq. (*c*) results in

$$\left[3 + 3\left(\frac{a}{b}\right)^4 + 2\left(\frac{a}{b}\right)^2\right]a_{11} + 2a_{12} + 2\left(\frac{a}{b}\right)^4 a_{21} + 2a_{13} + 2\left(\frac{a}{b}\right)^4 a_{31} = \frac{p_0 a^4}{4\pi^4 D}$$

$$2a_{11} + \left[3 + 48\left(\frac{a}{b}\right)^4 + 8\left(\frac{a}{b}\right)^2\right]a_{12} + 2a_{13} + 32\left(\frac{a}{b}\right)^2 a_{22} = \frac{p_0 a^4}{4\pi^4 D}$$

$$2\left(\frac{a}{b}\right)^4 a_{11} + \left[48 + 3\left(\frac{a}{b}\right)^4 + 8\left(\frac{a}{b}\right)^2\right]a_{21} + 2\left(\frac{a}{b}\right)^4 a_{31} + 32a_{22} = \frac{p_0 a^4}{4\pi^4 D}$$

$$32a_{21} + 16\left[3 + 3\left(\frac{a}{b}\right)^4 + 2\left(\frac{a}{b}\right)^2\right]a_{22} + 32\left(\frac{a}{b}\right)^4 a_{12} = \frac{p_0 a^4}{4\pi^4 D}$$

$$2a_{11} + 2a_{12} + \left[3 + 243\left(\frac{a}{b}\right)^4 + 18\left(\frac{a}{b}\right)^2\right]a_{31} + 162a_{33} = \frac{p_0 a^4}{4\pi^4 D}$$

$$2\left(\frac{a}{b}\right)^4 a_{11} + 2\left(\frac{a}{b}\right)^4 a_{21} + \left[243 + 3\left(\frac{a}{b}\right)^4 + 18\left(\frac{a}{b}\right)^2\right]a_{31} + 162a_{33} = \frac{p_0 a^4}{4\pi^4 D}$$

$$162\left(\frac{a}{b}\right)^4 a_{13} + 162a_{31} + 81\left[3 + 3\left(\frac{a}{b}\right)^4 + 2\left(\frac{a}{b}\right)^2\right]a_{33} = \frac{p_0 a^4}{4\pi^4 D}$$

Simultaneous solution of the above, for a square plate ($a = b$) yields

$$a_{11} = 0.11774p_1 \qquad a_{12} = a_{21} = 0.01184p_1$$

$$a_{22} = 0.00189p_1 \qquad a_{13} = a_{31} = 0.00268p_1$$

$$a_{33} = 0.00020p_1$$

in which $p_1 = p_0 a^4/4D\pi^4$. Upon substituting these values into Eq. (3.58), the maximum deflection is found to be

$$w_{max} = 0.00126 \frac{p_0 a^4}{D}$$

This result is exactly the same value given by Eq. (*d*) of Sec. 3.7.

Example 3.9 A rectangular portion ($a \times b$) of a machine room floor has all its edges built in, and supports a load P acting at a location $x = x_1$, $y = y_1$ (Figs. 3.3 and 3.13). Determine the maximum deflection of the plate.

SOLUTION Applying Eq. (3.55), the work done by the loading is

$$W = P \sum_{m=1}^{\infty} \sum_{n=1}^{\infty} a_{mn} \left(1 - \cos \frac{2m\pi x_1}{a}\right)\left(1 - \cos \frac{2n\pi y_1}{b}\right) \tag{d}$$

The condition $\partial\Pi/\partial a_{mn} = 0$ together with Eqs. (3.59) and (*d*) now leads to

$$4\pi^4 abD \left\{ \left[3\left(\frac{m}{a}\right)^4 + 3\left(\frac{n}{b}\right)^4 + 2\left(\frac{m}{a}\right)^2\left(\frac{n}{b}\right)^2 \right] a_{mn} + \sum_{r=1}^{\infty} 2\left(\frac{m}{a}\right)^4 a_{mr} + \sum_{r=1}^{\infty} 2\left(\frac{n}{b}\right)^4 a_{rn} \right\} - P\left(1 - \cos \frac{2m\pi x_1}{a}\right)\left(1 - \cos \frac{2n\pi y_1}{b}\right) = 0 \tag{e}$$

valid for $r \neq n$ and $r \neq m$. Consider, for simplicity, the case of a square plate ($a = b$) with P acting at its center and retain seven parameters, a_{11}, a_{12}, a_{22}, a_{21}, a_{13}, a_{31}, and a_{33}. We then obtain the following values

$$a_{11} = 0.12662P_1 \qquad a_{12} = a_{21} = -0.00601P_1$$
$$a_{22} = 0.00301P_1 \qquad a_{13} = a_{31} = 0.00278P_1$$
$$a_{33} = 0.00015P_1$$

in which $P_1 = Pa^2/D\pi^4$. The maximum deflection occurring at the center of the plate is, from Eq. (3.58), found to be

$$w_{max} = 0.00543 \frac{Pa^2}{D}$$

This is 3 percent smaller than the value determined from the solution of Eq. (1.17). By retaining more parameters in the series, we expect to improve the result.

PROBLEMS

Secs. 3.1 to 3.3

3.1 A structural steel door 1.5 m long, 0.9 m wide, and 15 mm thick is subjected to a uniform pressure p_0. The material properties are $E = 200$ GPa, $\nu = 0.3$ and allowable yield strength $\sigma_{yp} = 240$ MPa. The plate is regarded as simply supported. Determine, using the Navier's approach and retaining only the first four terms in the series solution: (*a*) the limiting value of p_0 that can be applied to the plate without causing permanent deformation; (*b*) the maximum deflection w that would be produced when p_0 reaches its limiting value.

3.2 Determine the reactions for a simply supported rectangular plate with a concentrated load P at its center (Fig. 3.3). Take $m = n = 1$.

3.3 A rectangular flat portion of a wind tunnel, simply supported on all edges, is under a uniform pressure p_0, distributed over the subregion shown in Fig. 3.3. If $c = a/4$, $d = b/4$, $x_1 = a/2$, and $y_1 = b/2$, find the maximum deflection and maximum stress in the plate, by retaining the first two terms of the series solution.

3.4 A concentrated load P is applied at the center of a simply supported rectangular plate (Fig. 3.1*a*). Determine, if $m = n = 1$, $\nu = 0.3$, and $a = 2b$: (*a*) the maximum deflection; (*b*) the maximum stress in the plate.

3.5 Verify the result given by Eq. (*j*) of Sec. 3.3 by evaluating the limits of $[\sin(m\pi c/a)]/c$ and $[\sin(n\pi y/b)]/d$.

3.6 A simply supported plate is subjected to loads P_1 and P_2 at points ($x = a/2$, $y = a/4$) and ($x = a/2$, $y = 3a/4$), respectively (Fig. 3.1*a*). Obtain the center deflection w by applying Navier's method and superposition. Retain only the first two terms in the series solution and take $a = b$.

Secs. 3.4 to 3.7

3.7 Given a rectangular plate simply supported along its edges and subjected to a uniform load of intensity p_0, determine the value of maximum deflection if $b = 2a$. Compare the result with those given in Table 3.1.

3.8 Determine the expression for shear force Q_x for the plate described in Prob. 3.7.

3.9 Verify the results given by Eqs. (3.14) and (3.15) by assuming a solution of the form $f_m = K_m e^{\lambda_m y}$, where K_m and λ_m are constants.

3.10 Verify the result for w given by Eq. (*i*) of Example 3.4. Take the first two terms in the series solution.

3.11 Redo Example 3.6 if the edge $y = 0$ of the plate is built in. Assume that all other conditions are the same.

3.12 Verify the results for p_m given in Figs. B and C of Table 3.2.

3.13 Determine the equation of the elastic surface for a plate with two opposite edges simply supported, the third edge free, and the fourth edge clamped (Fig. 3.6).

3.14 A water level control structure consists of a vertically positioned simply supported plate. The structure is filled with water up to the upper edge level at $x = 0$ (Fig. A of Table 3.2). Show that by taking only the first term of the series solution, the values for the deflection w and the bending moment M_x at the center, for $a = b/3$, are:

$$w = 0.00614\frac{p_0 a^4}{D} \qquad M_x = 0.06178 p_0 a^2 \qquad \left(x = \frac{a}{2},\ y = 0\right) \tag{P3.14}$$

3.15 Find the equation of the elastic surface for the plate loaded as shown in Fig. C of Table 3.2, if $a = b$ and $x_1 = a/2$.

3.16 Determine the equation of the elastic surface for the plate loaded as shown in Fig. B of Table 3.2. Let $x_1 = e = a/2$, $a = b$, and $m = 1$ to compare the result for the maximum deflection with that listed in Table 3.1.

3.17 A square steel plate 10 mm thick is built in at the bottom of a ship drawing 7 meters of water.

The allowable stress for the plate is 100 MPa. Determine the maximum plate dimension and the maximum deflection. Specific weight of salt water $\gamma = 10.054\ \text{kN/m}^3$. Take $E = 200$ GPa.

3.18 A rectangular plate, with two opposite sides $y = \pm b/2$ fixed and the sides $x = 0$ and $x = a$ simply supported, is subjected to a hydrostatic loading as shown in Fig. A of Table 3.2. By employing the method of superposition derive an expression for the reactional moments M_y along the clamped edges $y = \pm b/2$.

Secs. 3.8 to 3.12

3.19 Employ the strip method to determine the center deflection and the maximum bending moments for a plate with two opposite edges simply supported and the two edges clamped (Fig. 3.9*a*). Compare the results with those obtained in Example 3.7.

3.20 Employ the strip method to obtain the approximate center deflection and center bending moment for the simply supported rectangular plate under hydrostatic loading as shown in Fig. A of Table 3.2.

3.21 Simply supported silo flooring is loaded as shown in Fig. 3.11. Referring to the equations derived in Sec. 3.9, determine the deflection at the center of plate 1 and plate 2, for $a = b$. Take only the first term of the series solution.

3.22 Derive an expression for the moment $f(y)$ distributed along the midsupport of the continuous plate shown in Fig. 3.11. Take $a = b$.

3.23 A uniformly loaded airport platform is supported by equally spaced columns $(a = b)$ (Fig. 3.12*a*). Obtain the maximum deflection and maximum bending moment $M_{x,\max}$ in each panel, for $m = 2$. Compare the results with those given in Table 3.5.

3.24 Determine the value of maximum stress $\sigma_{x,\max}$ at the support locations of a uniformly loaded large plate carried by equally spaced columns $(a = b)$ (Fig. 3.12*a*).

3.25 Determine the required thickness of a plate platform under uniform loading p_0 supported by columns at a and $b = 2a$ distance apart (Fig. 3.12*a*). The plate is fabricated of a material of tensile yield strength σ_{yp}. Design the platform, selecting a factor of safety N, according to: (*a*) maximum principal stress theory; (*b*) maximum principal strain theory.

3.26 Derive the equation of the elastic surface of a simply supported rectangular plate on an elastic foundation (Fig. 3.3) if the plate experiences a central point load P. Use the equilibrium approach of Sec. 3.11.

3.27 Obtain the maximum stress σ_x in the plate loaded as described in Prob. 3.26 by retaining in the series solution the first term only. Take $b = 4a$ and $\nu = 0.3$.

3.28 A uniformly loaded rectangular steel plate rests freely on an elastic foundation of modulus k. What should be the value of k in order to limit the maximum deflection to 0.006 m. Let $E = 200$ GPa, $t = 3$ mm, $a = 0.8$ m, $b = 1{,}4$ m, and $p_0 = 20$ MPa. Retain only the first term of the series solution. The coordinates are placed as shown in Fig. 3.1*a*.

3.29 Redo Prob. 3.26 by employing the Ritz method.

3.30 A uniformly loaded rectangular plate has two opposite sides built in and the other two sides simply supported. Assume a solution of the form

$$w = \sum_{m=1}^{\infty} \sum_{n=1}^{\infty} a_{mn}\left(1 - \cos\frac{2m\pi x}{a}\right)\sin\frac{n\pi y}{b}$$

and determine the coefficients a_{mn} employing the Ritz method.

3.31 A uniformly loaded rectangular plate has its edges $y = 0$ and $y = b$ simply supported, the side $x = 0$ clamped, and the side $x = a$ free. Assume a solution of the form

$$w = C\alpha^2 \sin \pi\lambda \qquad \alpha = \frac{x}{a},\ \lambda = \frac{y}{b}$$

where C is an undetermined coefficient. Apply the Ritz method to derive an expression for the deflection surface. Evaluate the deflection at the middle of the free edge, if $a = b$ and $\nu = 0.3$. The coordinates are placed as shown in Fig. 3.1*a*.

CHAPTER

FOUR

PLATES OF VARIOUS GEOMETRICAL FORMS

4.1 INTRODUCTION

Structural members designed to resist lateral loading are generally circular or rectangular in shape, but in some situations other geometrical forms are used. As previously discussed, the bending problem of plates of any shape is solved if we derive the displacement function $w(x, y)$ which satisfies $\nabla^4 w = p/D$ and the specified conditions at the boundaries. We now deal with the problem of plate boundaries of geometrical form different from those previously treated. From the examples worked out so far, it becomes evident that a rigorous solution of the deflection for a plate with more complicated shape is likely to be very difficult. In developing approximate formulas for the deflections of polygonal plates, the membrane analogy (Sec. 1.6) has proved very valuable.

When the uniformity of the cross-sectional area of the plate is interrupted as in the case of a shipdeck or airplane fuselage with holes or windows, a perturbation in stress takes place. Determination of this disturbed stress distribution is of considerable practical importance and is discussed briefly in Sec. 4.5.

4.2 METHOD OF IMAGES

Certain problems of plates can be treated by arbitrary extension of the plate and/or introducing fictitious forces to produce the deflection forms sought.

Consider, as an example, the bending of an isosceles right triangular plate with simply supported edges under a concentrated load P acting at arbitrary point $A(x_1, y_1)$. The plate is bounded as shown by solid lines OBC in Fig. 4.1.

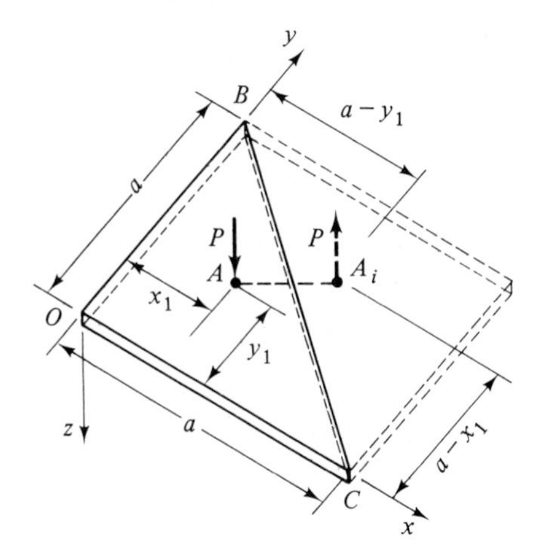

Figure 4.1

We begin the solution by utilizing Eq. (3.11) derived in Sec. 3.3. For this purpose it is assumed that the triangular plate is one half of a simply supported square plate subjected to the forces P and $-P$ at points A and $A_i[(a - y_1), (a - x_1)]$, respectively (Fig. 4.1). Point A_i is the *mirror* or *image point* of A with respect to diagonal BC. Owing to the loading described, the square plate deflects in such a way that diagonal BC is a *nodal line* or *fictitious support*. Hence, the deflection of OBC of the square plate is identical with that of a simply supported triangular plate OBC. The deflection owing to the force P, from Eq. (3.11), is

$$w_1 = \frac{4Pa^2}{\pi^4 D} \sum_{m=1}^{\infty} \sum_{n=1}^{\infty} \frac{\sin(m\pi x_1/a)\sin(n\pi y_1/a)}{(m^2 + n^2)^2} \sin\frac{m\pi x}{a} \sin\frac{n\pi y}{a} \tag{4.1}$$

Upon substitution of $-P$ for P, $(a - y_1)$ for x_1, and $(a - x_1)$ for y_1 in Eq. (4.1), we obtain the deflection due to the force $-P$ at A_i:

$$w_2 = -\frac{4Pa^2}{\pi^4 D} \sum_{m=1}^{\infty} \sum_{n=1}^{\infty} (-1)^{m+n} \frac{\sin(m\pi x_1/a)\sin(n\pi y_1/a)}{(m^2 + n^2)^2} \sin\frac{m\pi x}{a} \sin\frac{n\pi y}{b} \tag{4.2}$$

The deflection surface of the triangular plate is then

$$w = w_1 + w_2 \tag{4.3}$$

where w_1 and w_2 are given by Eqs. (4.1) and (4.2) respectively. By applying Eq. (4.3) together with the principle of superposition, the deflection may be determined of an isosceles right triangular plate for any kind of loading. The method described above, often referred to as the *method of images*, was introduced by Nadai.[13]

In the particular case of a triangular plate under a uniformly distributed load of intensity $p_0 = P/dx_1\,dy_1$, we find, from Eq. (4.3), after integration over the area of the plate, that

$$w = \frac{16p_0 a^4}{\pi^6 D}\left[\sum_{m=1,3,\ldots}^{\infty}\sum_{n=2,4,\ldots}^{\infty}\frac{n\sin(m\pi x/a)\sin(n\pi y/a)}{m(n^2-m^2)(m^2+n^2)^2} + \sum_{m=2,4,\ldots}^{\infty}\sum_{n=1,3,\ldots}^{\infty}\frac{m\sin(m\pi x/a)\sin(n\pi y/a)}{n(m^2-n^2)(m^2+n^2)^2}\right] \tag{4.4}$$

This series converges very rapidly. With the expression for the deflection of the plate determined, one can obtain the moments and the maximum stresses in the plate from Eqs. (1.10) and (1.12).

4.3 EQUILATERAL TRIANGULAR PLATE WITH SIMPLY SUPPORTED EDGES

We now treat the case of a simply supported equilateral triangular plate ABC, shown in Fig. 4.2. The equations of the boundaries are:

$$\begin{aligned} x + \frac{a}{3} &= 0 \qquad \text{on } BC \\ \frac{x}{\sqrt{3}} + y - \frac{2a}{3\sqrt{3}} &= 0 \qquad \text{on } AC \\ \frac{x}{\sqrt{3}} - y - \frac{2a}{3\sqrt{3}} &= 0 \qquad \text{on } AB \end{aligned} \tag{a}$$

Therefore, the function which vanishes at the boundary

$$\begin{aligned} w &= k\left(x + \frac{a}{3}\right)\left(\frac{x}{\sqrt{3}} + y - \frac{2a}{3\sqrt{3}}\right)\left(\frac{x}{\sqrt{3}} - y - \frac{2a}{3\sqrt{3}}\right) \\ &= k\left(\frac{x^3 - 3xy^2}{3} - \frac{a(x^3 + y^2)}{3} + \frac{4a^3}{81}\right) \end{aligned} \tag{b}$$

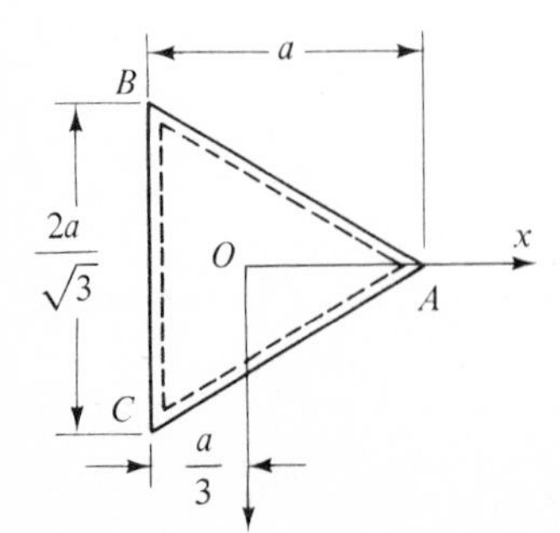

Figure 4.2

is the general expression for deflection of an equilateral triangular plate with simply supported edges. Here k is determined for two specific cases of loading as follows.

Equilateral triangular plate under uniform moment M_0 along its boundary It can be verified experimentally that the deflection surface of a plate loaded by uniform moment along its boundary, and the surface of a uniformly loaded membrane, uniformly stretched over the same triangular boundary, are identical. To arrive at an analytical solution, we introduce Eq (b) into Eq. (1.22b) and set $M = M_0$. It follows that

$$k = \frac{3M_0}{4aD}$$

The deflection is then

$$w = \frac{M_0}{4aD}\left[x^3 - 3xy^2 - a(x^2 + y^2) + \tfrac{4}{27}a^3\right] \tag{4.5}$$

and the center displacement is

$$w_0 = \frac{M_0 a^2}{27D} \qquad (x = y = 0) \tag{c}$$

The expressions for the moments are determined by means of Eqs. (4.5) and (1.10) as follows:

$$\begin{aligned} M_x &= \tfrac{1}{2}M_0\left[1 + \nu - (1 - \nu)\frac{3x}{a}\right] \\ M_y &= \tfrac{1}{2}M_0\left[1 + \nu + (1 - \nu)\frac{3x}{a}\right] \\ M_{xy} &= -\frac{3(1 - \nu)M_0 y}{2a} \end{aligned} \tag{4.6}$$

The largest moment takes place at the corners and acts on the planes bisecting the angles. For $\nu = 0.3$, it is given by

$$M_{y,\max} = 1.35M_0 \qquad (y = \tfrac{2}{3}a) \tag{d}$$

The corresponding stress $\sigma_{y,\max} = 8.10M_0/t^2$.

Equilateral triangular plate under uniform load p_0 In this case an expression for deflection of the form

$$w = \frac{p_0}{64aD}\left[x^3 - 3xy^2 - a(x^2 + y^2) + \tfrac{4}{27}a^3\right]\left(\tfrac{4}{9}a^2 - x^2 - y^2\right) \tag{4.7}$$

satisfies Eq. (1.17) and the conditions of the simply supported edges. Therefore Eq. (4.7) is the solution. By employing Eqs. (1.10) the bending moments may then be obtained. It can be shown that the center moments are

$$M_x = M_y = (1 + \nu)\frac{p_0 a^2}{54} \tag{4.8}$$

The largest moment takes place on the planes bisecting the angles of the triangle. For example, at points along the x axis (Fig. 4.2), for $\nu = 0.3$, we have

$$\begin{aligned} M_{x,\,\max} &= 0.0248 p_0 a^2 \qquad (x = -0.062a,\ y = 0) \\ M_{y,\,\max} &= 0.0259 p_0 a^2 \qquad (x = 0.129a,\ y = 0) \end{aligned} \tag{e}$$

The maximum stress is given by $\sigma_{y,\,\max} = 0.155 p_0 a^2/t^2$.

4.4 ELLIPTICAL PLATES

In this section, consideration is given the bending of an elliptic plate with semi-axes a and b, subjected to a uniform load of intensity p_0 (Fig. 4.3). Two situations are treated below.

Uniformly loaded elliptic plate with clamped edge The appropriate boundary conditions are

$$w = 0 \qquad \frac{\partial w}{\partial n} = 0 \qquad \left(\text{for } \frac{x^2}{a^2} + \frac{y^2}{b^2} = 1\right) \tag{a}$$

where n is the normal to the plate boundary. These conditions are satisfied if we assume for deflection, the expression:

$$w = k\left(1 - \frac{x^2}{a^2} - \frac{y^2}{b^2}\right)^2 \tag{b}$$

in which k is a constant. Substituting Eq. (b) into Eq. (1.17) and setting $p = p_0$, we have

$$k = \frac{p_0}{8D}\frac{a^4 b^4}{3a^4 + 2a^2 b^2 + 3b^4} \tag{c}$$

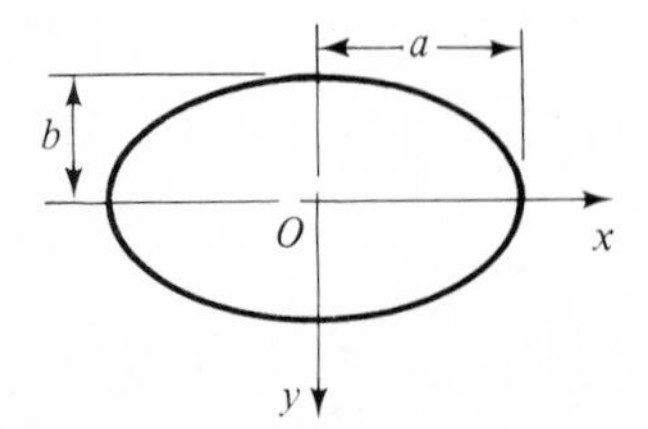

Figure 4.3

The required solution is then

$$w = \frac{p_0}{8D} \frac{a^4 b^4}{3a^4 + 2a^2 b^2 + 3b^4} \left(1 - \frac{x^2}{a^2} - \frac{y^2}{b^2}\right)^2 \tag{4.9}$$

and the maximum deflection occurs at the center of the plate and is given by

$$w_{\max} = \frac{p_0}{8D[(3/b^4) + (2/a^2 b^2) + (3/a^4)]} \tag{4.10}$$

By substituting Eq. (b) into Eq. (1.10), the bending and twisting moments are obtained

$$\begin{aligned} M_x &= 4Dk\left[\frac{1}{a^2} - \frac{3x^2}{a^4} - \frac{y^2}{a^2 b^2} + \nu\left(\frac{1}{b^2} - \frac{3y^2}{b^4} - \frac{x^2}{a^2 b^2}\right)\right] \\ M_y &= 4Dk\left[\frac{1}{b^2} - \frac{3y^2}{b^4} - \frac{x^2}{a^2 b^2} + \nu\left(\frac{1}{a^2} - \frac{3x^2}{a^4} - \frac{y^2}{a^2 b^2}\right)\right] \\ M_{xy} &= -(1 - \nu)\frac{8Dk}{a^2 b^2} xy \end{aligned} \tag{4.11}$$

At the center ($x = 0$, $y = 0$), the moments are

$$\begin{aligned} M_x &= 4Dk\left(\frac{1}{a^2} + \frac{\nu}{b^2}\right) \\ M_y &= 4Dk\left(\frac{1}{b^2} + \frac{\nu}{a^2}\right) \end{aligned} \tag{d}$$

At the ends of the minor and major axes, we have

$$\begin{aligned} M_x &= -\frac{8Dk}{a^2} \qquad M_y = -\frac{8\nu Dk}{a^2} \qquad (x = \pm a,\ y = 0) \\ M_x &= -\frac{8\nu Dk}{b^2} \qquad M_y = -\frac{8Dk}{b^2} \qquad (x = 0,\ y = \pm b) \end{aligned} \tag{4.12}$$

where k is given in Eq. (c). It is observed from Eqs. (4.12) that the *maximum* moment occurs at the *extremity of the minor axis*.

For $a = b$ the solution derived in this section reduces to the results obtained for a circular plate with clamped edge (Sec. 2.4). In the extreme case, $a \gg b$, Eqs. (4.10) and the second of (d) appear as

$$w_{\max} = \frac{p_0 (2b)^4}{384D} \qquad M_y = \frac{p_0 (2b)^2}{24} \tag{e}$$

Equations (e) correspond to the deflection and moment at the center of a fixed-end strip of span $2b$ under uniformly distributed load.

Table 4.1

a/b	1.0	1.5	2.0	3.0	4.0
$w \cdot (Et^3/p_0 b^4)$	0.70	1.26	1.58	1.88	2.02
$M_x \cdot (1/p_0 b^2)$	0.206	0.222	0.210	0.188	0.184
$M_y \cdot (1/p_0 b^2)$	0.206	0.321	0.379	0.433	0.465

Uniformly loaded elliptic plate with simply supported edge The boundary conditions are now given by

$$w = 0 \qquad \frac{\partial^2 w}{\partial n^2} = 0 \qquad \left(\text{for } \frac{x^2}{a^2} + \frac{y^2}{b^2} = 1\right)$$

where, as before, n is a normal to the plate boundary. After routine but somewhat more lengthy calculations than in the case of a fixed plate, expressions for deflection and bending moments may be obtained. Table 4.1 provides, for $\nu = 0.3$, some final numerical results for the deflection and bending moments at the center of the plate.[11]

4.5 STRESS CONCENTRATION AROUND HOLES IN A PLATE

The bending theory of plates applies only to plates of constant cross-sectional area. If the cross section changes gradually, reasonably accurate results can also be expected. On the other hand, where abrupt variation in the cross section takes place, ordinary bending theory cannot predict the high values of stress which actually exist. The condition referred to occurs in plates with holes, notches, or fillets. In some instances, the stresses in these regions can be analyzed by applying special theories which take into account the shear deformation. The solutions more often involve considerable difficulty. However, it is more usual to rely upon experimental methods. The finite element approach (Sec. 5.7) is also very efficient for this purpose.

The ratio k_b of the actual maximum stress to the nominal stress in the minimum section is defined as the *stress concentration factor*. Hence,

$$\sigma_{\max} = k_b \sigma_n \tag{4.13}$$

where σ_n is obtained by applying the familiar flexure formula for beams. The technical literature contains extensive information on stress concentration factors presented in the form of graphs and tables.[7,14]

It is convenient to examine the stress concentration at the edge of a small hole in a very large plate (Example 4.1). The results obtained for this case have been proven applicable to plates of any shape.

Example 4.1 A large, thin plate containing a circular hole of radius a and width $2a$, small as compared with the overall dimensions, is subjected to

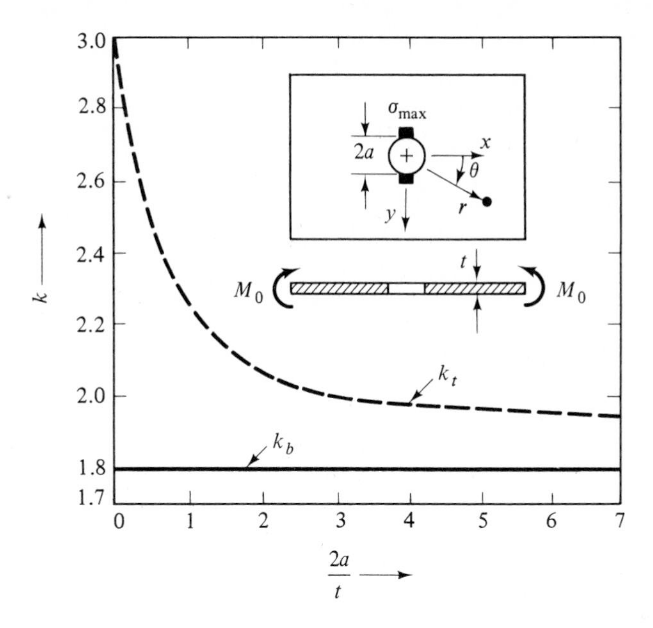

Figure 4.4

pure bending $M_x = M_0$ and $M_y = 0$ (Fig. 4.4). Determine the stress distribution around the hole.

SOLUTION We place the origin of coordinates at the center of the hole. In the absence of any hole, one has, from polar form of Eq. (1.30)

$$w_1 = -\frac{M_0 r^2}{4D(1-\nu^2)}[1 - \nu + (1+\nu)\cos 2\theta] \tag{a}$$

At $r = a$ the radial moment and effective transverse force owing to the M_0, obtained by means of Eqs. (*a*), (2.2), and (2.4), are:

$$M_{r1} = \tfrac{1}{2}M_0(1 + \cos 2\theta)$$
$$V_{r1} = \frac{1}{a}M_0 \cos 2\theta \tag{b}$$

When a hole of radius a is drilled through the plate, we superimpose on the original state of stress an additional state of stress so that: the combined radial moments and forces equal to zero at $r = a$; the superimposed stresses *only* are zero at $r = \infty$. Guided by the expression for w_1, we assume the additional deflection of the form[11]

$$w_2 = -\frac{M_0 a^2}{2D}\left[A \ln r + \left(B - C\frac{a^2}{r^2}\right)\cos 2\theta\right] \tag{c}$$

thus satisfying both conditions stated above and Eq. (2.6). The moment and effective transverse force corresponding to Eq. (*c*) at $r = a$ are determined as follows

$$\begin{aligned} M_{r2} &= -\tfrac{1}{2}M_0\{(1-\nu)A + [4\nu B - 6(1-\nu)C]\cos 2\theta\} \\ V_{r2} &= \frac{1}{a}M_0[2(3-\nu)B + 6(1-\nu)C]\cos 2\theta \end{aligned} \tag{d}$$

The conditions along the boundary of the hole are represented by

$$M_{r1} + M_{r2} = 0 \qquad V_{r1} + V_{r2} = 0 \qquad (r = a) \tag{e}$$

Substitution of Eqs. (*b*) and (*d*) into Eqs. (*e*) leads to the values of the constants. The final deflection expression $(w_1 + w_2)$ is then obtained. It follows that the tangential moment and shear force at the periphery of the circle $(r = a)$ are

$$\begin{aligned} M_\theta &= M_0\left[1 - \frac{2(1+\nu)}{3+\nu}\cos 2\theta\right] \\ Q_\theta &= \frac{4M_0}{(3+\nu)a}\sin 2\theta \end{aligned} \tag{4.14}$$

From Eqs. (4.14), it is evident that the maximum values of stress resultants occur at $\theta = \pi/2$ and $\theta = \pi/4$, respectively:

$$M_{\theta,\,max} = \frac{5+3\nu}{3+\nu}M_0 \qquad Q_{\theta,\,max} = \frac{4}{(3+\nu)a}M_0 \tag{4.15}$$

The stress concentration factor is therefore

$$k_b = \frac{5+3\nu}{3+\nu} \tag{4.16}$$

where reduction in the cross-sectional area due to small hole has been neglected. For $\nu = 1/3$, Eq. (4.16) yields $k_b = 1.80$.

Upon introducing Eqs. (4.15) into Eqs. (1.12) and (1.20) and setting $\nu = 1/3$, the maximum stresses are found to be $\sigma_{max} = 10.8M_0/t^2$ and $\tau_{max} = 1.8M_0/at$. The ratio of these stresses is

$$\frac{\tau_{max}}{\sigma_{max}} = \frac{t}{6a} \tag{f}$$

When, for example, the hole width is equal to the plate thickness the above quotient is 1/3. For $2a < t$, the ratio τ_{max}/σ_{max} is larger. The influence of shear upon the plate deformation is thus pronounced.

In Fig. 4.4, the dashed line represents a plot of the true stress concentration factor k_t, obtained from Reisner's theory which takes the shear deformation into account.[11] In both cases ν is taken to be 1/3. It is observed from the graphs that the stress concentration factor determined using the thin plate theory is quite crude.

The case of plates with noncircular holes may also be treated by the use of the bending theory, as illustrated in the example. For the reasons cited above, the solutions offer unacceptably poor accuracy.

PROBLEMS

Secs. 4.1 to 4.3

4.1 A simply supported wing panel in the form of an isosceles right triangle is subjected to a uniform load of intensity p_0 (Fig. 4.1). Determine, by retaining only the first term of the series solution, at point $A(x = a/4,\ y = a/4)$: (*a*) the deflection; (*b*) the bending moments.

4.2 Consider an equilateral triangular plate with simply supported edges under a uniform moment M_0 along its boundary (Fig. 4.2). Taking $\nu = 1/3$, find: (*a*) the twisting moment on the side AC; (*b*) the concentrated reactions at the corners.

Secs. 4.4 to 4.5

4.3 A manhole cover consists of a cast-iron elliptical plate 1.0 m long, 0.5 m wide, and 0.02 m thick. Determine the maximum uniform loading the plate can carry when the allowable stress is limited to 20 MPa. Use $\nu = 0.3$. Assume that (*a*) the plate is simply supported at the edge and (*b*) the plate is built in at the edge.

4.4 Consider two plates, one having a circular shape of radius c, the other an elliptical shape with semiaxes $a = 2c$ and $b = c$, both clamped at the edge and under a uniform load of intensity p_0. Determine the ratio of the maximum deflection and the maximum bending moment for the elliptical plate to those for the circular plate. Take $\nu = 0.3$.

4.5 An elliptical plate built in at the edge is subjected to a linearly varying pressure $p = p_0 x$ (Fig. 4.3). Given an expression for the deflection of the form

$$w = \frac{p_0 x}{24D} \frac{[1 - (x/a)^2 - (y/b)^2]^2}{(5/a^4) + (1/b^4) + (2/a^2b^2)} \tag{P4.5}$$

determine whether the above represents a possible solution. By means of Eq. (P4.5), obtain the bending moment M_x and the stress σ_x (at $x = a,\ y = 0$) for $a = 2b$.

4.6 Redo Prob. 4.4 if the edges of the plates are simply supported.

CHAPTER

FIVE

PLATE BENDING BY NUMERICAL METHODS

5.1 INTRODUCTION

In the previous chapters, equilibrium and energy methods are employed in the solution of a number of plate-bending problems. In some cases, however, these analytical solutions are not always possible, and one must resort to approximate numerical methods. The use of numerical approaches enables the engineer to expand his or her ability to solve design problems of practical significance. Treated are real shapes and loadings, as distinct from the somewhat limited variety of shapes and loadings amenable to simple analytical solution.

Among the most important of the numerical approaches are the method of *finite differences* (Secs. 5.2 to 5.6) and the *finite element method* (Secs. 5.7 to 5.10). Both techniques eventually require the solution of a system of linear algebraic equations (App. B). Such calculations are commonly performed by means of a digital computer employing matrix methods. A less formal method, leading to an improved understanding of how the physical behavior relates to the numerical analysis, was presented in Sec. 2.12.

5.2 FINITE DIFFERENCES

The method of finite differences replaces the plate differential equation and the expressions defining the boundary conditions with equivalent difference equations. The solution of the bending problem thus reduces to the simultaneous solution of a set of algebraic equations, written for every nodal point within the plate. This section is concerned with the fundamentals of finite differences.[15,16]

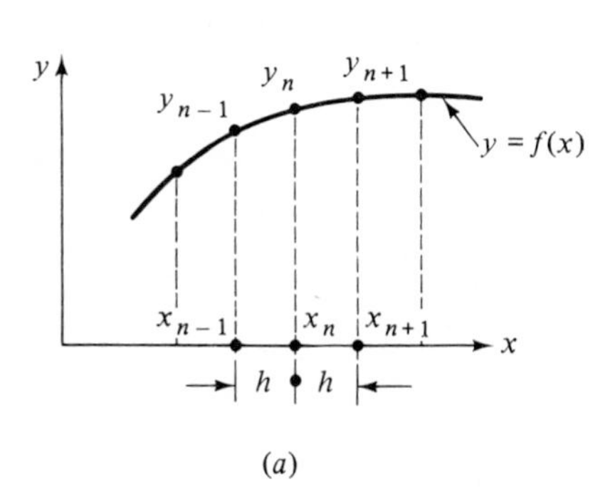

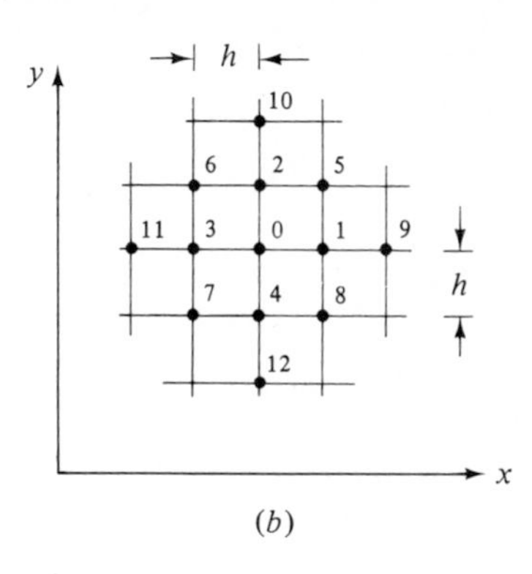

Figure 5.1

The finite difference expressions may be obtained from the definition of the first derivative with respect to x of a continuous function $y = f(x)$ (Fig. 5.1*a*):

$$\left(\frac{dy}{dx}\right)_n = \lim_{\Delta x \to 0} \frac{y_{n+1} - y_n}{\Delta x}$$

The subscript n denotes any arbitrary point on the curve. Over an increment $\Delta x = h$, the above expression represents an approximation to the derivative given by

$$\left(\frac{dy}{dx}\right)_n \approx \frac{\Delta y_n}{h} = \frac{y_{n+1} - y_n}{h}$$

Here Δy_n is called the *first forward difference* of y at point x_n,

$$\Delta y_n = y_{n+1} - y_n \approx h\left(\frac{dy}{dx}\right)_n \tag{5.1}$$

The *first backward difference* at n, ∇y_n, is

$$\nabla y_n = y_n - y_{n-1} \approx h\left(\frac{dy}{dx}\right)_n \tag{5.2}$$

The central differences contain nodal points symmetrically located with respect to x_n. The *first central difference* δy_n is thus

$$\delta y_n = \tfrac{1}{2}(y_{n+1} - y_{n-1}) \approx h\left(\frac{dy}{dx}\right)_n \tag{5.3}$$

A procedure identical to that used above will yield the higher-order derivatives. We shall hereafter consider only the central differences. The second derivative can be written using the difference representation of the first derivative:

$$h^2\left(\frac{d^2y}{dx^2}\right)_n \approx \Delta(\nabla y_n) = \nabla(\Delta y_n) = \delta^2 y_n$$

The *second central difference* at x_n, after substitution of Eqs. (5.1) and (5.2) into the above, is

$$\delta^2 y_n = \Delta y_n - \Delta y_{n-1} = (y_{n+1} - y_n) - (y_n - y_{n-1})$$

$$= y_{n+1} - 2y_n + y_{n-1} \approx h^2\left(\frac{d^2y}{dx^2}\right)_n \tag{5.4}$$

The *third* and the *fourth central differences* are

$$\delta^3 y_n = \delta(\delta^2 y_n) = \delta(y_{n+1} - 2y_n + y_{n-1}) = \delta y_{n+1} - 2\,\delta y_n + \delta y_{n-1}$$

$$= \tfrac{1}{2}(y_{n+2} - y_n) - (y_{n+1} - y_{n-1}) + \tfrac{1}{2}(y_n - y_{n-2})$$

$$= \tfrac{1}{2}(y_{n+2} - 2y_{n+1} + 2y_{n-1} - y_{n-2}) \approx h^3\left(\frac{d^3y}{dx^3}\right)_n \tag{a}$$

and

$$\delta^4 y_n = \delta^2(\delta^2 y_n) = \delta^2(y_{n+1} - 2y_n + y_{n-1}) = \delta^2 y_{n+1} - 2\,\delta^2 y_n + \delta^2 y_{n-1}$$

$$= (y_{n+2} - 2y_{n+1} + y_n) - 2(y_{n+1} - 2y_n + y_{n-1}) + (y_n - 2y_{n-1} + y_{n-2})$$

$$= y_{n+2} - 4y_{n+1} + 6y_n - 4y_{n-1} + y_{n-2} \approx h^4\left(\frac{d^4y}{dx^4}\right)_n \tag{b}$$

We now discuss the case of the function $w(x, y)$ of two variables. For the purpose of illustration consider a rectangular plate. By taking $\Delta x = \Delta y = h$, the plate is divided into a square *mesh* (Fig. 5.1*b*). Equations (5.3) and (5.4) lead to

$$\frac{\partial w}{\partial x} \approx \frac{1}{h}\,\delta_x w \qquad \frac{\partial w}{\partial y} \approx \frac{1}{h}\,\delta_y w$$

$$\frac{\partial^2 w}{\partial x^2} \approx \frac{1}{h^2}\,\delta_x^2 w \qquad \frac{\partial^2 w}{\partial y^2} \approx \frac{1}{h^2}\,\delta_y^2 w \qquad \frac{\partial^2 w}{\partial x\,\partial y} \approx \frac{1}{h}\,\delta_x\left(\frac{\partial w}{\partial y}\right)$$

Here the subscripts x and y designate the directions in which the differences are taken. The above expressions, based upon the definition of the partial derivatives, are written at a point 0 as follows

$$\frac{\partial w}{\partial x} \approx \frac{1}{2h}[w(x+h, y) - w(x-h, y)] = \frac{1}{2h}(w_1 - w_3)$$

$$\frac{\partial w}{\partial y} \approx \frac{1}{2h}[w(x, y+h) - w(x, y-h)] = \frac{1}{2h}(w_2 - w_4) \tag{c}$$

and

$$\frac{\partial^2 w}{\partial x^2} \approx \frac{1}{h^2}[w(x+h, y) - 2w(x, y) + w(x-h, y)] = \frac{1}{h^2}(w_1 - 2w_0 + w_3)$$

$$\frac{\partial^2 w}{\partial y^2} \approx \frac{1}{h^2}(w_2 - 2w_0 + w_4) \tag{d}$$

$$\frac{\partial^2 w}{\partial x\,\partial y} \approx \frac{1}{h^2}\,\delta_x(\delta_y w) = \frac{1}{2h^2}(\delta_x w_2 - \delta_x w_4) = \frac{1}{4h^2}(w_5 - w_6 + w_7 - w_8)$$

The finite difference approximation of the Laplace operator at the point 0 is thus

$$\nabla^2 w = \frac{1}{h^2}(w_1 + w_2 + w_3 + w_4 - 4w_0) \tag{5.5}$$

The formulas for approximating the higher-order partial derivatives are developed by applying Eqs. (*a*) and (*b*) in the x and y directions, respectively:

$$\begin{aligned}
\frac{\partial^3 w}{\partial x^3} &\approx \frac{1}{h^3}\,\delta_x^3 w = \frac{1}{2h^3}(w_9 - 2w_1 + 2w_3 - w_{11}) \\
\frac{\partial^4 w}{\partial x^4} &\approx \frac{1}{h^4}\,\delta_x^2 w = \frac{1}{h^4}(w_9 - 4w_1 + 6w_0 - 4w_3 + w_{11}) \\
\frac{\partial^3 w}{\partial y^3} &\approx \frac{1}{h^3}\,\delta_y^3 w = \frac{1}{2h^3}(w_{10} - 2w_2 + 2w_4 - w_{12}) \\
\frac{\partial^4 w}{\partial y^4} &\approx \frac{1}{h^4}\,\delta_y^4 w = \frac{1}{h^4}(w_{10} - 4w_2 + 6w_0 - 4w_4 + w_{12})
\end{aligned} \tag{e}$$

The mixed derivatives are also represented by

$$\begin{aligned}
\frac{\partial^3 w}{\partial x\,\partial y^2} &\approx \frac{1}{h^3}\,\delta_x(\delta_y^2 w) = \frac{1}{h^3}\,\delta_x(w_2 - 2w_0 + w_4) \\
&= \frac{1}{h^3}(\delta_x w_2 - 2\,\delta_x w_0 + \delta_x w_4) \\
&= \frac{1}{2h^3}(w_5 - w_6 - 2w_1 + 2w_3 + w_8 - w_7) \\
\frac{\partial^3 w}{\partial x^2\,\partial y} &\approx \frac{1}{h^3}\,\delta_y(\delta_x^2 w) = \frac{1}{2h^3}(w_5 + w_6 - 2w_2 + 2w_4 - w_8 - w_7) \\
\frac{\partial^4 w}{\partial x^2\,\partial y^2} &\approx \frac{1}{h^4}\,\delta_x^2(\delta_y^2 w) = \frac{1}{h^4}\,\delta_x^2(w_2 - 2w_0 + w_4) \\
&= \frac{1}{h^4}[w_5 + w_6 + w_7 + w_8 + 4w_0 - 2(w_1 + w_2 + w_3 + w_4)]
\end{aligned} \tag{f}$$

Having available the various derivatives as difference approximations, one can readily obtain the finite difference equivalent of the plate equations. For reference purposes, some useful finite difference operators are represented in the form of *coefficient patterns* in Table 5.1. It is noted that the *center point* in each case is the node about which each operator is written.

Similar formulas can be derived when the nodal points are *not evenly spaced*.

Table 5.1 Coefficient patterns for some finite difference operators

Operator	Pattern	Operator	Pattern
$2h\left(\frac{d}{dx}\right) \approx$	(y, x axes) $\begin{matrix} -1 & 0 & 1 \end{matrix}$	$h^2\left(\frac{d^2}{dx^2}\right) \approx$	$\begin{matrix} 1 & -2 & 1 \end{matrix}$
$h^3\left(\frac{d^3}{dx^3}\right) \approx$	$\begin{matrix} -1 & 2 & 0 & -2 & 1 \end{matrix}$	$h^4\left(\frac{d^4}{dx^4}\right) \approx$	$\begin{matrix} 1 & -4 & 6 & -4 & 1 \end{matrix}$
$h^2\nabla^2 \approx$	$\begin{matrix} & 1 & \\ 1 & -4 & 1 \\ & 1 & \end{matrix}$	$h^3\left(\frac{\partial^3}{\partial x\,\partial y^2}\right) \approx$	$\begin{matrix} -1 & 0 & 1 \\ 2 & 0 & -2 \\ -1 & 0 & 1 \end{matrix}$
$h^4\left(\frac{\partial^4}{\partial x^2\,\partial y^2}\right) \approx$	$\begin{matrix} 1 & -2 & 1 \\ -2 & 4 & -2 \\ 1 & -2 & 1 \end{matrix}$	$\frac{h^2}{D}(-M_x) \approx$	$\begin{matrix} & \nu & \\ 1 & -2-2\nu & 1 \\ & \nu & \end{matrix}$
$\frac{2h^3}{D}(-V_x) \approx$	$\begin{matrix} & & 0 & & \\ & \nu-2 & 0 & 2-\nu & \\ -1 & 6-2\nu & 0 & 2\nu-6 & 1 \\ & \nu-2 & 0 & 2-\nu & \\ & & 0 & & \end{matrix}$	$h^4\nabla^4 \approx$	$\begin{matrix} & & 1 & & \\ & 2 & -8 & 2 & \\ 1 & -8 & 20 & -8 & 1 \\ & 2 & -8 & 2 & \\ & & 1 & & \end{matrix}$

In the case of a rectangular mesh with $\Delta x = h$ and $\Delta y = k$, one need *only* to replace h by k in the foregoing derivatives of w with respect to y. That is, for example:

$$
\begin{aligned}
&\frac{\partial w}{\partial x} \approx \frac{1}{2h}(w_1 - w_3) \qquad \frac{\partial w}{\partial y} \approx \frac{1}{k^2}(w_2 - w_4) \\
&\frac{\partial^2 w}{\partial x^2} \approx \frac{1}{h^2}(w_1 - 2w_0 + w_3) \qquad \frac{\partial^2 w}{\partial y^2} \approx \frac{1}{k^2}(w_2 - 2w_0 + w_4) \\
&\frac{\partial^2 w}{\partial x\,\partial y} \approx \frac{1}{h}\delta_x\left(\frac{1}{k}\delta_y w\right) = \frac{1}{2hk}(\delta_x w_2 - \delta_x w_4) \\
&\qquad = \frac{1}{4hk}(w_5 - w_6 + w_7 - w_8)
\end{aligned}
\tag{g}
$$

The operator $\nabla^2 w$ given by Eq. (5.5) becomes then

$$
\nabla^2 w \approx \frac{1}{h^2}(w_1 - 2w_0 + w_3) + \frac{1}{k^2}(w_2 - 2w_0 + w_4) \tag{5.6}
$$

Unless otherwise specified, we shall refer to equidistance ($\Delta x = \Delta y = h$) nodal points.

Difference operators in cartesian coordinates x and y are well adapted to the solution of problems involving rectangular domains. When the plate has a curved or irregular boundary, special operators must be used at nodal points adjacent to the boundary (Sec. 5.4). One of the noncartesian meshes most commonly employed to cover the polygonal and irregular boundaries is the *triangular mesh* (Sec. 5.6). If the plate shape is a parallelogram, it may often be more accurate and convenient to use coordinates parallel to the edges of the plate, *skew coordinates*. The *polar mesh* is used in connection with shapes having some degree of axisymmetry (Sec. 5.5). The finite difference operators in any coordinate set are developed through transformation of the equations that relate the x and y coordinates to that set. In all cases, the procedure for determining the deflections and the moments is the same as described in the following section.

5.3 SOLUTION OF THE FINITE DIFFERENCE EQUATIONS

We are now in a position to transform the differential equation of the bent plate into an algebraic equation. Let us write this equation for an interior point, such as 0 in Fig. 5.1*b*. Referring to the operator ∇^4 of Table 5.1, one finds that the difference equation corresponding to Eq. (1.17) is

$$[w_9 + w_{10} + w_{11} + w_{12} + 2(w_5 + w_6 + w_7 + w_8) \\ - 8(w_1 + w_2 + w_3 + w_4) + 20w_0]\frac{1}{h^4} = \frac{p_0}{D} \qquad (5.7)$$

Similar expressions are written for every nodal point within the plate. At the same time, the boundary conditions must also be converted into finite difference form. The set of difference equations is then solved to yield the deflection.

As an alternate approach to the plate-bending problem, it is shown in Sec. 1.6 that Eq. (1.17) can be replaced by two second-order equations (1.22). The application of the operator ∇^2 of Table 5.1 to the latter equations at point 0 leads to

$$(M_1 + M_2 + M_3 + M_4 - 4M_0)\frac{1}{h^2} = -p_0$$

$$(w_1 + w_2 + w_3 + w_4 - 4w_0)\frac{1}{h^2} = -\frac{M_0}{D} \qquad (5.8a,b)$$

Identical equations are written for the remaining nodes within the plate. Solution of the problem now requires the determination of those values of M and w satisfying the system of algebraic equations and the given boundary conditions. In the case of a *simply supported* plate, M and w are zero at the boundary, and one can solve the first set of equations *independently* of the second system to find all the values of M within the boundary. The second set is then solved. For plates with edges *clamped or free, or for mixed* boundary conditions, it is necessary to solve both sets *simultaneously.* Note that for plates with mixed boundary conditions, the values of M may be different at the edges. The deflection w may then be obtained more advantageously by direct application of Eq. (5.7) instead of from Eqs. (5.8).

With the values of M and w available at all nodes, one can derive expressions for the moments and shear forces from Eqs. (1.10) and Eqs. (*b*) of Sec. 1.6. These are, at point 0:

$$M_x = \frac{D}{h^2}[2w_0 - w_1 - w_3 + \nu(2w_0 - w_2 - w_4)]$$

$$M_y = \frac{D}{h^2}[2w_0 - w_2 - w_4 + \nu(2w_0 - w_1 - w_3)] \tag{5.9}$$

$$M_{xy} = \frac{D(1-\nu)}{4h^2}(-w_5 + w_6 - w_7 + w_8)$$

and

$$Q_x = \frac{1}{2h}(M_1 - M_3) \qquad Q_y = \frac{1}{2h}(M_2 - M_4) \tag{5.10}$$

The stresses are now readily determined through the application of Eqs. (1.12) and (1.20).

The finite difference method can best be illustrated by reference to numerical examples.

Example 5.1 Use finite difference techniques to analyze the bending of a square plate ($a \times a$) with simply supported edges subjected to a uniformly distributed load of p_0.

SOLUTION The domain is divided into a number of small squares, 16 for example. We thus have $h = a/4$. In labeling nodal points, it is important to take into account any conditions of symmetry which may exist. This has been done in Fig. 5.2*a*. It is seen that only one-eighth of the surface need then be considered, this indicated in the figure by the shaded area. From the boundary conditions at points located on the boundary, M and w are zero. Application of the operator ∇^2 of Table 5.1 to $\nabla^2 M = -p$ at the nodes 1, 2, and 3 results in

$$2M_2 - 4M_1 = -\frac{p_0 a^2}{16}$$

$$2M_1 + M_3 - 4M_2 = -\frac{p_0 a^2}{16}$$

$$4M_2 - 4M_3 = \frac{p_0 a^2}{16}$$

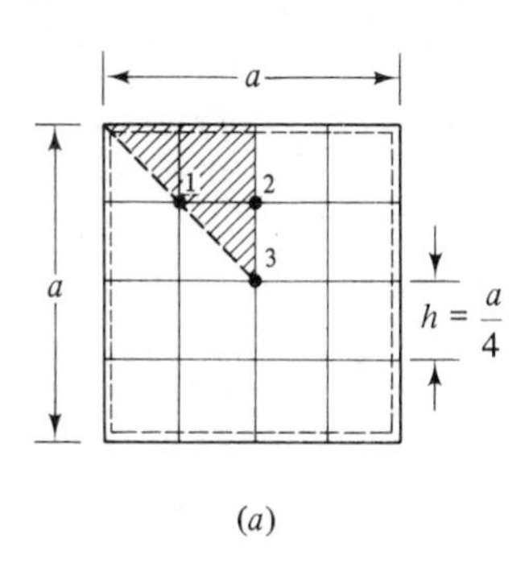

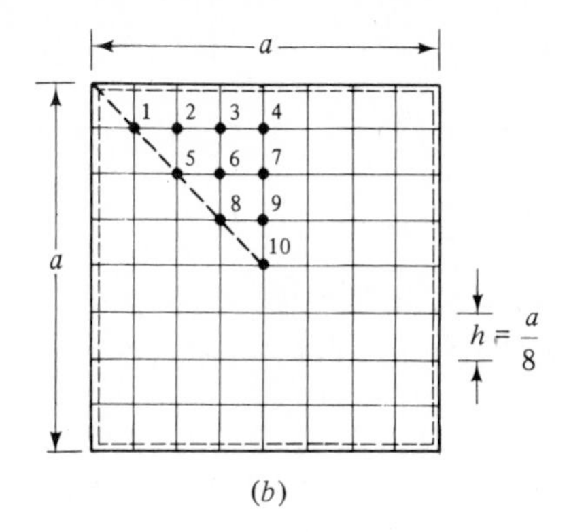

Figure 5.2

Simultaneous solution yields

$$M_1 = \tfrac{11}{256}p_0a^2 \qquad M_2 = \tfrac{7}{128}p_0a^2 \qquad M_3 = \tfrac{9}{128}p_0a^2 \tag{a}$$

Similarly, $\nabla^2 w = -M/D$ together with Eqs. (*a*), leads to

$$2w_2 - 4w_1 = -\frac{11p_0h^4}{16D}$$

$$w_1 + w_3 - 4w_2 = -\frac{7p_0h^4}{8D}$$

$$4w_2 - 4w_3 = -\frac{9p_0h^4}{8D}$$

From the above, one obtains

$$w_1 = 0.00214\frac{p_0a^4}{D} \qquad w_2 = 0.00293\frac{p_0a^4}{D} \qquad w_3 = 0.00403\frac{p_0a^4}{D}$$

Deflection w_3 at the center of the plate is 0.79 percent less than the "exact" value given in Example 3.1. The bending moment at the center of the plate is, from Eqs. (1.21) and (*a*) for $\nu = 0.3$

$$M_x = M_y = (1+\nu)\frac{M_3}{2} = 0.0457p_0a^2$$

This value is approximately 4.5 percent less than the "exact" value, $0.0479p_0a^2$.

By means of a finer network, one expects to improve the results. Let us subdivide the domain into 64 small squares, each with $h = a/8$. Taking into account the symmetry, we number the nodal points as shown in Fig. 5.2*b*.

The values of M and w are zero on the boundary. Writing the finite difference equations at points 1 through 10, we have 20 simultaneous equations for the 20 unknowns M and w at the inner nodal points. Solving these, we obtain

$$M_1 = 0.01778p_0a^2 \qquad M_6 = 0.05377p_0a^2$$

$$M_2 = 0.02774p_0a^2 \qquad M_7 = 0.05664p_0a^2$$

$$M_3 = 0.03291p_0a^2 \qquad M_8 = 0.06523p_0a^2$$

$$M_4 = 0.03452p_0a^2 \qquad M_g = 0.06888p_0a^2$$

$$M_5 = 0.04466p_0a^2 \qquad M_{10} = 0.07278p_0a^2$$

and

$$w_1 = 0.000663p_0a^4/D \qquad w_6 = 0.002733p_0a^4/D$$

$$w_2 = 0.001186p_0a^4/D \qquad w_7 = 0.002937p_0a^4/D$$

$$w_3 = 0.001515p_0a^4/D \qquad w_8 = 0.003507p_0a^4/D$$

$$w_4 = 0.001627p_0a^4/D \qquad w_9 = 0.003770p_0a^4/D$$

$$w_5 = 0.002134p_0a^4/D \qquad w_{10} = 0.004055p_0a^4/D$$

The deflection at the center w_{10} is now 0.12 percent less than the exact value. For the center, $M_x = M_y = 0.0473p_0a^2$, which is 1.22 percent less than the "exact" value. Thus for the situation described a small number of subdivisions of the plate yields results of acceptable accuracy for practical applications. On the basis of the results for $h = a/4$ and $h = a/8$, a still better approximation can be obtained by applying extrapolation approaches[15] or by reducing the net size h further.

To calculate the moments at any other location we apply finite difference form of Eqs. (1.10). For example, at point 7, for $\nu = 0.3$, we have

$$(M_x)_7 = D[2w_7 - 2w_6 + \nu(2w_7 - w_4 - w_9)]\frac{64}{a^2} = 0.0353p_0a^2$$

$$(M_y)_7 = D[2w_7 - w_4 - w_9 + \nu(2w_7 - 2w_6)]\frac{64}{a^2} = 0.0384p_0a^2$$

In a like manner, the twisting moment at point 6 is found to be

$$(M_{xy})_6 = D(1 - \nu)(w_2 - w_4 - w_6 + w_9)\frac{16}{a^2} = 0.00668p_0a^2$$

We can obtain the shear forces at node 6, referring to Eq. (5.10), as follows

$$(Q_x)_6 = (M_7 - M_5)\frac{4}{a} = 0.0479p_0a$$

$$(Q_y)_6 = (M_3 - M_8)\frac{4}{a} = 0.1293p_0a$$

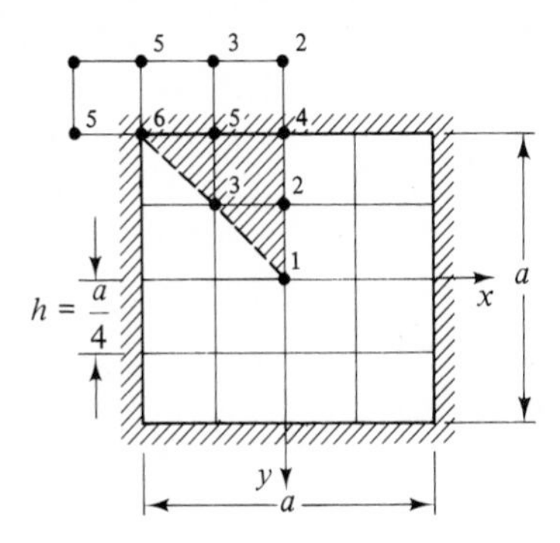

Figure 5.3

The support reactions are computed similarly through the use of Eqs. (1.23) and (1.24).

Example 5.2 Determine the deflection and moment at various points of a clamped-edge square plate of sides a under a uniformly distributed load p_0 per unit surface area. Take $h = a/4$ and employ two approaches of solution: (a) application of Eqs (1.22) and (b) application of Eq. (1.17).

SOLUTION The boundary conditions are (Fig. 5.3):

$$w = 0 \qquad \frac{\partial w}{\partial x} = \frac{\partial w}{\partial y} = 0 \qquad \left(x = y = \pm\frac{a}{2}\right) \tag{b}$$

That is, the deflection and slopes are zero at the points on the border. Referring to Eq. (5.3), for the first central difference, we obtain at the boundary

$$2hw'_n = w_{n+1} - w_{n-1} = 0 \tag{c}$$

where the prime denotes the first derivative with respect to x or y. Equations (b) and (c) indicate that the values of w (and M) at the nodes immediately outside the boundary are equal to the values of w (and M) at the points immediately inside the boundary along the same normal. Owing to symmetry, only the triangular shaded area is considered, with the corresponding labeling of nodes shown in Fig. 5.3.

(a) Applying the operator ∇^2 of Table 5.1 to $\nabla^2 M = -p_0$ at points 1, 2, and 3 respectively, we have

$$\begin{aligned} 4M_2 - 4M_1 &= -p_0 h^2 \\ 2M_3 + M_1 + M_4 - 4M_2 &= -p_0 h^2 \\ 2M_5 + 2M_2 - 4M_3 &= -p_0 h^2 \end{aligned}$$

and, referring to $\nabla^2 w = -M/D$

$$4w_2 - 4w_1 = -M_1 h^2/D$$
$$2w_3 + w_1 - 4w_2 = -M_2 h^2/D$$
$$2w_2 - 4w_3 = -M_3 h^2/D$$
$$2w_2 = -M_4 h^2/D$$
$$2w_3 = -M_5 h^2/D$$
$$0 = M_6$$

Solving the above expressions simultaneously, substituting $h = a/4$, we obtain

$$\begin{aligned} &M_1 = 0.03792 p_0 a^2 \qquad && w_1 = 0.00180 p_0 a^4/D \\ &M_2 = 0.02230 p_0 a^2 && w_2 = 0.00121 p_0 a^4/D \\ &M_3 = 0.01369 p_0 a^2 && w_3 = 0.00082 p_0 a^4/D \\ &M_4 = 0.03862 p_0 a^2 \\ &M_5 = 0.026159 p_0 a^2 \\ &M_6 = 0 \end{aligned} \tag{d}$$

It follows from Eqs. (1.21) that, for $\nu = 0.3$:

$$(M_x)_1 = (M_y)_1 = 0.02465 p_0 a^2$$
$$(M_x)_4 = (M_y)_4 = -0.02510 p_0 a^2$$

Compared with the series solution for this problem, the calculated deflection w_1 is 43 percent high, the numerical value of moments at point 1 is about 6 percent high, while at point 4 it is about 25 percent low. Similar calculations conducted for $n = 8$ result in a center deflection $w_1 = 0.00143 p_0 a^4/D$, about 13 percent higher than the series solution.

(*b*) Application of the operator ∇^4 of Table 5.1 to $\nabla^4 w = p_0/D$ at nodes 1, 2, and 3, respectively, leads to

$$-32w_2 + 8w_3 + 20w_1 = \frac{p_0}{D}\left(\frac{a}{4}\right)^4$$

$$-8w_1 - 16w_3 + 26w_2 = \frac{p_0}{4}\left(\frac{a}{4}\right)^4$$

$$2w_1 - 16w_2 + 24w_3 = \frac{p_0}{4}\left(\frac{a}{4}\right)^4$$

Solution of the above set yields values of w_1, w_2, and w_3 equal to those given by Eqs. (*d*). The accuracy may be improved by increasing the number of nodes or by applying an extrapolation procedure.

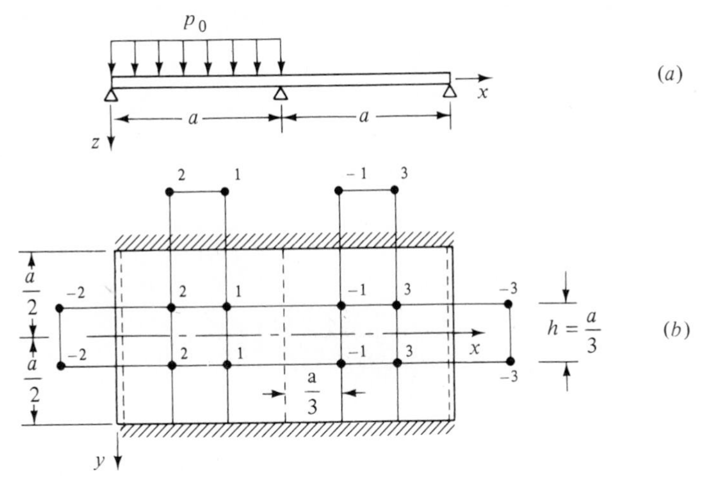

Figure 5.4

Example 5.3 A stockroom floor, half of which is carrying a uniform loading, is approximated by a continuous plate with opposite edges $y = \pm a/2$ clamped, remaining edges and the midspan simply supported (Fig. 5.4*a*). We wish to obtain deflections at points 1, 2, and 3 (Fig. 5.4b).

SOLUTION The boundary conditions are

$$w = 0 \qquad \frac{\partial^2 w}{\partial x^2} = 0 \qquad (x = 0,\ x = a,\ x = 2a)$$

$$w = 0 \qquad \frac{\partial w}{\partial y} = 0 \qquad \left(y = \frac{a}{2},\ y = -\frac{a}{2}\right) \tag{e}$$

Taking into account symmetry of the solution about the x axis and satisfying the finite difference expressions (Sec. 5.2) corresponding to Eqs. (*e*), we shall number the nodes as shown in the figure. The operator ∇^4 (Table 5.1) is applied to $\nabla^4 w = p_0/D$ at points 1, 2, and 3, respectively, to yield

$$2w_1 - 6w_2 = p_0 h^4/D$$

$$-6w_1 + 12w_2 = p_0 h^4/D$$

$$6w_2 + 12w_3 = 0$$

Simultaneous solution of the above, after setting $h = a/3$, results in

$$w_1 = w_2 = \frac{p_0 a^4}{486D} \qquad w_3 = -\frac{p_0 a^4}{972D}$$

By decreasing the size of the mesh, the accuracy of the solution can be improved.

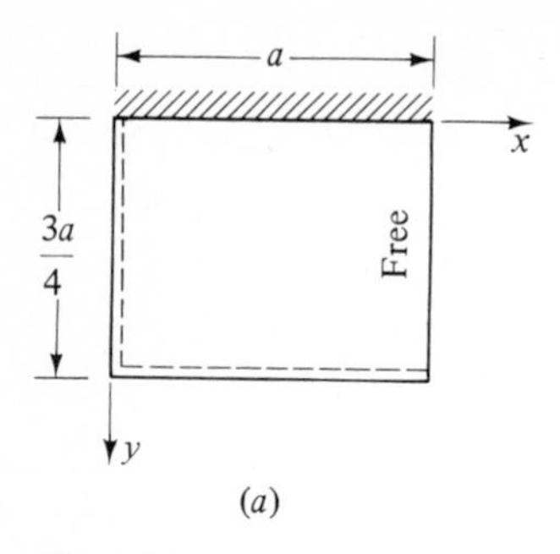

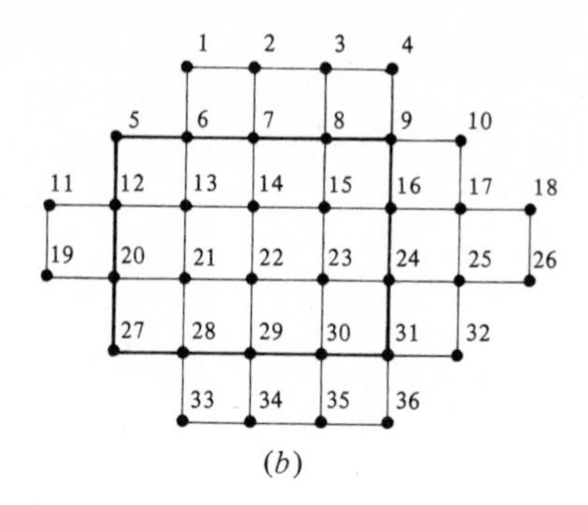

Figure 5.5

Example 5.4 Consider the case of a uniformly loaded rectangular plate with two adjacent edges simply supported, the third edge free, and the fourth edge built in (Fig. 5.5*a*). Use the finite difference method with $h = a/4$ to obtain w at various points.

SOLUTION The domain is divided into twelve squares of sides $a/4$ (Fig. 5.5*b*). At the nodes of the simply supported and the built-in edges, we have $w = 0$. The finite difference formula corresponding to Eq. (1.17) is applied to all interior nodes and to nodes 16 and 24 on the free edge, where the deflections are to be found. This application gives *eight* equations involving *thirty-six* nodes, indicated in the mesh on the figure. The boundary conditions for each edge, replaced by the finite difference forms, are listed in Table 5.2. It is seen

Table 5.2

Edge(s)	Conditions
A. Clamped at $y = 0$	$w = 0$: $w_5 = w_6 = w_7 = w_8 = w_9 = 0$ $\partial w/\partial x = 0$: $w_{13} = w_1$, $w_{14} = w_2$, $w_{15} = w_3$, $w_{16} = w_4$ Continuity of zero curvature ($\partial^2 w/\partial x^2 = 0$) along x axis: $w_8 = -w_{10} = 0$
B. Simply supported at $x = 0$ and $y = 3a/4$	$w = 0$: $w_{12} = w_{20} = w_{27} = w_{28} = w_{29} = w_{30} = w_{31} = 0$ $\partial^2 w/\partial x^2 = 0$: $w_{13} = -w_{11}$, $w_{21} = -w_{19}$ $\partial^2 w/\partial y^2 = 0$: $w_{21} = -w_{33}$, $w_{22} = -w_{34}$, $w_{23} = -w_{35}$, $w_{24} = -w_{36}$ Continuity of zero curvature ($\partial^2 w/\partial x^2 = 0$) along edge $y = 3a/4$: $w_{30} = -w_{32} = 0$
C. Free at $x = a$	$V_x = 0$, at nodes 16 and 24, respectively: $w_{18} - w_{14} + (\nu - 2)(w_8 - w_{25} + w_{23} - w_{10}) + (6 - 2\nu)(w_{15} - w_{17}) = 0$ $w_{26} - w_{22} + (\nu - 2)(w_{15} + w_{35} - w_{17} - w_{32}) + (6 - 2\nu)(w_{23} - w_{25}) = 0$ $M_x = 0$, at nodes 16 and 24, respectively: $w_{15} + w_{17} + \nu(w_9 + w_{24}) - 2(1 + \nu)w_{16} = 0$ $w_{23} + w_{25} + \nu(w_{16} + w_{31}) - 2(1 + \nu)w_{24} = 0$

from the table that all edge conditions are represented by a total of *twenty-eight* expressions. We thus have a total of thirty-six independent equations in terms of the thirty-six values of the deflections corresponding to each node.

Writing the finite difference expressions for Eq. (1.17) by applying the operator ∇^4 of Table 5.1 at nodes 13, 14, 15, 16, 21, 22, 23, and 24, and inserting the w's given in cases A and B of Table 5.2, we have eight simultaneous equations containing twelve unknown values of w at nodal points 13 through 18 and 21 through 26. Case C of Table 5.2 provides four more equations in terms of the same unknown values of w. The resulting twelve independent expressions are represented in the following matrix form (for $\nu = 0.3$):

$$\begin{bmatrix} 20 & -8 & 1 & 0 & 0 & 0 & -8 & 2 & 0 & 0 & 0 & 0 \\ -8 & 21 & -8 & 1 & 0 & 0 & 2 & -8 & 2 & 0 & 0 & 0 \\ 1 & -8 & 21 & -8 & 1 & 0 & 0 & 2 & -8 & 2 & 0 & 0 \\ 0 & 1 & -8 & 21 & -8 & 1 & 0 & 0 & 2 & -8 & 2 & 0 \\ -8 & 2 & 0 & 0 & 0 & 0 & 18 & -8 & 1 & 0 & 0 & 0 \\ 2 & -8 & 2 & 0 & 0 & 0 & -8 & 19 & -8 & 1 & 0 & 0 \\ 0 & 2 & -8 & 2 & 0 & 0 & 1 & -8 & 19 & -8 & 1 & 0 \\ 0 & 0 & 2 & -8 & 2 & 0 & 0 & 1 & -8 & 19 & -8 & 1 \\ 0 & -1 & 5.4 & 0 & -5.4 & 1 & 0 & 0 & -1.7 & 0 & 1.7 & 0 \\ 0 & 0 & -1.7 & 0 & 1.7 & 0 & 0 & -1 & 5.4 & 0 & -5.4 & 1 \\ 0 & 0 & 1 & -2.6 & 1 & 0 & 0 & 0 & 0 & 0.3 & 0 & 0 \\ 0 & 0 & 0 & 0.3 & 0 & 0 & 0 & 0 & 1 & -2.6 & 1 & 0 \end{bmatrix} \begin{Bmatrix} w_{13} \\ w_{14} \\ w_{15} \\ w_{16} \\ w_{17} \\ w_{18} \\ w_{21} \\ w_{22} \\ w_{23} \\ w_{24} \\ w_{25} \\ w_{26} \end{Bmatrix} = \frac{p_0 h^4}{D} \begin{Bmatrix} 1 \\ 1 \\ 1 \\ 1 \\ 1 \\ 1 \\ 1 \\ 1 \\ 0 \\ 0 \\ 0 \\ 0 \end{Bmatrix}$$

The above equations are solved to yield

$$\begin{array}{lll} w_{13} = 0.25819N & w_{14} = 0.38943N & w_{15} = 0.45037N \\ w_{16} = 0.51951N & w_{17} = 0.79839N & w_{18} = 1.07433N \\ w_{21} = 0.30383N & w_{22} = 0.46598N & w_{23} = 0.54558N \\ w_{24} = 0.63989N & w_{25} = 0.96226N & w_{26} = 2.27753N \end{array}$$

where $N = p_0 h^4/D$ and $h = a/4$. From these values, we determine the bending moment at any node in the plate as illustrated in the previous examples.

5.4 PLATES WITH CURVED BOUNDARIES

We now treat the bending problems of *simply supported* plates having *curved* or *irregular* boundaries. Dividing a portion of such a boundary into a square mesh (Fig. 5.6*a*) shows that the ∇^2 operator (Table 5.1) does not apply to point 0 because of the unequal lengths of arms 01, 02, 03, and 04. When at least one of the arms is not equal to h, the operator pattern is referred to as an *irregular star*. There are available various approaches to such situations. The one that will be discussed assumes that it is required to develop an irregular star using the actual boundary points, rather than those falling "outside," associated with the continued regular mesh.

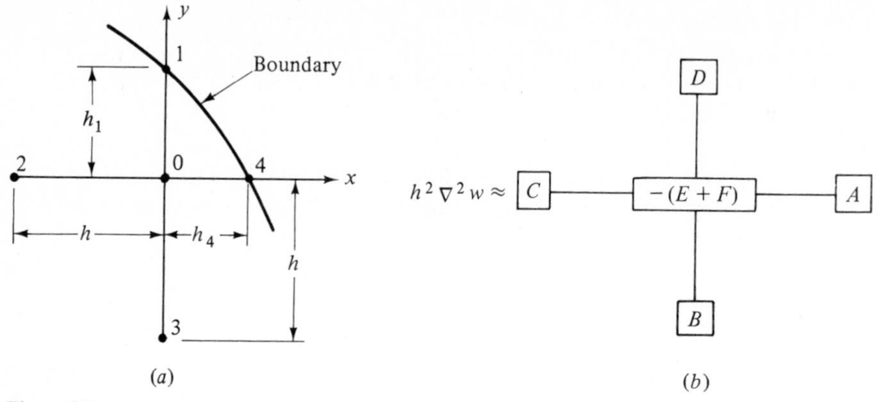

Figure 5.6

It is assumed that in the region immediately surrounding the node 0, the expression for w can be approximated by

$$w(x, y) = w_0 + a_1 x + a_2 y + a_3 x^2 + a_4 y^2 \tag{a}$$

This equation, referring to Fig. 5.6*a*, gives the following expressions for w at points 1, 2, 3, and 4:

$$\begin{aligned} w_1 &= w_0 + a_2 h_1 + a_4 h_1^2 \\ w_2 &= w_0 - a_1 h + a_3 h^2 \\ w_3 &= w_0 - a_2 h + a_4 h^2 \\ w_4 &= w_0 + a_1 h_4 + a_3 h_4^2 \end{aligned}$$

from which

$$\begin{aligned} a_1 &= \frac{h^2(w_4 - w_0) + h_4^2(w_0 - w_2)}{hh_4(h + h_4)} \qquad & a_2 &= \frac{h^2(w_1 - w_0) + h_1^2(w_0 - w_3)}{hh_1(h + h_1)} \\ a_3 &= \frac{h(w_4 - w_0) - h_4(w_0 - w_2)}{hh_4(h + h_4)} \qquad & a_4 &= \frac{h(w_1 - w_0) - h_1(w_0 - w_3)}{hh_1(h + h_1)} \end{aligned} \tag{b}$$

At point 0 ($x = 0$, $y = 0$), Eq. (*a*) gives

$$\left(\frac{\partial^2 w}{\partial x^2}\right)_0 = 2a_3 \qquad \left(\frac{\partial^2 w}{\partial y^2}\right)_0 = 2a_4 \tag{c}$$

Next, we introduce Eqs. (*b*) into Eqs. (*c*). Adding the resulting expressions, the operator $\nabla^2 w$ is obtained in the form

$$h^2\left(\frac{\partial^2 w}{\partial x^2} + \frac{\partial^2 w}{\partial y^2}\right)_0 \approx \frac{2w_1}{\alpha_1(1+\alpha_1)} + \frac{2w_2}{1+\alpha_4} + \frac{2w_3}{1+\alpha_1} + \frac{2w_4}{\alpha_4(1+\alpha_4)} - \left(\frac{2}{\alpha_1} + \frac{2}{\alpha_4}\right)w_0 \tag{5.11a}$$

in which $\alpha_i = h_i/h$, $i = 1, 4$. For convenience, the above may be written

$$h^2(\nabla^2 w)_0 \approx Aw_1 + Bw_2 + Cw_3 + Dw_4 - (E + F)w_0 \tag{5.11b}$$

The pattern associated with Eq. (5.11) is shown in Fig. 5.6*b*.

The case where all of the arms h_i, $i = 1, 2, 3, 4$, are smaller than h may also be treated by following a procedure identical with that described in this section. In so doing, we obtain:

$$h^2(\nabla^2 w)_0 \approx \frac{2w_1}{\alpha_1(\alpha_1+\alpha_3)} + \frac{2w_2}{\alpha_2(\alpha_2+\alpha_4)} + \frac{2w_3}{\alpha_3(\alpha_1+\alpha_3)} + \frac{2w_4}{\alpha_4(\alpha_2+\alpha_4)} - \left(\frac{2}{\alpha_1\alpha_3} + \frac{2}{\alpha_2\alpha_4}\right)w_0 \quad (5.12)$$

The finite difference equivalent forms of the Laplace operator, Eqs. (5.11) and (5.12), may be applied in the same manner as the standard form given by Eq. (5.5).

For situations involving other than simply supported plates with curved boundaries, a somewhat different approach for approximating $\nabla^2 w$ is used. In any case, the development of the finite difference expression that replaces $\nabla^4 w$ near the curved boundary is not simple.

Example 5.5 Determine the deflection and the bending moments at the center of a simply supported elliptical steel plate under uniform load of intensity p_0 (Fig. 5.7). Let $a = 0.15$ *m*, $b = 0.1$ m, and $h = 0.05$ m. Compare the results with those given in Table 4.1.

SOLUTION Owing to the symmetry, it is sufficient to find a solution in one-fourth of the entire domain. The equation of the ellipse together with the given data yields $h_1 = 0.044$ m, $h_2 = 0.0245$ m, and $h_3 = 0.03$ m. At nodes 1, 2, 3, and 4 the standard finite difference expression applies, while at 5 and 6 we employ the operator of Fig. 5.6*b*. Note that $w = 0$ and $M = 0$ at the boundary. We can thus write six finite difference equations corresponding to Eq. (1.22*a*), presented in matrix form as follows

$$\begin{bmatrix} -4 & 2 & 0 & 2 & 0 & 0 \\ 1 & -4 & 1 & 0 & 2 & 0 \\ 0 & 1 & -4 & 0 & 0 & 2 \\ 1 & 0 & 0 & -4 & 2 & 0 \\ 0 & 1.06 & 0 & 1 & -4.26 & 1 \\ 0 & 0 & 1.34 & 0 & 1.25 & -7.41 \end{bmatrix} \begin{Bmatrix} M_1 \\ M_2 \\ M_3 \\ M_4 \\ M_5 \\ M_6 \end{Bmatrix} = -p_0 h^2 \begin{Bmatrix} 1 \\ 1 \\ 1 \\ 1 \\ 2 \\ 1 \end{Bmatrix} \quad (d)$$

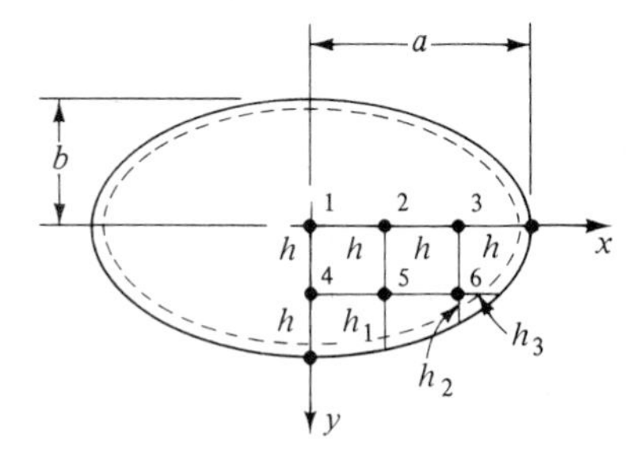

Figure 5.7

which when solved yields

$$\begin{aligned} M_1 &= 1.384 p_0 h^2 \qquad M_2 = 1.230 p_0 h^2 \qquad M_3 = 0.769 p_0 h^2 \\ M_4 &= 1.038 p_0 h^2 \qquad M_5 = 0.884 p_0 h^2 \qquad M_6 = 0.423 p_0 h^2 \end{aligned} \tag{e}$$

Similarly, application of Eq. (1.22b) at nodes 1 through 6, together with the values of moment given by Eqs. (*e*), results in six equations from which

$$\begin{aligned} w_1 &= 1.520 \frac{p_0 h^4}{D} \qquad w_2 = 1.282 \frac{p_0 h^4}{D} \qquad w_3 = 0.674 \frac{p_0 h^4}{D} \\ w_4 &= 1.066 \frac{p_0 h^4}{D} \qquad w_s = 0.853 \frac{p_0 h^4}{D} \qquad w_6 = 0.323 \frac{p_0 h^4}{D} \end{aligned} \tag{f}$$

The bending moments at the center, by means of Eqs. (*f*) and (5.9), are therefore

$$\begin{aligned} M_x &= \frac{D}{h^2}[2w_1 - 2w_2 + \nu(2w_1 - 2w_4)] = 0.187 p_0 b^2 \\ M_y &= \frac{D}{h^2}[2w_1 - 2w_4 + \nu(2w_1 - 2w_2)] = 0.263 p_0 b^2 = M_{\max} \end{aligned} \tag{g}$$

The value of $M_{\max}$ is 18 percent less than the result listed in Table 4.1. The maximum deflection occurs at the center and, from Eqs. (*f*), with $\nu = 0.3$, we obtain

$$w_1 = 1.038 \frac{p_0 b^4}{Et^3} \tag{h}$$

This is 17.6 percent less than the value furnished in Table 4.1.

5.5 THE POLAR MESH

It is convenient to employ a *polar mesh* (Fig. 5.8*a*) to cover the domains of the circular plates. The laplacian, in terms of the polar coordinates r and θ, is, from Eq. (2.1):

$$\nabla^2 w = \frac{\partial^2 w}{\partial r^2} + \frac{1}{r}\frac{\partial w}{\partial r} + \frac{1}{r^2}\frac{\partial^2 w}{\partial \theta^2} \tag{a}$$

Using the notation of Fig. 5.8*a* and by means of Eqs. (5.3) and (5.4), we have

$$\frac{\partial^2 w}{\partial r^2} \approx \frac{1}{h^2}(w_2 - 2w_0 + w_4) \qquad \frac{\partial w}{\partial r} \approx \frac{1}{2h}(w_2 - w_4)$$

$$\frac{\partial^2 w}{\partial \theta^2} \approx \frac{1}{\phi^2}(w_1 - 2w_0 + w_3)$$

where

$$h = \nabla r \qquad \phi = \Delta\theta$$

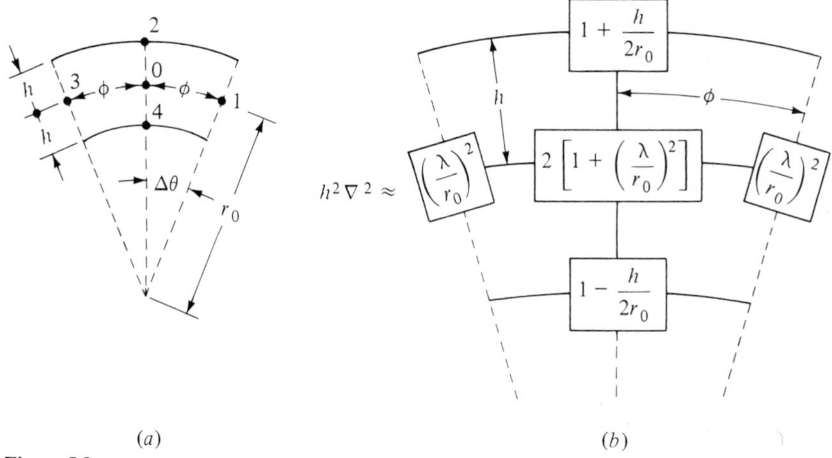

Figure 5.8

The pattern for ∇^2 in polar coordinates is presented by Fig. 5.8*b*, in which $\lambda = h/\phi$.

Other operator patterns may be developed in a like manner. Referring to the operator of Fig. 5.8*b*, the governing equations (1.22) are replaced by the finite differences in terms of the nodal values of a polar mesh.

5.6 THE TRIANGULAR MESH

Some plates have boundary configurations that can be covered with ease by the use of a *triangular mesh* shown in Fig. 5.9*a*. In this case, it is necessary to express the governing plate equations in terms of the *triangular coordinates*, q_1, q_2, and q_3. Referring to Fig. 5.9*b* the equations relating the cartesian coordinates to the triangular coordinates are:

$$
\begin{aligned}
x &= q_1 + q_2 \cos\alpha + q_3 \cos\beta \\
y &= q_2 \sin\alpha + q_3 \sin\beta
\end{aligned}
\tag{5.13}
$$

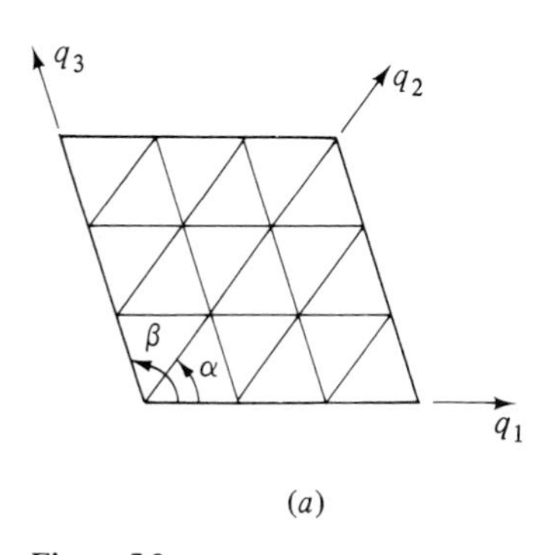

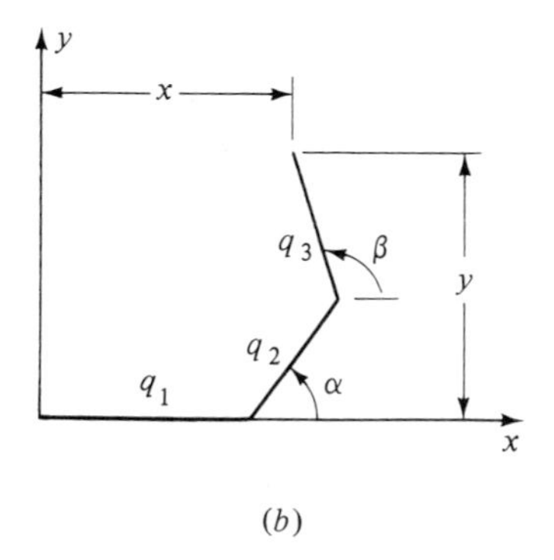

Figure 5.9

In which α and β are the angles between q_1 and q_2, and q_3 and q_1, respectively. The following partial derivatives are obtained from the above

$$\frac{\partial x}{\partial q_1} = 1 \qquad \frac{\partial x}{\partial q_2} = \cos\alpha \qquad \frac{\partial x}{\partial q_3} = \cos\beta$$
$$\frac{\partial y}{\partial q_1} = 0 \qquad \frac{\partial y}{\partial q_2} = \sin\alpha \qquad \frac{\partial y}{\partial q_3} = \sin\beta \tag{a}$$

When an expression for $w(x, y)$ is given, Eqs. (5.13) and (a) can be employed to define the corresponding expression $w(q_1, q_2, q_3)$ and its derivatives with respect to the triangular coordinates. The chain rule, together with Eqs. (a), yields

$$\frac{\partial w}{\partial q_1} = \frac{\partial w}{\partial x}\frac{\partial x}{\partial q_1} + \frac{\partial w}{\partial y}\frac{\partial y}{\partial q_1} = \frac{\partial w}{\partial x}$$
$$\frac{\partial w}{\partial q_2} = \frac{\partial w}{\partial x}\cos\alpha + \frac{\partial w}{\partial y}\sin\alpha \tag{b}$$
$$\frac{\partial w}{\partial q_3} = \frac{\partial w}{\partial x}\cos\beta + \frac{\partial w}{\partial y}\sin\beta$$

The second partial derivatives of the deflection, written in the matrix form, are then (Prob. 5.13):

$$\begin{Bmatrix} \partial^2 w/\partial q_1^2 \\ \partial^2 w/\partial q_2^2 \\ \partial^2 w/\partial q_3^2 \end{Bmatrix} = \begin{bmatrix} 1 & 0 & 0 \\ \cos^2\alpha & 2\sin\alpha\cos\alpha & \sin^2\alpha \\ \cos^2\beta & 2\sin\beta\cos\alpha & \sin^2\beta \end{bmatrix} \begin{Bmatrix} \partial^2 w/\partial x^2 \\ \partial^2 w/\partial x\,\partial y \\ \partial^2 w/\partial y^2 \end{Bmatrix} \tag{c}$$

Upon introducing the first of Eqs. (c) into the second and the third, and eliminating $\partial^2 w/\partial x\,\partial y$, we obtain

$$\frac{\partial^2 w}{\partial y^2} = R\left[2\,\frac{\partial^2 w}{\partial q_1^2}\cos\alpha\cos\beta\sin(\beta-\alpha) - \frac{\partial^2 w}{\partial q_2^2}\sin 2\beta + \frac{\partial^2 w}{\partial q_3^2}\sin 2\alpha\right]$$

and thus

$$\nabla^2 w = \frac{\partial^2 w}{\partial x^2} + \frac{\partial^2 w}{\partial y^2} = \frac{\partial^2 w}{\partial q_1^2} + \frac{\partial^2 w}{\partial y^2}$$
$$= R\left[\frac{\partial^2 w}{\partial q_1^2}\sin 2(\beta-\alpha) - \frac{\partial^2 w}{\partial q_2^2}\sin 2\beta + \frac{\partial^2 w}{\partial q_3^2}\sin 2\alpha\right] \tag{5.14}$$

where $R = \frac{1}{2}\sin\alpha\sin\beta\sin(\beta-\alpha)$.

In the case of a commonly used equilateral triangular mesh, where $\alpha = 60°$, and $\beta = 120°$, Eq. (5.14) reduces to

$$\nabla^2 w = \frac{2}{3}\left(\frac{\partial^2 w}{\partial q_1^2} + \frac{\partial^2 w}{\partial q_2^2} + \frac{\partial^2 w}{\partial q_3^2}\right) \tag{5.15}$$

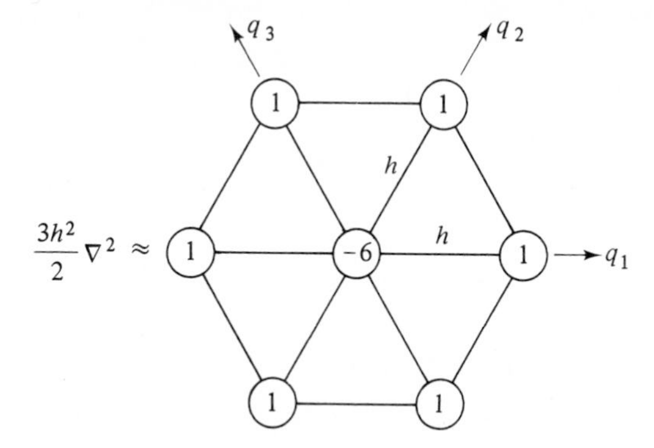

Figure 5.10

The finite difference operator pattern for the above, obtained by application of the operator for the second derivative (Table 5.1) in the q_1, q_2, q_3 directions, is given by Fig. 5.10. Other finite difference operators may be developed similarly. The finite difference equivalent of Eq. (1.22) can then be readily written in terms of the triangular coordinates.

Example 5.6 Determine the center deflection of a simply supported skew plate under uniform loading of intensity p_0 (Fig. 5.11). Use $h = a/4$.

SOLUTION The domain is divided into 32 small triangles. In labeling nodal points, the condition of symmetry is taken into account as shown in the figure. Note that $M = 0$ and $w = 0$ at the boundary. Operator ∇^2 of Fig. 5.10 is applied to Eqs. (1.22) at nodal points 1, 2, 3, and 4, resulting in the following two sets of expressions:

$$\begin{bmatrix} -6 & 4 & 2 & 0 \\ 1 & -5 & 1 & 1 \\ 1 & 2 & -6 & 0 \\ 0 & 2 & 0 & -6 \end{bmatrix} \begin{Bmatrix} M_1 \\ M_2 \\ M_3 \\ M_4 \end{Bmatrix} = -\tfrac{3}{2} p_0 h^2 \begin{Bmatrix} 1 \\ 1 \\ 1 \\ 1 \end{Bmatrix} \tag{d}$$

and

$$\begin{bmatrix} -6 & 4 & 2 & 0 \\ 1 & -5 & 1 & 1 \\ 1 & 2 & -6 & 0 \\ 0 & 2 & 0 & -6 \end{bmatrix} \begin{Bmatrix} w_1 \\ w_2 \\ w_3 \\ w_4 \end{Bmatrix} = -\frac{3}{2} \frac{h^2}{D} \begin{Bmatrix} M_1 \\ M_2 \\ M_3 \\ M_4 \end{Bmatrix} \tag{e}$$

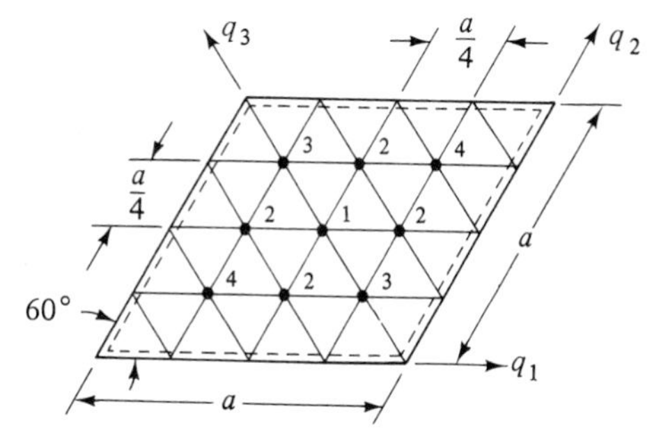

Figure 5.11

From the above: $w_1 = 0.00283 p_0 a^4/D$. By increasing the number of nodes, we expect to improve the result. For example, selecting $h = a/8$, it can be shown that[15] the center deflection $w_1 = 0.00262 p_0 a^4/D$. The "exact" coefficient for the solution is 0.00256.

5.7 PROPERTIES OF A FINITE ELEMENT

The powerful finite element method has developed simultaneously with the widespread use of digital computers and the increasing emphasis upon numerical methods. This approach permits the prediction of stress and deflection in a plate with a degree of ease and precision never before possible. In the finite element method, the plate is *discretized* into a finite number of elements (usually triangular or rectangular in shape), connected at their nodes and along hypothetic interelement boundaries. Hence equilibrium and compatibility must be satisfied at each node and along the boundaries between elements. There are a number of finite element methods. We here discuss only the commonly used finite displacement approach wherein the governing set of algebraic equations is expressed in terms of unknown nodal displacements. The literature associated with this method is voluminous.[17,18]

To begin with, a number of basic quantities relevant to an individual *finite element* of an isotropic plate is defined. The derivations are based on the assumptions of *small deflection* theory, described in Sec. 1.2. Illustrated in Chap. 6 is the determination of stiffness matrices for plate elements with nonisotropic properties, after derivation of the stress-strain relations for the orthotropic materials. The plate, in general, may have any nonuniform shape and loading.

Consider the thin plate of Fig. 5.12*a* which is replaced by an assembly of triangular finite elements indicated by the dashed lines. The properties belonging to a discrete element will be designated by *e*.

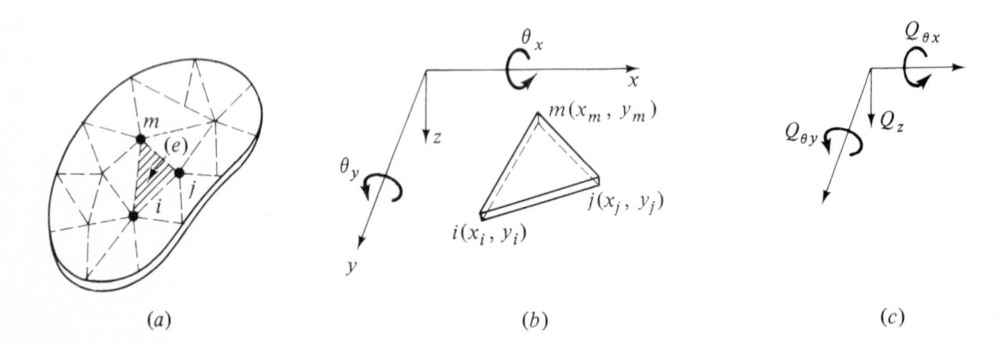

Figure 5.12

Displacement matrix The nodal displacements $\{\delta\}_e$ are related to the displacements within the element by means of a *displacement function* $\{w\}_e$. The latter is expressed in the following general form

$$\{w\}_e = [P]\{\delta\}_e \tag{5.16}$$

where the braces indicate a *column* matrix and the matrix $[P]$ is a function of position, to be later determined for a specific element. This matrix is often referred to as *shape function.* It is, of course, desirable that a displacement function be chosen such that the true displacement field be as closely represented as possible.

Strain, stress, and elasticity matrices Referring to Eqs. (1.3), we define, for commonality in finite element analysis of all types of problems, a *generalized "strain"-displacement* matrix of the form

$$\begin{Bmatrix} \varepsilon_x \\ \varepsilon_y \\ \gamma_{xy} \end{Bmatrix}_e = \left\{ -\frac{\partial^2 w}{\partial x^2}, \; -\frac{\partial^2 w}{\partial y^2}, \; -2\frac{\partial^2 w}{\partial x \, \partial y} \right\} \tag{5.17a}$$

or

$$\{\varepsilon\}_e = [B]\{\delta\}_e \tag{5.17b}$$

in which $[B]$ is also yet to be determined. The *stress-generalized "strain"* relationship, from Eqs. (1.8) is as follows

$$\begin{Bmatrix} \sigma_x \\ \sigma_y \\ \tau_{xy} \end{Bmatrix}_e = \frac{Ez}{1 - \nu^2} \begin{bmatrix} 1 & \nu & 0 \\ \nu & 1 & 0 \\ 0 & 0 & (1-\nu)/2 \end{bmatrix} \{\varepsilon\}_e \tag{5.18a}$$

Concisely,

$$\{\sigma\}_e = z[D^*]\{\varepsilon\}_e \tag{5.18b}$$

Moments are connected to stresses by Eqs. (1.9):

$$\{M\}_e = \begin{Bmatrix} M_x \\ M_y \\ M_{xy} \end{Bmatrix}_e = \int_{-t/2}^{t/2} z\{\sigma\}_e \, dz \tag{5.19}$$

Substitution of Eqs. (5.18) into the above yields the following moment-generalized "strain" relations

$$\{M\}_e = \left(\int_{/t/2}^{t/2} z^2[D^*] \, dz \right) \{\varepsilon\}_e \tag{5.20a}$$

or

$$\{M\}_e = [D]\{\varepsilon\}_e \tag{5.20b}$$

The *elasticity matrix* for an isotropic plate is therefore

$$[D] = \frac{t^3}{12}[D^*] = \frac{Et^3}{12(1-\nu^2)}\begin{bmatrix} 1 & \nu & 0 \\ \nu & 1 & 0 \\ 0 & 0 & (1-\nu)/2 \end{bmatrix} \tag{5.21}$$

The stress-strain relations, and thus elasticity matrix, differ for anisotropic materials (Sec. 6.7).

Owing to the many causes (e.g., shrinkage, temperature changes) some *initial strains* may be produced in the plate. In the case of a transversely loaded and heated plate, for instance, the stress and moment matrices become, respectively

$$\{\sigma\}_e = [D^*](\{\varepsilon\} - \{\varepsilon_0\})_e \tag{5.18c}$$

and

$$\{M\}_e = [D](\{\varepsilon\} - \{\varepsilon_0\})_e \tag{5.20c}$$

wherein $\{\varepsilon_0\}_e$ is the thermal strain matrix. The thermal stress problems are treated in Chap. 9.

5.8 FORMULATION OF THE FINITE ELEMENT METHOD

A convenient approach for derivation of the finite element governing expressions and the characteristics is based upon the principle of potential energy. The variation in the potential energy $\Delta\Pi$ of the entire plate shown in Fig. 5.12*a*, from Eq. (1.42), is

$$\Delta\Pi = \sum_1^n \iint_A (M_x\,\Delta\varepsilon_x + M_y\,\Delta\varepsilon_y + M_{xy}\,\Delta\gamma_{xy})\,dx\,dy - \sum_1^n \iint_A (p\,\Delta w)\,dx\,dy = 0 \tag{5.22a}$$

where n, A, and p represent the number of uniform thickness elements comprising the plate, surface area of an element, and the lateral load per unit surface area, respectively. Expression (5.22*a*) may be rewritten as follows

$$\sum_1^n \iint_A (\{\Delta\varepsilon\}_e^T\{M\}_e - p\,\Delta w)\,dx\,dy = 0 \tag{5.22b}$$

in which superscript T denotes the transpose of a matrix. Introduction of Eqs. (5.16), (5.17), and (5.20) into Eq. (5.22b) yields

$$\sum_1^n \iint_A \{\Delta\delta\}_e^T([k]_e\{\delta\}_e - \{Q\}_e) = 0 \tag{a}$$

The element *stiffness* matrix $[k]_e$ equals

$$[k]_e = \iint_A [B]^T[D][B]\,dx\,dy \tag{5.23}$$

The element *nodal force* matrix $\{Q\}_e$, due to initial strain and transverse load, is

$$\{Q\}_e = \iint_A [B]^T[D]\{\varepsilon_0\}\,dx\,dy + \iint_A [P]^T p\,dx\,dy \tag{5.24}$$

Since the changes in $\{\delta\}_e$ are independent and arbitrary, Eq. (a) leads to the expression

$$[k]_e\{\delta\}_e = \{Q\}_e \tag{5.25}$$

for the element nodal force equilibrium.

We now assemble Eqs. (a) to obtain

$$\{\Delta\delta\}^T([K]\{\delta\} - \{Q\}) = 0 \tag{b}$$

The above must be valid for all $\{\Delta\delta\}$. This yields the following *governing* equations for the entire plate:

$$[K]\{\delta\} = \{Q\} \tag{5.26}$$

where

$$[K] = \sum_1^n [k]_e \qquad \{Q\} = \sum_1^n \{Q\}_e \tag{5.27}$$

We observe that the *plate stiffness matrix* $[K]$ and the *plate nodal force* matrix $\{Q\}$ are determined by superposition of all element stiffness and nodal force matrices, respectively.

The *general procedure* for solving a plate- (or shell-) bending problem by the finite element method is summarized as follows:

(1) Determine $[k]_e$ from Eq. (5.23) in terms of the given element properties. Generate $[K] = \Sigma[k]_e$.
(2) Determine $\{Q\}_e$ from Eq. (5.24) in terms of the applied loading. Generate $\{Q\} = \Sigma\{Q\}_e$.
(3) Determine the nodal displacements from Eq. (5.26) by satisfying the boundary conditions: $\{\delta\} = [K]^{-1}\{Q\}$.

Then, determine element moment from $\{M\}_e = [D]\{\delta\}_e$, and the element stress from Eqs. (5.18) or (1.12).

This procedure will be better understood when the characteristics of a certain element are derived. Two most commonly used plate-bending elements, each requiring a different type of general displacement function, are discussed in the following sections.

5.9 TRIANGULAR FINITE ELEMENT

The triangular element can easily accommodate irregular boundaries and can be graduated in size to permit small elements in regions of stress concentration. Because of this, it is used extensively in the finite element approach. Consider as the finite element model a triangular plate element *ijm* coinciding with the *xy* plane (Fig. 5.12*b*). Note the *counterclockwise* numbering convention of the nodes. Each nodal displacement of the element has three components: a deflection in the *z* direction (w), a rotation about the x axis (θ_x), and a rotation about the y axis (θ_y). Rotations are related to the slopes as follows

$$\theta_x = \frac{\partial w}{\partial y} \qquad \theta_y = \frac{\partial w}{\partial x} \tag{5.28}$$

The positive directions of the rotations are determined by the *right-hand rule* as shown in the figure.

Displacement function The nodal displacement matrix for the element is

$$\{\delta\}_e = \begin{Bmatrix} \delta_i \\ \delta_j \\ \delta_m \end{Bmatrix} = \{w_i, \theta_{xi}, \theta_{yi}, w_j, \theta_{xj}, \theta_{yj}, w_m, \theta_{xm}, \theta_{ym}\} \tag{5.29}$$

The displacement function, defining the displacement at any point within the element *ijm*, is chosen to be a modified third-order polynomial of the form

$$w_e = a_1 + a_2 x + a_3 y + a_4 x^2 + a_5 xy + a_6 y^2 + a_7 x^3 + a_8(x^2 y + xy^2) + a_9 y^3 \tag{5.30}$$

leading to a reasonably simple theoretical development. Observe that the number of terms in the above is identical with the number of nodal displacements of the element. This function preserves the continuity of displacements but not the slopes along the element surfaces. For practical engineering purposes, however, in most cases the accuracy of the solution based upon Eq. (5.30) is acceptable. A displacement function of eighteen order, corresponding to a triangle with six nodes, leads to improved results, but the analysis becomes more involved than described here.

When coefficients a_1 through a_9 are known, Eq. (5.30) will provide the displacement at all locations in the plate. The nodal displacements can be written as follows:

$$\begin{Bmatrix} w_i \\ \theta_{xi} \\ \theta_{yi} \\ w_j \\ \theta_{xj} \\ \theta_{yj} \\ w_m \\ \theta_{xm} \\ \theta_{ym} \end{Bmatrix} = \begin{bmatrix} 1 & x_i & y_i & x_i^2 & x_i y_i & y_i^2 & x_i^3 & (x_i^2 y_i + x_i y_i^2) & y_i^3 \\ 0 & 0 & 1 & 0 & x_i & 2y_i & 0 & (x_i^2 + 2x_i y_i) & 3y_i^2 \\ 0 & 1 & 0 & 2x_i & y_i & 0 & 3x_i^2 & (2x_i y_i + y_i^2) & 0 \\ 1 & x_j & y_j & x_j^2 & x_j y_j & y_j^2 & x_j^2 & (x_j^2 y_j + x_j y_j^2) & y_j^3 \\ 0 & 0 & 1 & 0 & x_j & 2y_j & 0 & (x_j^2 + 2x_j y_j) & 3y_j^2 \\ 0 & 1 & 0 & 2x_j & y_j & 0 & 3x_j^2 & (2x_j y_j + y_j^2) & 0 \\ 1 & x_m & y_m & x_m^2 & x_m y_m & y_m^2 & x_m^3 & (x_m^2 y_m + x_m y_m^2) & y_m^3 \\ 0 & 0 & 1 & 0 & x_m & 2y_m & 0 & (x_m^2 + 2x_m y_m) & 3y_m^2 \\ 0 & 1 & 0 & 2x_m & y_m & 0 & 3x_m^2 & (2x_m y_m + y_m^2) & 0 \end{bmatrix} \begin{Bmatrix} a_1 \\ a_2 \\ a_3 \\ a_4 \\ a_5 \\ a_6 \\ a_7 \\ a_8 \\ a_9 \end{Bmatrix} \tag{5.31a}$$

or concisely,

$$\{\delta\}_e = [C]\{a\} \tag{5.31b}$$

From the foregoing, the solution for the unknown constants is

$$\{a\} = [C]^{-1}\{\delta\}_e \tag{5.32}$$

Equations (5.31) show that matrix $[C]$ is dependent upon the coordinate dimensions of the nodal points. The displacement function may now be written in the form of Eq. (5.16):

$$\{w\}_e = [P]\{\delta\}_e = [L][C]^{-1}\{\delta\}_e \tag{5.33}$$

where

$$[L] = [1, x, y, x^2, xy, y^2, x^3, x^2y + xy^2, y^3] \tag{5.34}$$

Substitution of Eq. (5.30 into Eq. (5.17) gives

$$\{\varepsilon\}_e = \begin{bmatrix} 0 & 0 & 0 & -2 & 0 & 0 & -6x & -2y & 0 \\ 0 & 0 & 0 & 0 & 0 & -2 & 0 & -2x & -6y \\ 0 & 0 & 0 & 0 & -2 & 0 & 0 & -4(x+y) & 0 \end{bmatrix} \{a_1, a_2, \ldots, a_9\} \tag{5.35a}$$

or

$$\{\varepsilon\}_e = [H]\{a\} \tag{5.35b}$$

One can determine the generalized "strain"-displacement matrix, by introducing Eq. (5.32) into (5.35)

$$\{\varepsilon\}_e = [B]\{\delta\}_e = [H][C]^{-1}\{\delta\}_e$$

We thus have

$$[B] = [H][C]^{-1} \tag{5.36}$$

The stiffness matrix Upon substituting $[B]$ from Eq. (5.36) into Eq. (5.23) one has

$$[k]_e = [[C]^{-1}]^T\left(\iint [H]^T[D][H]\,dx\,dy\right)[C]^{-1} \tag{5.37}$$

Here the matrices $[H]$, $[D]$, and $[C]$ are given by Eqs. (5.35), (5.21), and (5.31), respectively. After multiplying the matrices under the integral sign, the integrations can be performed to determine the element stiffness matrix.

External nodal forces The nodal forces owing to transverse surface loading may be obtained from Eq. (5.24) or by physical intuition. The element nodal force matrix is represented by

$$\{Q\}_e = \begin{Bmatrix} Q_i \\ Q_j \\ Q_m \end{Bmatrix} = \{Q_{zi}, Q_{\theta xi}, Q_{\theta yi}, Q_{zj}, Q_{\theta xj}, Q_{\theta yj}, Q_{zm}, Q_{\theta xm}, Q_{\theta ym}\} \tag{5.38}$$

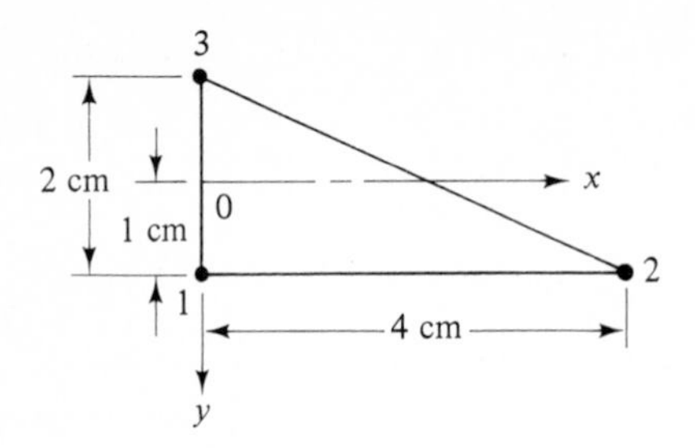

Figure 5.13

where Q_z, $Q_{\theta x}$, and $Q_{\theta y}$ denote the lateral force in the z direction, the moment per unit length about the x axis, and the moment per unit length about the y axis (Fig. 5.12c), respectively.

The determination of the element nodal forces is demonstrated by the following longhand solution.

Example 5.7 The element 123, shown in Fig. 5.13, represents a portion of a thin elastic plate which is under uniform loading of intensity p_0. Determine the nodal force matrix. Assume that the weight of element is negligible.

SOLUTION Referring to the figure (where $i = 1$, $j = 2$, $m = 3$), matrix $[C]$, defined by Eq. (5.31) is obtained as

$$[C] = \begin{bmatrix} 1 & 0 & 1 & 0 & 0 & 1 & 0 & 0 & 1 \\ 0 & 0 & 1 & 0 & 0 & 2 & 0 & 0 & 3 \\ 0 & 1 & 0 & 0 & 1 & 0 & 0 & 1 & 0 \\ 1 & 4 & 1 & 16 & 4 & 1 & 64 & 20 & 1 \\ 0 & 0 & 1 & 0 & 4 & 2 & 0 & 24 & 3 \\ 0 & 1 & 0 & 8 & 1 & 0 & 48 & 9 & 0 \\ 1 & 0 & -1 & 0 & 0 & 1 & 0 & 0 & -1 \\ 0 & 0 & 1 & 0 & 0 & -2 & 0 & 0 & 3 \\ 0 & 1 & 0 & 0 & -1 & 0 & 0 & 1 & 0 \end{bmatrix}$$

From the above we have

$$[[C]^{-1}]^T = \begin{bmatrix} 0.5 & 0 & 0.75 & -0.187 & 0 & 0 & 0.031 & 0 & 0.25 \\ -0.25 & 0.042 & -0.25 & 0.042 & 0 & 0.25 & 0 & -0.042 & 0.25 \\ 0 & 0.583 & 0 & -0.417 & 0.5 & 0 & 0.062 & -0.083 & 0 \\ 0 & 0 & 0 & 0.187 & 0 & 0 & -0.031 & 0 & 0 \\ 0 & -0.042 & 0 & -0.042 & 0 & 0 & 0 & 0.042 & 0 \\ 0 & 0 & 0 & -0.25 & 0 & 0 & 0.062 & 0 & 0 \\ 0.5 & 0 & -0.75 & 0 & 0 & 0 & 0 & 0 & 0.25 \\ 0.25 & 0 & -0.25 & 0 & 0 & -0.25 & 0 & 0 & 0.25 \\ 0 & 0.417 & 0 & -0.083 & -0.5 & 0 & 0 & 0.083 & 0 \end{bmatrix}$$

The nodal forces due to uniform surface loading p_0 are given by Eqs. (5.24) and (5.33)

$$\{Q\}_e = [[C]^{-1}]^T p_0 \iint [L]^T \, dx \, dy \tag{a}$$

Substituting Eq. (5.34) and the limits of integrations into Eq. (*a*), one has

$$\{Q\}_e = [[C]^{-1}]^T p_0 \int_{-1}^{1} \int_{0}^{2y+2} \{1, x, y, x^2, xy, y^2, x^3, x^2y + xy^2, y^3\}\, dx\, dy$$

This expression is readily integrated and then multiplied by $[[C]^{-1}]^T$ to yield

$$\{Q\}_e = p_0\{1.60, -0.49, 0.89, 120, -0.31, -1.07, 1.20, 0.53, 0.71\}$$

The element stiffness matrix $[k]_e$ may be determined similarly (see Prob. 5.15).

It is clear that the finite element method, even in the simplest of cases, requires considerable algebra. For any problem of practical significance, the digital computer must be employed to perform the necessary matrix algebra. Compared with other methods, however, the finite element approach offers a distinct advantage in the treatment of plates having irregular shapes, nonuniform load, isotropic, or anisotropic materials.

5.10 RECTANGULAR FINITE ELEMENT

Let us now consider the *rectangular element* shown in Fig. 5.14. To ensure at least an approximate fulfillment of the continuity of slopes, three nodal displacement components described in Sec. 5.9 are taken into account.

Displacement function The element nodal displacement matrix is represented by

$$\{\delta\}_e = \{w_i, \theta_{xi}, \theta_{yi}, w_j, \theta_{xj}, \theta_{yj}, w_m, \theta_{xm}, \theta_{ym}, w_n, \theta_{xn}, \theta_{yn}\} \tag{5.39}$$

The following polynomial expression for the displacement of the element *ijmn* is selected[17]

$$w_e = a_1 + a_2 + a_3 y + a_4 x^2 + a_5 xy + a_6 y^2 + a_7 x^3 + a_8 x^2 y + a_9 xy^2 + a_{10} y^3 + a_{11} x^3 y + a_{12} xy^3 \tag{5.40}$$

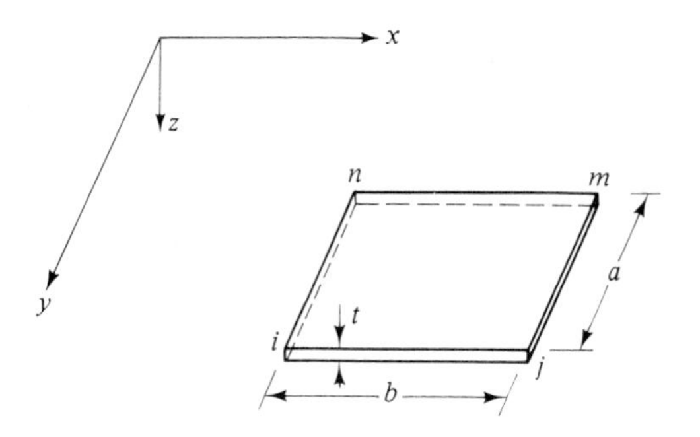

Figure 5.14

Nodal displacements, upon introduction of Eqs. (5.40) and (5.28) into Eq. (5.39), are next found. In concise form, these are

$$\{\delta\}_e = [C]\{a\} \tag{5.41}$$

in which $[C]$, a 12×12 matrix, depends upon the nodal coordinates as in Eqs. (5.31). Inversion of the above provides the values of the unknown coefficients $a_1, \ldots, a_{12}$:

$$\{a\} = [C]^{-1}\{\delta\}_e \tag{5.42}$$

The displacement function is expressed by Eq. (5.33). However, the matrix $[L]$ in that equation now has the form

$$[L] = [1, x, y, x^2, xy, y^2, x^3, x^2y, xy^2, y^3, x^3y, xy^3]$$

Inserting Eqs. (5.41) into Eq. (5.17), one has

$$\{\varepsilon\}_e = \begin{bmatrix} 0 & 0 & 0 & -2 & 0 & 0 & -6x & -2y & 0 & 0 & -6xy & 0 \\ 0 & 0 & 0 & 0 & 0 & -2 & 0 & 0 & -2x & -6y & 0 & -6xy \\ 0 & 0 & 0 & 0 & -2 & 0 & 0 & -4x & -4y & 0 & -6x^2 & -6y^2 \end{bmatrix} \{a_1, \ldots, a_{12}\} \tag{5.43a}$$

or

$$\{\varepsilon\}_e = [H]\{a\} \tag{5.43b}$$

As before, the generalized "strain"-displacement matrix is found upon inserting Eqs. (5.42) into Eqs. (5.43).

The stiffness matrix The element stiffness matrix $[k]_e$ is obtained by introducing matrices $[H]$, $[D]$, and $[C]$ given by Eqs. (5.43), (5.21), and (5.41) into Eq. (5.37). It follows that

$$[k]_e = \frac{Et^3}{180ab(1 - \nu^2)}[R]\left\{[k_1] + [k_2] + \nu[k_3] + \frac{1 - \nu}{2}[k_4]\right\}[R] \tag{5.44}$$

Explicit expressions of bending-stiffness coefficients $[k_1]$ to $[k_4]$ and the matrix $[R]$ are given[17,18] in Table 5.3. These coefficients in general provide rapid convergence and satisfactory accuracy.

External nodal forces The element nodal force matrix is

$$\{Q\}_e = \begin{Bmatrix} Q_i \\ Q_j \\ Q_m \\ Q_n \end{Bmatrix}$$

where the Q's are defined in the previous section.

The unknown displacements, strains, and stresses may now be calculated employing the general procedure of Sec. 5.8.

Table 5.3 Stiffness coefficients in Eqs. (5.44) and (6.36) for a rectangular plate element (Fig. 5.14).

$$[k_1] = \left(\frac{b}{a}\right)^2 \begin{bmatrix} 60 & & & & & & & & & & & \\ 0 & 0 & & & & & & & & & & \\ 30 & 0 & 20 & & & & & & & & & \\ 30 & 0 & 15 & 60 & & & & & & & & \\ 0 & 0 & 0 & 0 & 0 & & & & \text{Symmetric} & & & \\ 15 & 0 & 10 & 30 & 0 & 20 & & & & & & \\ -60 & 0 & -30 & -30 & 0 & -15 & 60 & & & & & \\ 0 & 0 & 0 & 0 & 0 & 0 & 0 & 0 & & & & \\ 30 & 0 & 10 & 15 & 0 & 5 & -30 & 0 & 20 & & & \\ -30 & 0 & -15 & -60 & 0 & -30 & 30 & 0 & -15 & 60 & & \\ 0 & 0 & 0 & 0 & 0 & 0 & 0 & 0 & 0 & 0 & 0 & \\ 15 & 0 & 5 & 30 & 0 & 10 & -15 & 0 & 10 & -30 & 0 & 20 \end{bmatrix}$$

$$[k_3] = \begin{bmatrix} 30 & & & & & & & & & & & \\ -15 & 0 & & & & & & & & & & \\ 15 & -15 & 0 & & & & & & & & & \\ -30 & 0 & -15 & 30 & & & & & & & & \\ 0 & 0 & 0 & 15 & 0 & & & & \text{Symmetric} & & & \\ -15 & 0 & 0 & 15 & 15 & 0 & & & & & & \\ -30 & 15 & 0 & 30 & 0 & 0 & 30 & & & & & \\ 15 & 0 & 0 & 0 & 0 & 0 & -15 & 0 & & & & \\ 0 & 0 & 0 & 0 & 0 & 0 & -15 & 15 & 0 & & & \\ 30 & 0 & 0 & -30 & -15 & 0 & -30 & 0 & 15 & 30 & & \\ 0 & 0 & 0 & -15 & 0 & 0 & 0 & 0 & 0 & 15 & 0 & \\ 0 & 0 & 0 & 0 & 0 & 0 & 15 & 0 & 0 & -15 & -15 & 0 \end{bmatrix}$$

$$[k_2] = \left(\frac{a}{b}\right)^2 \begin{bmatrix} 60 & & & & & & & & & & & \\ -30 & 20 & & & & & & & & & & \\ 0 & 0 & 0 & & & & & & & & & \\ -60 & 30 & 0 & 60 & & & & & & & & \\ -30 & 10 & 0 & 30 & 20 & & & & \text{Symmetric} & & & \\ 0 & 0 & 0 & 0 & 0 & 0 & & & & & & \\ 30 & -15 & 0 & -30 & -15 & 0 & 60 & & & & & \\ -15 & 10 & 0 & 15 & 5 & 0 & -30 & 20 & & & & \\ 0 & 0 & 0 & 0 & 0 & 0 & 0 & 0 & 0 & & & \\ -30 & 15 & 0 & 30 & 15 & 0 & -60 & 30 & 0 & 60 & & \\ -15 & 5 & 0 & 15 & 10 & 0 & -30 & 10 & 0 & 30 & 20 & \\ 0 & 0 & 0 & 0 & 0 & 0 & 0 & 0 & 0 & 0 & 0 & 0 \end{bmatrix}$$

$$[k_4] = \begin{bmatrix} 84 & & & & & & & & & & & \\ -6 & 8 & & & & & & & & & & \\ 6 & 0 & 8 & & & & & & & & & \\ -84 & 6 & -6 & 84 & & & & & & & & \\ -6 & -2 & 0 & 6 & 8 & & & & \text{Symmetric} & & & \\ -6 & 0 & -8 & 6 & 0 & 8 & & & & & & \\ -84 & 6 & -6 & 84 & 6 & 6 & 84 & & & & & \\ 6 & -8 & 0 & -6 & 2 & 0 & -6 & 8 & & & & \\ 6 & 0 & -2 & -6 & 0 & 2 & -6 & 0 & 8 & & & \\ 84 & -6 & 6 & -84 & -6 & -6 & -84 & 6 & 6 & 84 & & \\ 6 & 2 & 0 & -6 & -8 & 0 & -6 & -2 & 0 & 6 & 8 & \\ -6 & 0 & 2 & 6 & 0 & -2 & 6 & 0 & -8 & -6 & 0 & 8 \end{bmatrix}$$

$$[R] = \begin{bmatrix} [r] & [0] & [0] & [0] \\ [0] & [r] & [0] & [0] \\ [0] & [0] & [r] & [0] \\ [0] & [0] & [0] & [r] \end{bmatrix} \quad \text{where} \quad [r] = \begin{bmatrix} 1 & 0 & 0 \\ 0 & b & 0 \\ 0 & 0 & a \end{bmatrix}$$

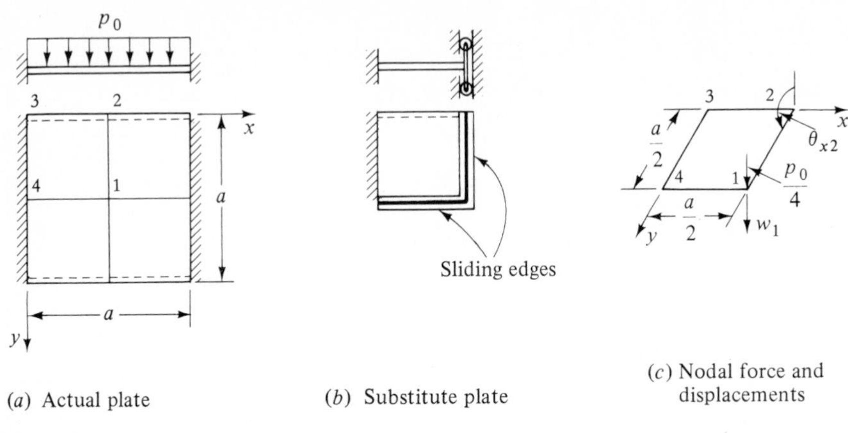

(*a*) Actual plate (*b*) Substitute plate (*c*) Nodal force and displacements

Figure 5.15

Example 5.8 Consider a square plate of sides a with two opposite edges $x = 0$ and $x = a$ simply supported and the remaining edges clamped (Fig. 5.15*a*). Compute maximum value of w if the plate is subjected to uniformly distributed load of intensity p_0. Take $a = 2$ m and $\nu = 0.3$.

SOLUTION Symmetry in deflection dictates that only one quarter-plate need be analyzed, provided that *sliding-edge* conditions (1.28) are introduced along the lines of symmetry. The *substitute plate* is shown in Fig. 5.15*b*. For the sake of simplicity in calculations, we employ only one element per quarter-plate.

A concentrated load $p_0(1 \times 1)/4 = p_0/4$ is assigned to node 1. The boundary constraints permit only a lateral displacement w_1 at node 1 and a rotation θ_{x2} at node 2 (Fig. 5.15*c*). Nodal force and the displacement matrices are

$$\{Q\}_e = \{p_0/4, 0, 0, 0, 0, 0, 0, 0, 0, 0, 0, 0\}$$

$$\{\delta\}_e = \{w_1, 0, 0, 0, \theta_{x2}, 0, 0, 0, 0, 0, 0, 0\}$$

The force-displacement relationship (5.26), together with the values of stiffness coefficients (Table 5.3), is then readily reduced to the form

$$\begin{Bmatrix} p_0/4 \\ 0 \end{Bmatrix} = \frac{Et^3}{180(1-0.9)}$$

$$\begin{bmatrix} (60 + 60 + 0.3 \times 30 + 0.35 \times 84) & (0 - 30 + 0 - 0.35 \times 6) \\ (0 - 30 + 0 - 0.35 \times 6) & (0 + 20 + 0 + 0.35 \times 8) \end{bmatrix} \begin{Bmatrix} w_1 \\ \theta_{x2} \end{Bmatrix}$$

From the above, we obtain

$$w_1 = w_{\max} = 0.3617 \frac{p_0}{Et^3}$$

The "exact" solution of this problem (see Example 3.7) is $0.3355p_0/Et^3$.

When the quarter-plate is divided into 4, 16, and 25 elements, the results are, respectively:

$$w_{\max} = 0.3512p_0/Et^3$$

$$w_{\max} = 0.3397p_0/Et^3$$

$$w_{\max} = 0.3378p_0/Et^3$$

It is apparent that the accuracy of the solution increases as the mesh is refined.

PROBLEMS

Secs. 5.1 to 5.3

5.1 Verify that the effective shear forces are represented, as finite difference approximations at point 0 (Fig. 5.1*b*), in the form

$$\begin{aligned} V_x &= -\frac{D}{2h^3}[w_9 - w_{11} - 2(3-\nu)(w_1 - w_3) + (2-\nu)(w_5 - w_6 - w_7 + w_8)] \\ V_y &= -\frac{D}{2h^3}[w_{10} - w_{12} - 2(3-\nu)(w_2 - w_4) + (2-\nu)(w_5 + w_6 - w_7 + w_8)] \end{aligned} \tag{P5.1}$$

Referring to Table 5.1, check the correctness of the result for V_x.

5.2 Determine a finite difference expression corresponding to $\nabla^4 w = p/D$ at a nodal point 0 (Fig. 5.1*b*), for a rectangular mesh. Take $\Delta x = h$ and $\Delta y = k$.

5.3 Determine the finite difference equivalent of M_x, M_y, and V_x at a nodal point 0 (Fig. 5.1*b*) for a rectangular mesh with $\Delta x = h$ and $\Delta y = k$.

5.4 Calculate the maximum stress at the nodal point 22 for the plate shown in Fig. 5.5.

5.5 Swept wing of an aircraft is approximated by a simply supported skew plate subjected to uniform loading p_0 (Fig. P5.5). Determine the deflection w at the nodal points 1 through 5.

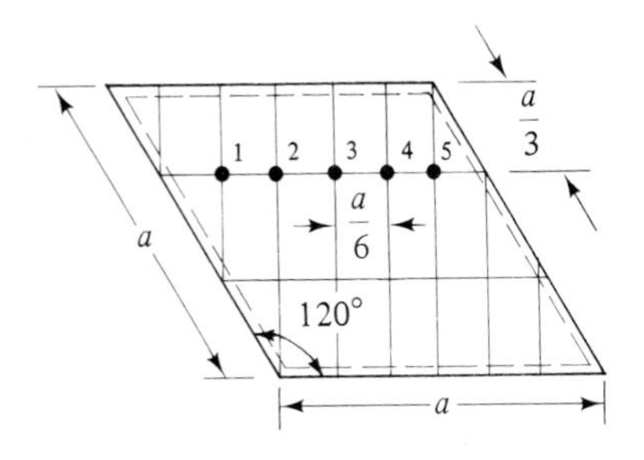

Figure P5.5

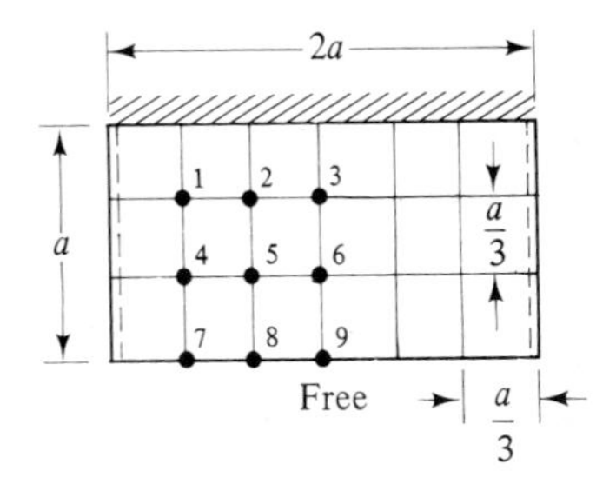

Figure P5.6

5.6 Consider a uniformly loaded plate with two opposite sides simply supported, the third side clamped, and the fourth side free (Fig. P5.6). Determine: (*a*) the displacement w at the nodal points 1 through 9; (*b*) the bending moments at nodal point 9.

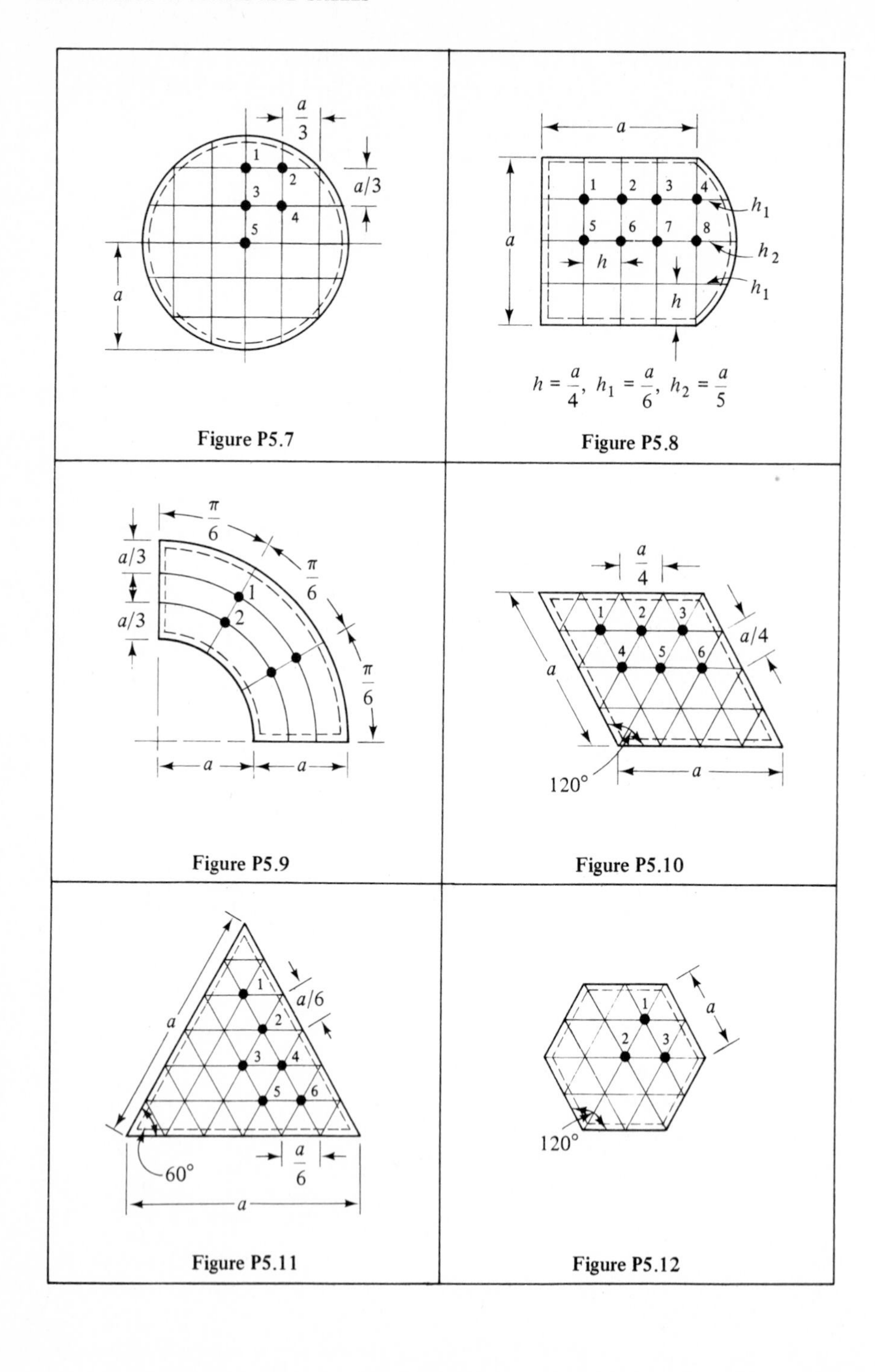

Figure P5.7

Figure P5.8

Figure P5.9

Figure P5.10

Figure P5.11

Figure P5.12

Secs. 5.4 to 5.6

5.7 through 5.12 Each of the variously shaped plates shown in the figures is simply supported at all edges and carries a uniform loading of intensity p_0. Determine the deflection w at the nodal points labeled on the mesh configurations describing the plates.

5.13 Verify the results given by Eqs. (*c*) of Sec. 5.6 using the chain rule.

5.14 Redo Prob. 5.9 if the inner edge is built in and the remaining edges are simply supported.

Secs. 5.7 to 5.10

5.15 Develop the stiffness matrix of the finite element, described in Example 5.7, in terms of E and ν.

5.16 Derive the matrix $[C]$ given by Eq. (5.41) for the rectangular element of sides a and b (Fig. 5.14), by locating the origin of xyz at the node n.

5.17 Calculate maximum deflection of a simply supported and uniformly loaded 2 m × 2 m square plate, using only one element per quarter-plate. Compare the result with that given in Table 3.1.

CHAPTER

SIX

ORTHOTROPIC PLATES

6.1 INTRODUCTION

The plates analyzed thus far have been assumed to be composed of a single homogeneous and isotropic material. However, plates of anisotropic materials have important applications owing to their exceptionally high bending stiffness. A nonisotropic or anisotropic material displays direction-dependent properties.[19] Simplest among them are those in which the material properties differ in *two* mutually perpendicular directions. A material so described is *orthotropic*, e.g., wood. A number of manufactured materials are approximated as orthotropic.[20] Examples include corrugated and rolled metal sheet, fillers in sandwich plate construction, plywood, fiber reinforced composites, reinforced concrete, and gridwork. The latter consists of two systems of equally spaced parallel ribs (beams), mutually perpendicular, and attached rigidly at the points of intersection.

Presented in this chapter are the fundamental equations for the small-deflection theory of bending of thin orthotropic plates. Orthotropic properties of several commonly employed materials are discussed as are applications to various orthotropic plates. Topics on plates made from two or more different materials and numerical methods are also included.

6.2 BASIC RELATIONSHIPS

Solution of the bending problem of orthotropic plates requires reformulation of the Hooke's law. Equation (1.7) now assumes the following form:

$$
\begin{aligned}
\sigma_x &= E_x + E_{xy}\varepsilon_y \\
\sigma_y &= E_y\varepsilon_y + E_{xy}\varepsilon_x \\
\tau_{xy} &= G\gamma_{xy}
\end{aligned}
\tag{6.1}
$$

in which the moduli, E_x, E_y, E_{xy}, and G are all independent of one another. In particular, Eq. (1.6) is *no longer applicable*, as this relation involving E, ν, and G relies upon material isotropy. An alternate representation of Eqs. (6.1) is

$$
\begin{aligned}
\sigma_x &= \frac{E'_x}{1-\nu_x\nu_y}(\varepsilon_x + \nu_y\varepsilon_y) \\
\sigma_y &= \frac{E'_y}{1-\nu_x\nu_y}(\varepsilon_y + \nu_x\varepsilon_x) \\
\tau_{xy} &= G\gamma_{xy}
\end{aligned}
\tag{6.2}
$$

Here ν_x, ν_y, and E'_x, E'_y are the effective Poisson's ratios and effective moduli of elasticity, respectively. Subscripts x and y relate to the directions. The shear modulus of elasticity G is the *same* for both isotropic and orthotropic materials. Clearly, the two sets of elastic constants in Eqs. (6.1) and (6.2) are connected by

$$
\begin{aligned}
E_x &= \frac{E'_x}{1-\nu_x\nu_y} \\
E_y &= \frac{E'_y}{1-\nu_x\nu_y} \\
E_{xy} &= \frac{E'_x\nu_y}{1-\nu_x\nu_y} = \frac{E'_y\nu_x}{1-\nu_x\nu_y}
\end{aligned}
\tag{6.3}
$$

The strain-displacement relations (1.3) are based upon purely geometrical considerations and are *unchanged* for orthotropic plates. The same is true for the conditions of equilibrium of Sec. 1.5. The stresses are obtained by introducing Eqs. (1.3) into (6.1):

$$
\begin{aligned}
\sigma_x &= -z\left(E_x\frac{\partial^2 w}{\partial x^2} + E_{xy}\frac{\partial^2 w}{\partial y^2}\right) \\
\sigma_y &= -z\left(E_y\frac{\partial^2 w}{\partial y^2} + E_{xy}\frac{\partial^2 w}{\partial x^2}\right) \\
\tau_{xy} &= -2Gz\frac{\partial^2 w}{\partial x\,\partial y}
\end{aligned}
\tag{6.4}
$$

The formulas for the moments, inserting the foregoing into Eqs. (1.9), and integrating the resulting expression, are

$$M_x = -\left(D_x \frac{\partial^2 w}{\partial x^2} + D_{xy} \frac{\partial^2 w}{\partial y^2}\right)$$

$$My = -\left(D_y \frac{\partial^2 w}{\partial y^2} + D_{xy} \frac{\partial^2 w}{\partial x^2}\right) \qquad (6.5)$$

$$M_{xy} = -2G_{xy} \frac{\partial^2 w}{\partial x\, \partial y}$$

where

$$D_x = \frac{t^3 E_x}{12} \qquad D_y = \frac{t^3 E_y}{12} \qquad D_{xy} = \frac{t^3 E_{xy}}{12} \qquad G_{xy} = \frac{t^3 G}{12} \qquad (6.6)$$

The expressions for D_x, D_y, D_{xy}, and G_{xy} represent the *flexural rigidities* and the *torsional rigidity* of an orthotropic plate, respectively. We can obtain the vertical shear forces in the plate by substituting Eqs. (6.5) into Eqs. (*b*) and (*c*) of Sec. 1.5:

$$Q_x = -\frac{\partial}{\partial x}\left(D_x \frac{\partial^2 w}{\partial x^2} + H \frac{\partial^2 w}{\partial y^2}\right)$$

$$Q_y = -\frac{\partial}{\partial y}\left(Dy \frac{\partial^2 w}{\partial y^2} + H \frac{\partial^2 w}{\partial x^2}\right) \qquad (6.7)$$

where

$$H = D_{xy} + 2G_{xy} \qquad (6.8)$$

The governing differential equation of deflection for an orthotropic plate, through the use of Eqs. (1.15) and (6.5), is expressed in the form

$$D_x \frac{\partial^4 w}{\partial x^4} + 2H \frac{\partial^4 w}{\partial x^2\, \partial y^2} + D_y \frac{\partial^4 w}{\partial y^4} = p \qquad (6.9)$$

which is solved for w, upon satisfying the given boundary conditions, as illustrated in Secs. 6.4 and 6.5.

6.3 DETERMINATION OF RIGIDITIES

Discussed briefly in this section is the determination of the rigidities of orthotropic materials used in plate structures. Practical considerations often lead to assumptions, with regard to material properties, resulting in approximate expressions for the elastic constants. The accuracy of these approximations is generally the most significant factor in the orthotropic plate problem.

The orthotropic plate moduli and Poisson's ratios

$$E'_x,\ E'_y,\ \nu_x,\ \nu_y,\ G \qquad (6.10)$$

are obtained by tension and shear tests, as in the case of isotropic materials. The plate rigidities are calculated from Eqs. (6.3), (6.6), and (6.8):

$$D_x = \frac{t^3 E'_x}{12(1 - \nu_x \nu_y)} \qquad G_{xy} = \frac{t^3 G}{12}$$

$$D_y = \frac{t^3 E'_y}{12(1 - \nu_x \nu_y)} \qquad H = D_{xy} + 2G_{xy} \tag{6.11}$$

$$D_{xy} = \frac{t^3 \nu_x \nu_y}{12(1 - \nu_x \nu_y)}$$

Table 6.1 Various orthotropic plates

Geometry	Rigidities
A. Reinforced concrete slab with x and y directed reinforcement steel bars 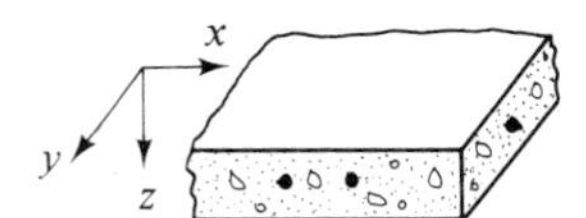	$D_x = \frac{E_c}{1-\nu_c^2}\left[I_{cx} + \left(\frac{E_s}{E_c} - 1\right)I_{sx}\right]$ $\quad D_y = \frac{E_c}{1-\nu_c^2}\left[I_{cy} + \left(\frac{E_s}{E_c} - 1\right)I_{sy}\right]$ $G_{xy} = \frac{1-\nu_c}{2}\sqrt{D_x D_y}$ $\quad H = \sqrt{D_x D_y}$ $\quad D_{xy} = \nu_c\sqrt{D_x D_y}$ ν_c: Poisson's ratio for concrete E_c, E_s: Elastic modulus of concrete and steel, respectively $I_{cx}(I_{sx})$, $I_{cy}(I_{sy})$: Moment of inertia of the slab (steel bars) about neutral axis in the section x = constant and y = constant, respectively
B. Plate reinforced by equidistant stiffeners 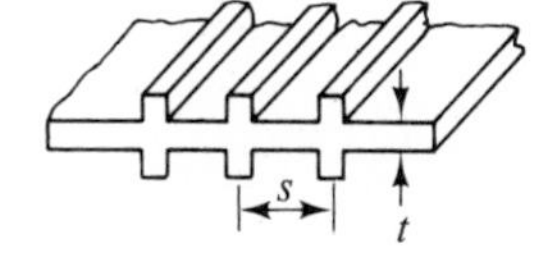	$D_x = H = \frac{Et^3}{12(1-\nu^2)}$ $\quad D_y = \frac{Et^3}{12(1-\nu^2)} + \frac{E'I}{s}$ E, E': Elastic modulus of plating and stiffeners, respectively ν: Poisson's ratio of plating s: Spacing between centerlines of stiffeners I: Moment of inertia of the stiffener cross section with respect to midplane of plating
C. Plate reinforced by a set of equidistant ribs 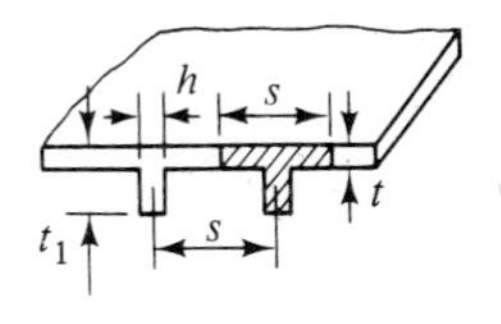	$D_x = \frac{Est^3}{12[s - h + h(t/t_1)^3]}$ $\quad D_y = \frac{EI}{s}$ $H = 2G'_{xy} + \frac{C}{s}$ $\quad D_{xy} = 0$ C: Torsional rigidity of one rib I: Moment of inertia about neutral axis of a T-section of width s (shown as shaded) G'_{xy}: Torsional rigidity of the plating E: Elastic modulus of the plating
D. Corrugated plate 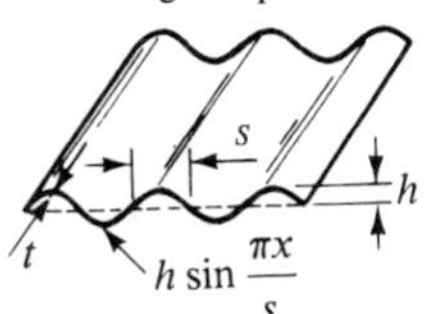	$D_x = \frac{s}{\lambda}\frac{Et^3}{12(1-\nu^2)}$ $\quad D_y = EI,\ H = \frac{\lambda}{a}\frac{Et^3}{12(1+\nu)}$ $\quad D_{xy} = 0$ where $\lambda = s\left(1 + \frac{\pi^2 h^2}{4s^2}\right)$ $\quad I = 0.5h^2 t\left[1 - \frac{0.81}{1 + 2.5(h/2s)^2}\right]$

When it is *not possible*, however, to determine the constants of (6.10) experimentally, one resorts to approximations derived using analytical techniques. The latter approaches consist of constructing an orthotropic plate with elastic properties equal to the average properties of components of the original plate.[21] Such a plate is termed an *equivalent* or *transformed* orthotropic plate. For example, in the case of a plate reinforced by ribs, the bending stiffness of the ribs and the plating are combined and taken to be uniform across the replacement model. Subsequently, the constants of (6.10) are approximated with Eqs. (6.11) giving the rigidities. For reference purposes, Table 6.1 presents the rigidities for some commonly encountered cases.[11,22]

It is noted that when $E'_x = E'_y = E$ (and hence $\nu_x = \nu_y = \nu$), Eqs. (6.3) become

$$E_x = E_y = \frac{E}{1-\nu^2} \qquad E_{xy} = \frac{\nu E}{1-\nu^2} \tag{6.12}$$

Consequently,

$$D_x = D_y = \frac{Et^3}{12(1-\nu^2)} \qquad G_{xy} = \frac{Et^3}{24(1+\nu)} \qquad H = \frac{Et^3}{12(1-\nu^2)} = D \tag{6.13}$$

It follows that Eq. (6.9), as expected, reduces to that of an isotropic plate [Eq. (1.17)].

6.4 RECTANGULAR ORTHOTROPIC PLATES

The general procedure for the determination of the deflection and stress in rectangular orthotropic plates is identical with that employed for isotropic plates. We now apply Navier's method (Sec. 3.2) to treat the case of a simply supported rectangular orthotropic plate under a nonuniform load $p(x, y)$ (Fig. 3.1).

Introduction of Eqs. (3.1) into (6.9) yields

$$\sum_{m=1}^{\infty}\sum_{n=1}^{\infty}\left\{a_{mn}\left(\frac{m^4\pi^4}{a^4}D_x + 2H\frac{m^2n^2\pi^4}{a^2b^2} + \frac{n^4\pi^4}{b^4}D_y\right) - p_{mn}\right\}\sin\frac{m\pi x}{a}\sin\frac{n\pi y}{b} = 0$$

Inasmuch as the above must be valid for all x and y, it follows that the terms in the brackets must be zero, leading to

$$a_{mn} = \frac{p_{mn}}{(m^4\pi^4/a^4)D_x + 2H(m^2n^2\pi^4/a^2b^2) + (n^4\pi^4/b^4)D_y} \tag{6.14}$$

The expression of the plate-deflection surface, by substitution of Eqs. (6.14) and (3.3) into (3.1*b*), is therefore

$$w = \frac{4}{ab}\sum_{m=1}^{\infty}\sum_{n=1}^{\infty}\int_0^b\int_0^a \frac{p(x,y)\sin(m\pi x/a)\sin(n\pi y/b)\,dx\,dy}{(m^4\pi^4/a^4)D_x + 2H(m^2n^2\pi^4/a^2b^2) + (n^4\pi^4/b^4)D_y} \times \sin\frac{m\pi x}{a}\sin\frac{n\pi y}{b} \tag{6.15}$$

For an isotropic material, from Eqs. (6.12) and (6.13), $D_x = D_y = H = D$, and the above coincides with Eq. (3.5).

In the particular case of a rectangular plate under a uniformly distributed load p_0, referring to Sec. 3.3 we can readily obtain from Eq. (6.15):

$$w = \frac{16p_0}{\pi^6} \sum_m^\infty \sum_n^\infty \frac{\sin (m\pi x/a) \sin (n\pi y/b)}{mn[(m^4/a^4)D_x + 2H(m^2n^2/a^2b^2) + (n^4/b^4)Dy]} \tag{6.16}$$

When the material is isotropic the above reduces to Eq. (3.6). For example, if the plate is made of a *reinforced concrete*, from Table 6.1, we have $H = \sqrt{D_x D_y}$. Based upon the notation

$$a_1 = a \sqrt[4]{\frac{D}{D_x}}, \qquad b_1 = b \sqrt[4]{\frac{D}{D_y}} \tag{a}$$

Eq. (6.16) becomes

$$w = \frac{16p_0}{D\pi^6} \sum_m^\infty \sum_n^\infty \frac{\sin (m\pi x/a) \sin (n\pi y/b)}{mn(m^2/a_1^2 + n^2/b_1^2)^2} \qquad (m, n = 1, 3, \ldots) \tag{b}$$

which is of the same form as Eq. (3.6). We are led to conclude that the center deflection of the reinforced concrete plate ($a \times b$) having rigidities D_x, D_y is equal to that of an isotropic plate ($a_1 \times b_1$) of rigidity D.

Having the expression for the deflection of the plate available, we can obtain the bending moments from Eqs. (6.5) and the stresses by applying the relationships of Sec. 1.4.

Deflection of orthotropic rectangular plates is also determined by following the same basic procedures that were prescribed in Chaps. 3 and 5 for the Lévy's solution and for the finite difference approach. Referring to Table 5.1 the pertinent coefficient pattern for finite difference expression of Eq. (6.9) of the orthotropic plate may readily be obtained. It is given in Fig. 6.1 for the case of evenly distributed nodes, i.e., for $\Delta x = \Delta y = h$.

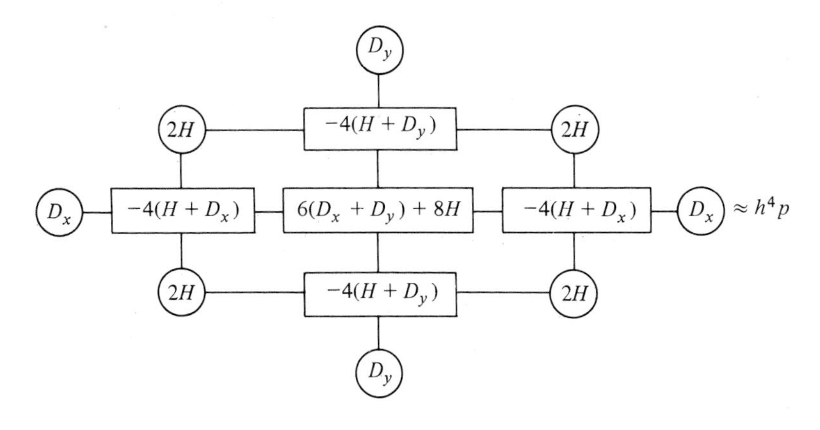

Figure 6.1

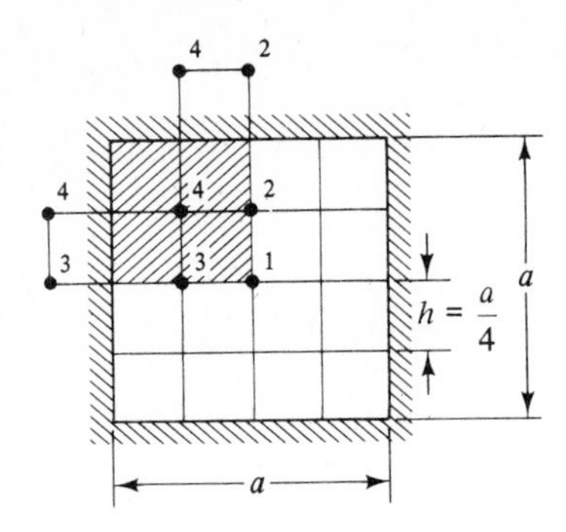

Figure 6.2

Example 6.1 A square orthotropic plate is subjected to a uniform loading of intensity p_0. Assume the plate edges are clamped and parallel with the principal directions of orthotropy. Determine the deflection, using the finite difference approach by dividing the domain into equal nets with $h = a/4$. Take $D_x = D_0$, $D_y = 0.5D_0$, $H = 1.248D_0$, and $\nu_x = \nu_y = 0.3$.

SOLUTION For this case, the governing expression for deflection, Eq. (6.9), appears

$$\frac{\partial^4 w}{\partial x^4} + 2.496 \frac{\partial^4 w}{\partial x^2\,\partial y^2} + 0.5 \frac{\partial^4 w}{\partial y^4} = \frac{p_0}{D_0} \tag{c}$$

Considerations of symmetry indicate that only one-quarter (shaded portion) of the plate need by analyzed (Fig. 6.2). The conditions that the slopes vanish at the edges are satisfied by numbering the nodes located outside the plate surface as shown in the figure (See Example 5.2). The values of w are zero on the boundary.

Applying Fig. 6.1 at the nodes 1, 2, 3, and 4, we obtain

$$\begin{bmatrix} 18.718 & -17.718 & -13.718 & 9.718 \\ -8.859 & 20.718 & 4.859 & -13.718 \\ -6.859 & 4.859 & 19.718 & -17.718 \\ 2.429 & -6.859 & -8.859 & 21.718 \end{bmatrix} \begin{Bmatrix} w_1 \\ w_2 \\ w_3 \\ w_4 \end{Bmatrix} = \frac{p_0 h^4}{D_0} \begin{Bmatrix} 1 \\ 1 \\ 1 \\ 1 \end{Bmatrix} \tag{d}$$

The simultaneous solution of Eqs. (d) results in

$$w_1 = 0.496 \frac{p_0 h^4}{D_0} \qquad w_2 = 0.339 \frac{p_0 h^4}{D_0}$$

$$w_3 = 0.359 \frac{p_0 h^4}{D_0} \qquad w_4 = 0.241 \frac{p_0 h^4}{D_0}$$

The center deflection, $w_1 = 0.0019 p_0 a^4/D_0$, is about 24 percent more than the "exact" value,[19] $0.00156 p_0 a^4/D_0$. By decreasing the size of the mesh increment, the accuracy of the solution can be improved.

6.5 ELLIPTIC AND CIRCULAR ORTHOTROPIC PLATES

Consider an elliptic orthotropic plate with semiaxes a and b, clamped at the edge and subjected to the uniformly distributed load p_0 (Fig. 4.3). Assume that the principal axes of the ellipse and the principal directions of the orthotropic material are parallel. The solution procedure follows a pattern similar to that described in Sec. 4.4. Thus, we let

$$w = k\left(1 - \frac{x^2}{a^2} - \frac{y^2}{b^2}\right)^2 \tag{a}$$

in which k is a constant to be determined. Substitution of the above into Eq. (6.9) leads to an expression which is satisfied when

$$k = \frac{p_0}{8} \frac{a^4 b^4}{3b^4 D_x + 2a^2 b^2 H + 3a^4 D_y} \tag{b}$$

The expression describing the deflected surface of the plate is then

$$w = \frac{p_0}{8} \frac{a^4 b^4}{3b^4 D_x + 2a^2 b^2 H + 3a^4 D_y}\left(1 - \frac{x^2}{a^2} - \frac{y^2}{b^2}\right)^2 \tag{6.17}$$

This equation satisfies the boundary conditions for an elliptic plate with fixed edge, presented by Eqs. (a) in Sec. 4.4 The maximum deflection occurs at the center of the plate and is given by

$$w_{\max} = \frac{p_0 a^4 b^4}{24b^4 D_x + 16a^2 b^2 H + 24a^4 D_y} \tag{6.18}$$

As anticipated, in the case of an isotropic plate, Eqs. (6.17) and (6.18) reduce to Eqs. (4.9) and (4.10), respectively. Expressions for the moments may then be obtained from Eqs. (6.5).

The result obtained above for an elliptic plate may readily be reduced to the case of a circular plate by setting $b = a$. For a built-in-edge orthotropic *circular plate* of radius a under uniform load, we have, from Eq. (6.17):

$$w = \frac{p_0}{64D_1}(a^2 - r^2)^2 \tag{6.19}$$

where

$$r = \sqrt{x^2 + y^2} \qquad D_1 = \tfrac{1}{8}(3D_x + 2H + 3D_y) \tag{c}$$

When $D_x = D_y = H = D$, Eq. (6.19) is identical with Eq. (2.14), the deflection formula for an isotropic circular plate.

The bending moments and the twisting moment are calculated by means of expression (6.19), which when introduced into Eqs. (6.5), yields

$$M_x = \frac{p_0}{16D_1}[(D_x + D_{xy})(a^2 - r^2) - 2(D_x x^2 + D_{xy} y^2)]$$

$$M_y = \frac{p_0}{16D_1}[(D_y + D_{xy})(a^2 - r^2) - 2(D_y y^2 + D_{xy} x^2)] \tag{6.20}$$

$$M_{xy} = \frac{p_0}{4D_1} D_{xy} xy$$

The stresses are then determined through application of the formulas given in Sec. 1.4.

6.6 MULTILAYERED PLATES

Structures composed of an arbitrary number of bounded layers (e.g., aircraft and marine windshields, and portions of space vehicles) can often be satisfactorily approximated by considering a laminated plate. Generally, in these structural assemblies, each layer may possess a different thickness, orientation of the principal axes, and the anisotropic properties. We shall treat only the places consisting of *isotropic* layers.

The strain-displacement relations (1.3), for the ith layer (Fig. 6.3), are expressed as follows

$$\varepsilon_x^{(i)} = -z_i \frac{\partial^2 w}{\partial x^2} \qquad \varepsilon_y^{(i)} = -z_i \frac{\partial^2 w}{\partial y^2} \qquad \gamma_{xy}^{(i)} = -2z_i \frac{\partial^2 w}{\partial x\, \partial y} \tag{6.21}$$

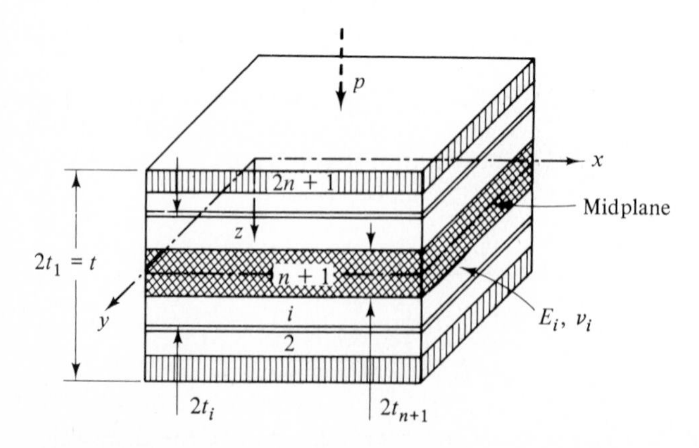

Figure 6.3

Hooke's law (1.7) now appears

$$\sigma_x^{(i)} = \frac{E_i}{1 - \nu_i^2}[\varepsilon_x^{(i)} + \nu_i \varepsilon_y^{(i)}]$$

$$\sigma_y^{(i)} = \frac{E_i}{1 - \nu_i^2}[\varepsilon_y^{(i)} + \nu_i \varepsilon_x^{(i)}] \tag{6.22}$$

$$\tau_{xy}^{(i)} = \frac{E_i}{2(1 + \nu_i)} \gamma_{xy}^{(i)}$$

Substituting strains defined by Eqs. (6.21) into the above, integrating over each layer, and summing the results, we obtain the stress resultants:

$$\begin{Bmatrix} M_x \\ M_y \\ M_{xy} \end{Bmatrix} = \sum_i \int_{z_{i-1}}^{z_i} \begin{Bmatrix} \sigma_x \\ \sigma_y \\ \tau_{xy} \end{Bmatrix}^{(i)} z\,dz \tag{6.23}$$

Stresses defined by Eqs. (1.8), for the ith layer, are

$$\sigma_x^{(i)} = -z_i \frac{E_i}{1 - \nu_i^2}\left(\frac{\partial^2 w}{\partial x^2} + \nu_i \frac{\partial^2 w}{\partial y^2}\right)$$

$$\sigma_y^{(i)} = -z_i \frac{E_i}{1 - \nu_i^2}\left(\frac{\partial^2 w}{\partial y^2} + \nu_i \frac{\partial^2 w}{\partial x^2}\right) \tag{6.24}$$

$$\tau_{xy}^{(i)} = -z_i \frac{E_i}{1 + \nu_i} \frac{\partial^2 w}{\partial x\, \partial y}$$

The general method of deriving the governing equation for multilayered plates follows a pattern identical with that described in Chap. 1. It can be shown that[23] the differential equation (1.17) now becomes

$$\nabla^4 w = \frac{p}{D_t} \tag{6.25}$$

where the D_t is the *transformed flexural rigidity* of laminated plates.

Layered plates of a symmetric structure about the midplane are of practical significance. For a plate of $2n + 1$ *symmetrical* isotropic layers (Fig. 6.3) the transformed flexural rigidity is given by

$$D_t = \frac{2}{3}\left[\sum_{i=1}^{n} \frac{E_i}{1 - \nu_i^2}(t_i^3 - t_{i+1}^3) + \frac{E_{n+1} t_{n+1}^3}{1 - \nu_{n+1}^2}\right] \tag{6.26}$$

If boundary conditions, transverse load p, and the isotropic material properties of each layer are known, Eq. (6.25) may be solved for $w(x, y)$. The stress components in the ith layer may then be computed from Eq. (6.24).

We observe that, upon introduction of transformed rigidity, solution of a multilayered plate problem reduces to that of a corresponding homogeneous plate. All analytical and numerical techniques are thus equally applicable to homogeneous and laminated plates.

6.7 THE FINITE ELEMENT SOLUTION

In Secs. 6.4 and 6.5, solutions of orthotropic plate problems were limited to simple cases in which there was uniformity of structural geometry and loading. In this section, the finite element approach of Chap. 5 is applied for computation of displacement and stress in an orthotropic plate of *arbitrary* shape and thickness, subjected to *nonuniform* loads.

For plates made of any nonisotropic material it is necessary to rederive the elasticity matrix $[D]$. When the principal directions of orthotropy are parallel to the directions of the x and y coordinates, the stress-generalized "strain" relationship is given by Eqs. (6.2). Written in the matrix form, they are as follows:

$$\begin{Bmatrix} \sigma_x \\ \sigma_y \\ \tau_{xy} \end{Bmatrix}_e = z \begin{bmatrix} \dfrac{E'_x}{1-\nu_x\nu_y} & \dfrac{\nu_y E'_x}{1-\nu_x\nu_y} & 0 \\ \dfrac{\nu_x E'_y}{1-\nu_x\nu_y} & \dfrac{E'_y}{1-\nu_x\nu_y} & 0 \\ 0 & 0 & G \end{bmatrix} \begin{Bmatrix} \varepsilon_x \\ \varepsilon_y \\ \gamma_{xy} \end{Bmatrix}_e \tag{6.27a}$$

or, succinctly

$$\{\sigma\}_e = z[D^*]\{\varepsilon\}_e \tag{6.27b}$$

The elasticity matrix, from Eq. (5.21), is therefore

$$[D] = \frac{t^3}{12} \begin{bmatrix} \dfrac{E'_x}{1-\nu_x\nu_y} & \dfrac{\nu_y E'_x}{1-\nu_x\nu_y} & 0 \\ \dfrac{\nu_x E'_y}{1-\nu_x\nu_y} & \dfrac{E'_y}{1-\nu_x\nu_y} & 0 \\ 0 & 0 & G \end{bmatrix} \tag{6.28}$$

The principal directions of orthotropy usually do *not coincide* with the x and y directions, however. Let us consider a plate in which x' and y' represent the principal directions of the material (Fig. 6.4). The stress and generalized

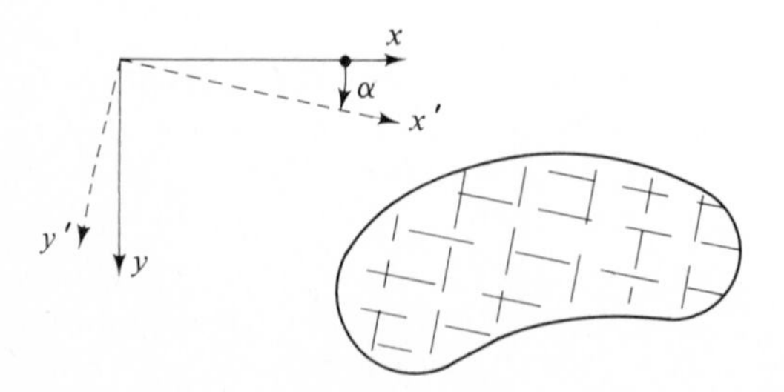

Figure 6.4

"strain," in the directions of these coordinates, are related by

$$\begin{Bmatrix} \sigma_{x'} \\ \sigma_{y'} \\ \tau_{x'y'} \end{Bmatrix} = z \begin{bmatrix} \dfrac{E'_{x'}}{1 - \nu_{x'}\nu_{y'}} & \dfrac{\nu_{y'}E'_{x'}}{1 - \nu_{x'}\nu_{y'}} & 0 \\ \dfrac{\nu_{x'}E'_{y'}}{1 - \nu_{x'}\nu_{y'}} & \dfrac{E'_{y'}}{1 - \nu_{x'}\nu_{y'}} & 0 \\ 0 & 0 & G \end{bmatrix} \begin{Bmatrix} \varepsilon_{x'} \\ \varepsilon_{y'} \\ \gamma_{x'y} \end{Bmatrix} \tag{6.29a}$$

or

$$\{\sigma'\} = z[D^{*\prime}]\{\varepsilon'\} \tag{6.29b}$$

Equations for transformation of the strain components ε_x, ε_y, γ_{xy} at any point of the plate, referring to Eqs. (P1.2), are written in the following matrix form

$$\begin{Bmatrix} \varepsilon_{x'} \\ \varepsilon_{y'} \\ \gamma_{x'y'} \end{Bmatrix} = \begin{bmatrix} \cos^2\alpha & \sin^2\alpha & \sin\alpha\cos\alpha \\ \sin^2\alpha & \cos^2\alpha & -\sin\alpha\cos\alpha \\ -2\sin\alpha\cos\alpha & 2\sin\alpha\cos\alpha & \cos^2\alpha - \sin^2\alpha \end{bmatrix} \begin{Bmatrix} \varepsilon_x \\ \varepsilon_y \\ \gamma_{xy} \end{Bmatrix} \tag{6.30a}$$

Concisely,

$$\{\varepsilon'\} = [T]\{\varepsilon\} \tag{6.30b}$$

where $[T]$ is called the strain transformation matrix. Similarly, the transformation relating stress components in x, y, z to those in x', y', z' is written as

$$\{\sigma\} = [T]^T\{\sigma'\} \tag{6.31}$$

Upon introducing Eqs. (6.29) together with Eqs. (6.30) into the above, we obtain

$$\{\sigma\} = z[T]^T[D^{*\prime}][T]\{\varepsilon\} = z[D^*]\{\varepsilon\} \tag{6.32}$$

in which

$$[D^*] = [T]^T[D^{*\prime}][T] \tag{6.33}$$

We thus have, from Eqs. (5.21) and (6.33), the expression

$$[D] = \frac{t^3}{12}[T]^T[D^{*\prime}][T] \tag{6.34}$$

for the elasticity matrix of the orthotropic plate in which the principal directions of orthotropy are not oriented along the x and y axes.

With Eq. (6.28) or (6.34), explicit expressions[18] of stiffness matrices for orthotropic plate elements may be evaluated as outlined in the preceding chapter. In the case of a *rectangular*, orthotropic plate element, we obtain

$$[k]_e = \frac{1}{15ab}[R]\{D_x[k_1] + D_y[k_2] + D_{xy}[k_3] + G_{xy}[k_4]\}[R] \tag{6.35}$$

The coefficients $[k_1]$ to $[k_4]$ and $[R]$ are listed in Table 5.3. For any particular orthotropic material, the appropriate values of the rigidities D_x, D_y, D_{xy}, and G_{xy} (Table 6.1) are specified.

The process of arriving at solutions for the orthotropic plates is identical to that described in Secs. 5.8 to 5.10.

PROBLEMS

Secs. 6.1 to 6.3

6.1 A plate is reinforced by single equidistant stiffeners (Table 6.1). Compute the rigidities. The plate and stiffeners are made of steel with $E = 200$ GPa, $\nu = 0.3$, $s = 200$ mm, $t = 20$ mm, and $I = 12 \times 10^{-7}$ m^4.

6.2 Determine the rigidities of an orthotropic steel bridge deck which may be approximated as a steel plate reinforced by a set of equidistant steel ribs (Table 6.1). Assume the following properties: $t = h = 10$ mm, $t_1 = 30$ mm, $s = 100$ mm, $\nu = 0.3$, and $E = 210$ GPa. Torsional rigidity of one rib is:[1] $C = JG = 0.246h^3(t_1 - t)G$.

6.3 Show that Eq. (1.34) for the strain energy appears in the following form in the case of orthotropic plates:

$$U = \frac{1}{2}\int_0^b \int_0^a \left[D_x\left(\frac{\partial^2 w}{\partial x^2}\right)^2 + 2D_{xy}\frac{\partial^2 w}{\partial x^2}\frac{\partial^2 w}{\partial y^2} + D_y\left(\frac{\partial^2 w}{\partial y^2}\right)^2 + 4G_{xy}\left(\frac{\partial^2 w}{\partial x\, \partial y}\right)^2\right] dx\, dy \qquad \text{(P6.3)}$$

Secs. 6.4 to 6.7

6.4 A rectangular building floor slab made of a reinforced concrete material is subjected to a concentrated center load P (Fig. 3.3). Determine expressions for: (a) the deflection surface; (b) the bending moment M_x. The plate edges can be assumed simply supported. Take $b = 2a$, $m = n = 1$, $I_{sx} = I_{cx}/2$, $I_{sy} = I_{cy}/2$, $t = 0.2$ m, $E_s = 200$ GPa, $\nu_c = 0.15$, and $E_c = 21.4$ GPa.

6.5 Determine the value of the largest deflection in the plate described in Prob. 6.4, if $a = b$. Retain the first two terms of the series solution.

6.6 A simply supported square plate of sides a is subjected to a uniform load p_0 (Fig. 3.1). The plate is constructed from the material described in Prob. 6.1. What should be the value of p_0 for an allowable deflection $w_{max} = 1$ mm. Retain only the first term of the series solution.

6.7 Determine, by taking $n = m = 1$, the center deflection of a simply supported square plate uniformly loaded by p_0. Assume the plate is constructed of the material described in Prob. 6.2.

6.8 Derive an expression for the deflection of an orthotropic clamped rectangular plate under a uniform load p_0 (Fig. 3.13). Use the Ritz method by retaining the first term of the series solution. Find the maximum deflection if $a = b$.

6.9 A steel clamped manhole cover, subjected to uniform load p_0, consists of a flat plate reinforced by equidistant steel stiffeners and is elliptical in form (Fig. 4.3). The material properties are given in Prob. 6.1 and $a = 2b = 4$ m. Compute the maximum displacement w, assuming that the principal x and y axes of the ellipse and the material coincide.

6.10 Redo Prob. 6.9 for a circular plate, $a = b$.

6.11 Derive expressions for the bending moments of an orthotropic elliptical plate with built-in edge.

6.12 A 5-mm thick large plate is fabricated of an orthotropic material having the properties:

$$E'_{x'} = 2E'_{y'} = 2.2G = 13.6 \text{ GPa} \qquad \nu_{x'} = 2\nu_{y'} = 0.2$$

The angle between the principal directions of the material (x', y') and the reference axes (x, y) is $\alpha = 30°$ (Fig. 6.4). Determine the elasticity matrix of the plate.

CHAPTER
SEVEN
PLATES UNDER COMBINED LATERAL AND DIRECT LOADS

7.1 INTRODUCTION

The classical stress analysis relations of the small deformation theory of plates resulting from a lateral loading have been developed in the preceding chapters. Attention will now be directed to situations in which lateral and *in-plane* or *direct* force systems act at a plate section. The latter forces are also referred to as the *membrane forces*. These forces may be applied directly at the plate edges, or they may arise as a result of temperature changes (Chap. 9).

To begin, the governing differential equations are modified to include the simultaneous action of the combined loading. This is followed by consideration of buckling stresses caused by in-plane compression, pure shear, and biaxial compression, upon application of equilibrium, energy, and finite difference methods, respectively.

The problem of plates with small initial curvature under the action of combined forces is next discussed. The chapter concludes with consideration of a plate bent into a simple surface of practical importance.

7.2 GOVERNING EQUATION FOR THE DEFLECTION SURFACE

The midplane is *strained* subsequent to combined loading, and assumption (2) of Sec. 1.2 is no longer valid. However, w is still regarded as *small* so that the remaining suppositions of Sec. 1.2 hold, and yet large enough so that the products of

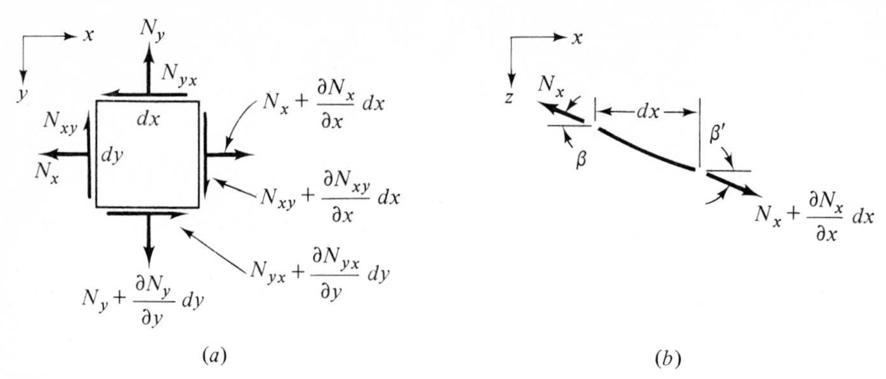

Figure 7.1

the in-plane forces or their derivatives and the derivatives of w are of the same order of magnitude as the derivatives of the shear forces (Q_x and Q_y). Thus, as before, the stress resultants are given by Eqs. (1.10) and (1.16).

Consider a plate element of sides dx and dy under the action of direct forces N_x, N_y, and $N_{xy} = N_{yx}$ which are functions of x and y only. Assume the body forces to be negligible. The top and front views of such an element are shown in Figs. 7.1a and b, respectively. The other resultants due to a lateral load, which also act on the element, are shown in Fig. 1.5. Referring to Fig. 7.1, from the equilibrium of $N_x\,dy$ forces, we obtain

$$\left(N_x + \frac{\partial N_x}{\partial x}\,dx\right)dy \cos \beta' - N_x\,dy \cos \beta \tag{a}$$

in which $\beta' = \beta + (\partial\beta/\partial x)\,dx$. Writing

$$\cos \beta = (1 - \sin^2 \beta)^{1/2} = 1 - \tfrac{1}{2}\sin^2 \beta + \cdots = 1 - \frac{\beta^2}{2} + \cdots$$

and noting that for β small, $\beta^2/2 \ll 1$ and $\cos \beta \approx 1$, and that likewise, $\cos \beta' \approx 1$, Eq. (a) reduces to $(\partial N_x/\partial x)\,dx\,dy$. The sum of the x components of $N_{xy}\,dx$ is treated in a similar way. The condition $\sum F_x = 0$ then leads to

$$\frac{\partial N_x}{\partial x} + \frac{\partial N_{xy}}{\partial y} = 0 \tag{7.1}$$

Furthermore, the condition $\sum F_y = 0$ results in

$$\frac{\partial N_{xy}}{\partial x} + \frac{\partial N_y}{\partial y} = 0 \tag{7.2}$$

To describe equilibrium in the z direction, it is necessary to consider the z components of the in-plane forces acting at each edge of the element. The z component of the normal forces acting on the x edges is equal to

$$-N_x\,dy \sin \beta + \left(N_x + \frac{\partial N_x}{\partial x}\,dx\right)dy \sin \beta' \tag{b}$$

Inasmuch as β and β' are small, $\sin\beta \approx \beta \approx \partial w/\partial x$ and $\sin\beta' \approx \beta'$, and hence

$$\beta' \approx \beta + \frac{\partial \beta}{\partial x}dx = \frac{\partial w}{\partial x} + \frac{\partial^2 w}{\partial x^2}dx$$

Neglecting higher-order terms, Eq. (b) is therefore

$$-N_x\,dy\frac{\partial w}{\partial x} + \left(N_x + \frac{\partial N_x}{\partial x}dx\right)dy\left(\frac{\partial w}{\partial x} + \frac{\partial^2 w}{\partial x^2}dx\right)$$

$$= N_x\frac{\partial^2 w}{\partial x^2}dx\,dy + \frac{\partial N_x}{\partial x}\frac{\partial w}{\partial x}dx\,dy$$

The z components of the shear forces N_{xy} on the x edges of the element are determined as follows. The slope of the deflection surface in the y direction on the x edges equals $\partial w/\partial y$ and $\partial w/\partial y + (\partial^2 w/\partial x\,\partial y)\,dx$. The z directed component of the shear forces is then

$$N_{xy}\frac{\partial^2 w}{\partial x\,\partial y}dx\,dy + \frac{\partial N_{xy}}{\partial x}\frac{\partial w}{\partial y}dx\,dy$$

An expression identical to the above is found for the z projection of shear forces N_{yx} acting on the y edges:

$$N_{yx}\frac{\partial^2 w}{\partial x\,\partial y}dx\,dy + \frac{\partial N_{yx}}{\partial y}\frac{\partial w}{\partial x}dx\,dy$$

For the forces in Figs. 7.1 and 1.5, from $\sum F_z = 0$, we thus have

$$\frac{\partial Q_x}{\partial x} + \frac{\partial Q_y}{\partial y} + p + N_x\frac{\partial^2 w}{\partial x^2} + N_y\frac{\partial^2 w}{\partial y^2} + 2N_{xy}\frac{\partial^2 w}{\partial x\,\partial y}$$

$$+ \left(\frac{\partial N_x}{\partial x} + \frac{\partial N_{yx}}{\partial y}\right)\frac{\partial w}{\partial x} + \left(\frac{\partial N_{xy}}{\partial x} + \frac{\partial N_y}{\partial y}\right)\frac{\partial w}{\partial y} = 0 \quad (c)$$

It is observed from Eqs. (7.1) and (7.2) that the terms within the parentheses in the above expression vanish. As the direct forces do not result in any moment along the edges of the element, Eqs. (b) and (c) of Sec. 1.5, and hence Eqs. (1.16), are unchanged.

Introduction of Eqs. (1.16) into Eq. (c) yields

$$\frac{\partial^4 w}{\partial x^4} + 2\frac{\partial^4 w}{\partial x^2\,\partial y^2} + \frac{\partial^4 w}{\partial y^4} = \frac{1}{D}\left(p + N_x\frac{\partial^2 w}{\partial x^2} + N_y\frac{\partial^2 w}{\partial y^2} + 2N_{xy}\frac{\partial^2 w}{\partial x\,\partial y}\right) \quad (7.3)$$

Expressions (7.1), (7.2), and (7.3) are the *governing differential equations* for a thin plate, subjected to *combined* lateral and direct forces. It is observed that Eq. (1.17) is now replaced by Eq. (7.3) to determine the deflection surface of the plate. Either Navier's or Lévy's method may be applied to obtain a solution.

Example 7.1 A rectangular plate with simply supported edges is subject to the action of combined uniform lateral load p_0 and uniform tension N (Fig. 7.2). Derive the equation of the deflection surface.

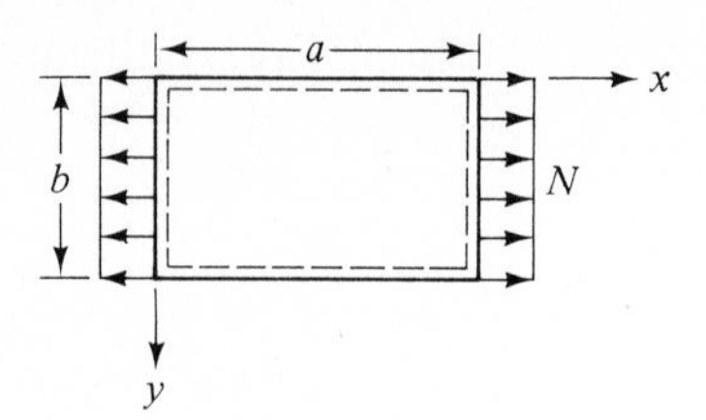

Figure 7.2

SOLUTION In this particular case, $N_x = N =$ constant and $N_y = N_{xy} = 0$, and hence Eqs. (7.1) and (7.2) are identically satisfied. The lateral load p_0 can be represented by (Sec. 3.3):

$$p = \frac{16p_0}{\pi^2} \sum_m^\infty \sum_n^\infty \frac{1}{mn} \sin \frac{m\pi x}{a} \sin \frac{n\pi y}{b} \qquad (m, n = 1, 3, \ldots)$$

Inserting the above in Eq. (7.3), we have

$$\frac{\partial^4 w}{\partial x^4} + 2\frac{\partial^4 w}{\partial x^2\, \partial y^2} + \frac{\partial^4 w}{\partial y^4} - \frac{N}{D}\frac{\partial^2 w}{\partial x^2} = \frac{16p_0}{\pi^2 D} \sum_m^\infty \sum_n^\infty \frac{1}{mn} \sin \frac{m\pi x}{a} \sin \frac{n\pi y}{b} \tag{7.4}$$

The conditions at the simply supported edges, expressed by Eqs. (*a*) of Sec. 3.2, are satisfied by assuming a deflection of the form given by Eq. (3.1*b*). When this is introduced into Eq. (7.4), we obtain

$$a_{mn} = \frac{16p_0}{\pi^6 Dmn\left[\left(\frac{m^2}{a^2} + \frac{n^2}{b^2}\right)^2 + \frac{N}{D}\left(\frac{m}{\pi a}\right)^2\right]} \qquad (m, n = 1, 3, \ldots)$$

The deflection is thus

$$w = \frac{16p_0}{\pi^6 D} \sum_m^\infty \sum_n^\infty \frac{\sin (m\pi x/a) \sin (n\pi y/b)}{mn\left[\left(\frac{m^2}{a^2} + \frac{n^2}{b^2}\right)^2 + \frac{N}{D}\left(\frac{m}{\pi a}\right)^2\right]} \tag{7.5}$$

Upon comparison of Eqs. (3.6) and (7.5), we are led to conclude that the presence of a tensile (compressive) force decreases (increases) the plate deflection.

7.3 COMPRESSION OF PLATES. BUCKLING

When a plate is compressed in its midplane, it becomes unstable and begins to *buckle* at a certain *critical value* of the in-plane force. Buckling of plates is qualitatively similar to column buckling.[1] However, a buckling analysis of the former case is not performed as readily as for the latter. Plate-buckling solutions using Eq. (7.3) usually involve considerable difficulty and subtlety,[24] and the conditions that result in the lowest *eigenvalue*, or the *actual buckling load*, are not

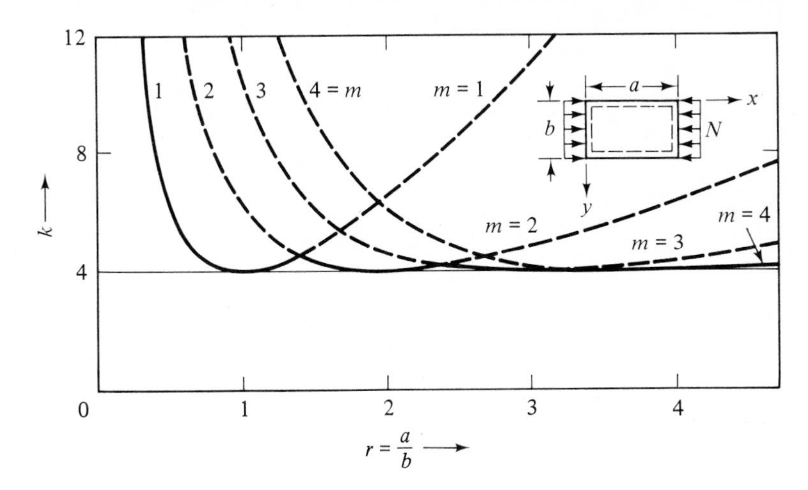

Figure 7.3

at all obvious in many situations. This is especially true in plates having other than simply supported edges. Often in these cases, the energy method (Sec. 7.4) is used to good advantage to obtain the approximate buckling loads.

For a plate, the in-plate load that results in an elastic *instability*, as in the case of a beam-column, is *independent* of the lateral loading. In the analysis of elastic stability, we thus take $p = 0$ in Eq. (7.3).

Thin plates or sheets, although quite capable of carrying tensile loadings, are poor in resisting compression. Usually, buckling or wrinkling phenomena observed in compressed plates (and shells) takes place rather suddenly and are very dangerous. Fortunately, there is close correlation between theory and experimental data concerned with buckling of plates subjected to various types of loads and edge conditions.

An illustration of plate behavior under compression loading is presented in the following example.

Example 7.2 A simply supported rectangular plate is subjected to uniaxial in-plane forces N (Fig. 7.3). Determine the buckling load.

SOLUTION For this case $N_x = -N =$ constant and $N_y = N_{xy} = 0$, and hence expressions (7.1) and (7.2) are satisfied identically. The governing equation for the displacement becomes

$$D\,\nabla^4 w + N\frac{\partial^2 w}{\partial x^2} = 0 \tag{7.6}$$

We assume the solution in the form

$$w = \sum_m^\infty \sum_n^\infty a_{mn} \sin\frac{m\pi x}{a} \sin\frac{n\pi y}{b} \qquad (m, n = 1, 2, \ldots) \tag{7.7}$$

which when substituted into Eq. (7.6) leads to

$$\sum_m^\infty \sum_n^\infty \left[D\pi^4 \left(\frac{m^2}{a^2} + \frac{n^2}{b^2} \right)^2 - N\pi^2 \frac{m^2}{a^2} \right] a_{mn} \sin \frac{m\pi x}{a} \sin \frac{n\pi y}{b} = 0$$

The nontrivial solution is

$$\pi^4 D \left(\frac{m^2}{a^2} + \frac{n^2}{b^2} \right)^2 - N\pi^2 \frac{m^2}{a^2} = 0$$

from which

$$N = \frac{\pi^2 a^2 D}{m^2} \left(\frac{m^2}{a^2} + \frac{n^2}{b^2} \right)^2 = \frac{\pi^2 D}{b^2} \left(\frac{mb}{a} + \frac{n^2 a}{mb} \right)^2 \tag{7.8}$$

Clearly, when N attains the value given by the right-hand side of Eq. (7.8), one has $a_{mn} \neq 0$ and hence $w \neq 0$, indicating plate buckling. It is observed from Eq. (7.8) that the minimum value of N occurs when $n = 1$. Thus, when the simply supported plate buckles, the *buckling mode* given by Eq. (7.8) can only be *one half-sine wave*, $\sin(\pi y/b)$, across the span, while several half-waves in the direction of compression can occur. The resulting expression is thus

$$N_{cr} = \frac{\pi^2 D}{b^2} \left(\frac{m}{r} + \frac{r}{m} \right)^2 = k^2 \frac{\pi^2 D}{b^2} \tag{7.9}$$

for the critical load. Here $k = [(m/r) + (r/m)]^2$ and $r = a/b$.

To ascertain the aspect ratio r at which the critical load is a minimum, we set

$$\frac{dN_{cr}}{dr} = \frac{2\pi^2 D}{b^2} \left(\frac{m}{r} + \frac{r}{m} \right) \left(-\frac{m^2}{r^2} + 1 \right) = 0$$

from which $m/r = 1$. This provides the following *minimum* value of the critical load:

$$N_{cr} = \frac{4\pi^2 D}{b^2} \tag{7.10}$$

The corresponding critical stress, N_{cr}/t, is given by

$$\sigma_{cr} = \frac{\pi^2 E}{3(1 - \nu^2)} \left(\frac{t}{b} \right)^2 \tag{7.11}$$

The variations of the buckling load factor k as functions of aspect ratio r for $m = 1, 2, 3, 4$ are sketched in Fig. 7.3. Clearly, for a specific m, the magnitude of k depends upon r only. Referring to the figure, the magnitude of N_{cr} and the number of half-waves m for any value of the aspect ratio r can readily be found. In the case of $r = 1.5$, for instance, from Fig. 7.3, $k = 4.34$ and $m = 2$. The corresponding critical load is $N_{cr} = 4.34\pi^2 D/b^2$ under which the plate will buckle into two half-waves in the direction of the loading.

It is also observed from Fig. 7.3 that a plate m times as long as it is wide will buckle in m half-sine waves. Thus, a *long plate* ($b \ll a$) with simply supported edges under a uniaxial compression *tends to buckle* into a number of *square cells* of side dimension b and its critical load for all practical purposes is given by Eq. (7.10).

7.4 APPLICATION OF THE ENERGY METHOD

The principle of minimum potential energy may be employed to analyze plates under the action of lateral and in-plane loading. We shall first develop expressions for the midsurface strains. Work done by the direct forces may then be evaluated, and the energy method applied readily to any particular problem.

Consider element $dx\,dy$ representing a point at the *midplane* of a plate (Fig. 7.4*a*). *Subsequent to the bending*, linear element AB is displaced to become $A'B'$. Inasmuch as midplane stressing does *not* occur, original element length dx remains unchanged, and its horizontal projection owing to the displacement w is

$$\left[dx^2 - \left(\frac{\partial w}{\partial x}\,dx\right)^2\right]^{1/2} = dx - \frac{1}{2}\left(\frac{\partial w}{\partial x}\right)^2 dx + \cdots \tag{a}$$

The *midplane* displacement per unit length (to a second approximation) in the x direction is therefore

$$\varepsilon_x = \frac{1}{2}\left(\frac{\partial w}{\partial x}\right)^2 \tag{7.12a}$$

Similarly, unit displacement of the midplane in the y direction equals

$$\varepsilon_y = \frac{1}{2}\left(\frac{\partial w}{\partial y}\right)^2 \tag{7.12b}$$

For the purpose of determining shear strains associated with plate bending, consider now two infinitesimal linear elements OA and OB in the x and y

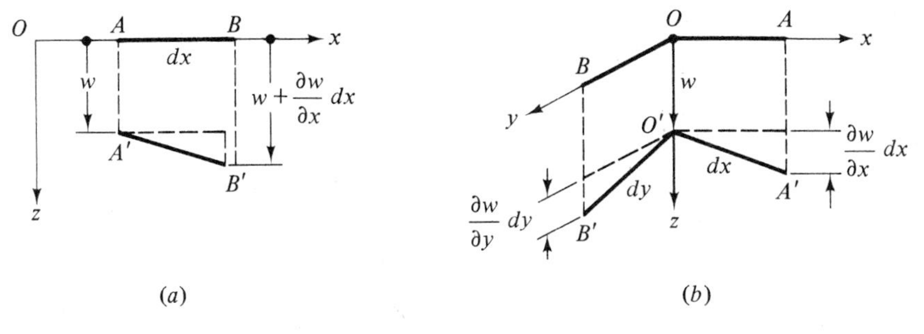

Figure 7.4

directions (Fig. 7.4*b*). Due to z displacement w, these elements move to the positions $O'A'$ and $O'B'$, having direction cosines l_1, m_1, n_1, and l_2, m_2, n_2, respectively, given by

$$l_1 = \frac{\left[dx^2 - \left(\frac{\partial w}{\partial x}dx\right)^2\right]^{1/2}}{dx} \approx 1 - \frac{1}{2}\left(\frac{\partial w}{\partial x}\right)^2 \qquad m_1 = 0 \qquad n_1 = \frac{\frac{\partial w}{\partial x}dx}{dx} = \frac{\partial w}{\partial x} \tag{b}$$

and

$$l_2 = 0 \qquad m_2 \approx 1 - \frac{1}{2}\left(\frac{\partial w}{\partial y}\right)^2 \qquad n_2 = \frac{\partial w}{\partial y} \tag{c}$$

Consider the following:

$$\gamma_{xy} = \frac{\pi}{2} - \measuredangle A'O'B' = \sin\left(\frac{\pi}{2} - \measuredangle A'O'B'\right) = \cos \measuredangle A'O'B' = l_1 l_2 + m_1 m_2 + n_1 n_2$$

This expression upon substitution of definitions (*b*) and (*c*), yields the midplane shear strain

$$\gamma_{xy} = \frac{\partial w}{\partial x}\frac{\partial w}{\partial y} \tag{7.12c}$$

Note that in previous discussion of plate bending, the strains given by Eqs. (7.12) were always omitted. They are now taken into account, since their products with direct forces may be of the same order of magnitude as the strain energy of bending. However, these strains are considered very small in comparison with

$$\varepsilon_x = \frac{1}{Et}(N_x - \nu N_y) \qquad \varepsilon_y = \frac{1}{Et}(N_y - \nu N_x) \qquad \gamma_{xy} = \frac{N_{xy}}{Gt}$$

caused by in-plane forces. It is further assumed that *direct forces are applied first (before lateral loads) and that they remain unchanged during plate bending*. The latter assumption is widely used in the classical treatment of plates as well as beams and shells.

The work done by the direct forces, due to displacement w only, then equals

$$W = \frac{1}{2}\iint_A \left[N_x\left(\frac{\partial w}{\partial x}\right)^2 + N_y\left(\frac{\partial w}{\partial y}\right)^2 + 2N_{xy}\frac{\partial w}{\partial x}\frac{\partial w}{\partial y}\right] dx\, dy \tag{7.13}$$

where A is the area of the plate. The plate-strain energy owing to bending is given by Eq. (1.34):

$$U = \frac{D}{2}\iint_A \left\{\left(\frac{\partial^2 w}{\partial x^2} + \frac{\partial^2 w}{\partial y^2}\right)^2 - 2(1-\nu)\left[\frac{\partial^2 w}{\partial x^2}\frac{\partial^2 w}{\partial y^2} - \left(\frac{\partial^2 w}{\partial x\,\partial y}\right)^2\right]\right\} dx\, dy \tag{7.14}$$

Application of the basic expressions derived in the preceding paragraphs is illustrated in the discussion of a plate-buckling problem which follows. The plate

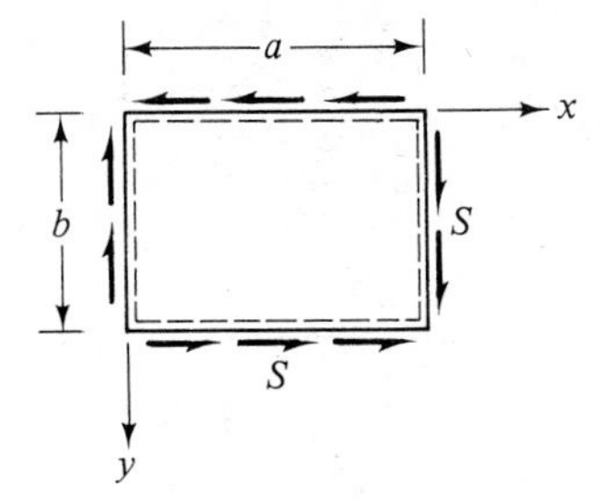

Figure 7.5

is taken to be subjected to *constant direct forces* (N_x, N_y, N_{xy}) *during bending just before it buckles*. This means that the magnitude of these in-plane forces is just equal to their critical values. Subsequently, it is assumed that the plate undergoes some small disturbances and buckling takes place. At the time of transition from one to the other of these equilibrium forms, no energy is gained or lost. Hence, the work done by the direct forces must be equal to the bending-strain energy stored in the plate. That is, buckling occurs if $U = W$. Thus, we have the expression

$$\Pi = U - W = 0 \tag{7.15}$$

for the potential energy of the plate.

Example 7.3 Determine the buckling load of a simply supported plate under the action of uniform shearing forces $N_{xy} = S$, $N_x = N_y = 0$, as shown in Fig. 7.5.

SOLUTION Assume that the deflection surface of the plate is described by an expression of the form:

$$w = \sum_m^\infty \sum_n^\infty a_{mn} \sin \frac{m\pi x}{a} \sin \frac{n\pi y}{b} \tag{d}$$

Clearly, the foregoing satisfies the boundary conditions at the simple supported plate edges. The work done by S during the buckling of the plate is, from Eq. (7.13),

$$W = S \int_0^b \int_0^a \frac{\partial w}{\partial x}\frac{\partial w}{\partial y}\, dx\, dy$$

Inserting Eq. (d) into this equation and observing that

$$\int_0^a \sin \frac{m\pi x}{a} \cos \frac{p\pi x}{a}\, dx = \begin{cases} 0 & (m \pm p \text{ is an even number}) \\ \dfrac{2a}{\pi}\dfrac{m}{m^2 - p^2} & (m \pm p \text{ is an odd number}) \end{cases}$$

we obtain

$$W = 4S \sum_m^\infty \sum_n^\infty \sum_p^\infty \sum_q^\infty a_{mn} a_{pq} \frac{mnpq}{(m^2 - p^2)(q^2 - n^2)} \tag{e}$$

wherein $m \pm p$ and $n \pm p$ are odd numbers.

The strain energy associated with the bending of the buckled plate is found upon substituting Eq. (*d*) into Eq. (7.14) and integrating the resulting expression. In so doing one has

$$U = \frac{\pi^4 ab}{8} D \sum_m^\infty \sum_n^\infty a_{mn}^2 \left(\frac{m^2}{a^2} + \frac{n^2}{b^2} \right)^2 \tag{f}$$

Equation (7.15) now becomes

$$\Pi = \frac{D\pi^4 ab}{8} \sum_m^\infty \sum_n^\infty a_{mn}^2 \left(\frac{m^2}{a^2} + \frac{n^2}{b^2} \right)^2$$

$$- 4S \sum_m^\infty \sum_n^\infty \sum_p^\infty \sum_q^\infty a_{mn} a_{pq} \frac{mnpq}{(m^2 - p^2)(q^2 - n^2)} = 0 \tag{7.16}$$

In order to ascertain the critical value of the shearing forces, the coefficient a_{mn} is to be found such that S is a minimum. This, it can be demonstrated (Sec. 1.9), is equivalent to requiring that Π be a minimum. Upon application of Eq. (1.43), we thus have

$$\frac{\pi^4 Dab}{4} a_{mn} \left(\frac{m^2}{a^2} + \frac{n^2}{b^2} \right)^2 - 8S \sum_p^\infty \sum_q^\infty a_{pq} \frac{mnpq}{(m^2 - p^2)(q^2 - n^2)} = 0 \tag{7.17}$$

in which p and q must be such that $m \pm p$, $n \pm q$ are odd numbers. Based upon the notation

$$r = \frac{a}{b} \qquad \lambda = \frac{\pi^4 D}{32 r b^2 S} \tag{7.18}$$

expression (7.17) assumes the following convenient form

$$\lambda a_{mn} \frac{(m^2 + n^2 r^2)^2}{r^2} - \sum_p^\infty \sum_q^\infty a_{pq} \frac{mnpq}{(m^2 - p^2)(q^2 - n^2)} = 0 \tag{7.19}$$

We thus have a system of linear equations in a_{mn}, of which an approximate solution may be found by retaining a finite number of parameters a_{mn}.

If, for example, only two parameters a_{11} and a_{22} are kept, we have, from Eq. (7.19):

$$\frac{\lambda(1 + r^2)^2}{r^2} a_{11} + \tfrac{4}{9} a_{22} = 0$$

$$\frac{16\lambda(1 + r^2)^2}{r^2} a_{22} + \tfrac{4}{9} a_{11} = 0$$

These equations have the following determinant

$$\begin{vmatrix} \dfrac{\lambda(1+r^2)^2}{r^2} & \dfrac{4}{9} \\ \dfrac{4}{9} & \dfrac{16\lambda(1+r^2)^2}{r^2} \end{vmatrix} = 0$$

with the solution

$$\lambda = \pm\frac{1}{9}\frac{r^2}{(1+r^2)^2}$$

The magnitude of the critical load, from Eq. (7.18), is thus

$$S_{cr} = \pm\frac{9\pi^4 D}{32b^2}\frac{(1+r^2)^2}{r^2} \tag{7.20}$$

This result differs by approximately 15 percent from the "exact" solution[24] for a square plate ($r = 1$). Accuracy may be improved by retaining additional parameters.

7.5 THE FINITE DIFFERENCE SOLUTION

The preceding sections have been concerned with rectangular plates subjected to uniform loads. The determination of the buckling load by means of analytical methods may be quite tedious and difficult, as was observed. In order to enable the stress analyst to cope with the numerous compressed plates of practical importance, numerical techniques must be relied upon. The following examples apply to the method of finite difference (Chap. 5) to the governing differential equations of a plate under in-plane loads.

Example 7.4 Compute the buckling load of a simply supported square plate under uniform compressive loading per unit of boundary length (Fig. 7.6*a*).

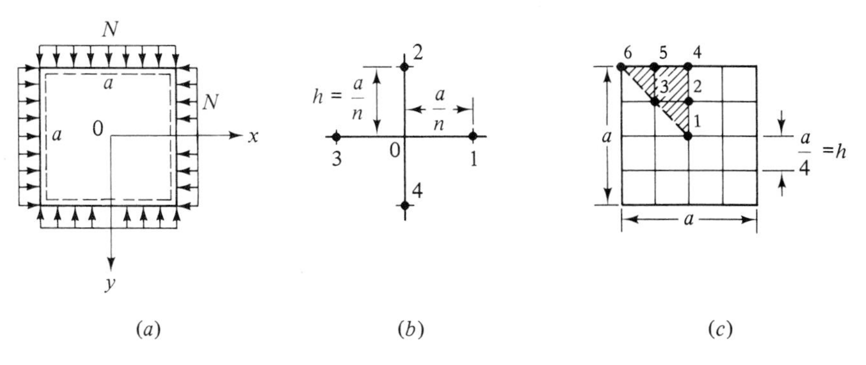

Figure 7.6

SOLUTION The plate is subjected to constant in-plane loads $N_x = N_y = -N$ and $p = N_{xy} = 0$. Upon substituting, expression (7.3) reduces to

$$\nabla^4 w + \frac{N}{D}\nabla^2 w = 0 \tag{7.21}$$

Along the edges, the conditions are described by:

$$\begin{aligned} w = 0 \qquad \frac{\partial^2 w}{\partial x^2} = 0 \qquad \left(x = \pm\frac{a}{2}\right) \\ w = 0 \qquad \frac{\partial^2 w}{\partial y^2} = 0 \qquad \left(y = \pm\frac{a}{2}\right) \end{aligned} \tag{a}$$

Hence

$$\nabla^2 w = 0 \qquad \text{(on the boundary)}$$

and the problem is reduced to finding the solution of

$$\nabla^2 M + \frac{N}{D} M = 0 \tag{7.22a}$$

$$\nabla^2 w = M \tag{7.22b}$$

under the condition that $w = 0$ and $M = 0$ on the boundary. For $w = 0$ along the boundary, it is observed that Eq. (7.22*b*), for the trivial solution $M = 0$, also leads to the trivial solution $w = 0$. It is concluded therefore that to treat the problem requires the solution of Eq. (7.22*a*) only.

The finite difference expression corresponding to Eq. (7.22*a*), referring to Fig. 7.6*b*, is written at node 0 as follows:

$$M_1 + M_2 + M_3 + M_4 + \left(\frac{Na^2}{n^2 D} - 4\right) M_0 = 0 \tag{7.23}$$

where n represents the number of divisions of the sides. For convenience, in the computations we shall denote

$$\lambda = \frac{Na^2}{n^2 D} - 4 = \frac{K}{n^2} - 4 \tag{7.24}$$

with

$$K = \frac{Na^2}{D} \tag{7.25}$$

The plate is now divided into a number of small squares, e.g., 16. Owing to conditions of symmetry one need only treat one-eighth of the plate, shown by the shaded area in Fig. 7.6*c*. On applying Eq. (7.23) at nodes 1, 2,

3 with $M_4 = M_5 = M_6 = 0$ along the boundary, three equations can be written. These are represented in the following matrix form:

$$\begin{bmatrix} \lambda & 4 & 0 \\ 1 & \lambda & 2 \\ 0 & 2 & \lambda \end{bmatrix} \begin{Bmatrix} M_1 \\ M_2 \\ M_3 \end{Bmatrix} = 0$$

The above leads to a nontrivial solution if the determinant is set equal to zero. In so doing, we obtain $\lambda = -2.8284$ and hence $K = 18.75$. The buckling load, from Eq. (7.25), is thus

$$N_{\text{cr}} = 18.75 \frac{D}{a^2}$$

This result is smaller than the "exact" solution[24] by 5.1 percent. By increasing the number of subdivisions, the accuracy may be improved.

Example 7.5 A portion of a missile launcher support fixture, approximated by a square plate with opposite edges $y = \pm a/2$ built-in and the other sides simply supported (Fig. 7.7a), carries uniform compressive loads N. Determine the critical buckling load.

SOLUTION We have $N_x = -N$, and $N_y = N_{xy} = p = 0$. Equation (7.3) then becomes

$$\nabla^4 w + \frac{N}{D} \frac{\partial^2 w}{\partial x^2} = 0 \tag{7.26}$$

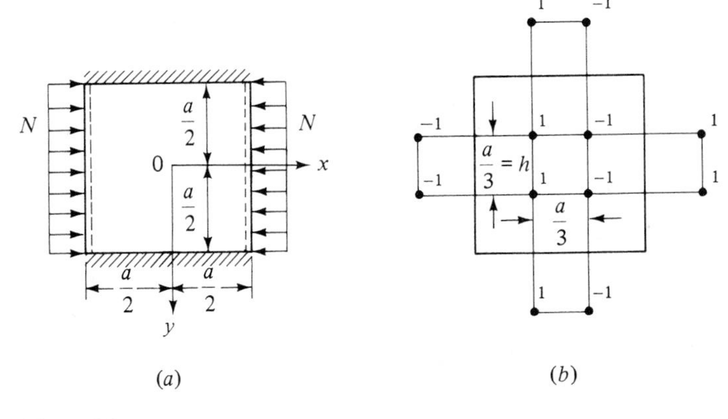

Figure 7.7

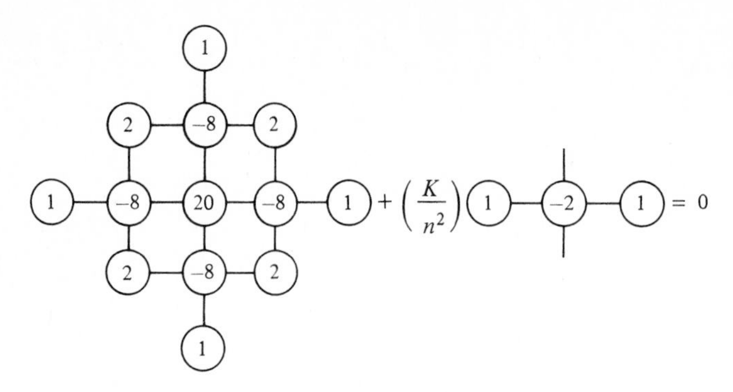

Figure 7.8

The boundary conditions are

$$w = 0 \qquad \frac{\partial^2 w}{\partial x^2} = 0 \qquad \left(x = \pm\frac{a}{2}\right)$$

$$w = 0 \qquad \frac{\partial w}{\partial y} = 0 \qquad \left(y = \pm\frac{a}{2}\right) \tag{b}$$

As before, employing $h = a/n$ and $K = Na^2/D$, we obtain the coefficient pattern for the finite difference expression of Eq. (7.26), Fig. 7.8.

Proceeding with the finite difference solution, the plate is subdivided into nine small squares. *Antisymmetrical deflections in the direction of the compression result in a smaller buckling load.* The finite difference equivalents of Eqs. (*b*) are fulfilled by numbering the nodes located outside of the boundary as shown in Fig. 7.7*b*. Note that the values of w are zero along the boundary. On applying Fig. 7.8 at point 1, we have

$$(-w_1 + w_1 + 0 + 0) + 2(0 + 0 + 0 - w_1) - 8(0 + 0 + w_1 - w_1)$$
$$+ 20w_1 + \frac{K}{9}(0 - 2w_1 - w_1) = 0$$

from which

$$w_1\left(18 - \frac{K}{3}\right) = 0$$

or $K = 54 = 5.471\pi^2$. Thus,

$$N_{cr} = \frac{5.471\pi^2 D}{a^2} \tag{c}$$

It can be shown that using $n = 4$ and the antisymmetrical deflections, the critical load is (Prob. 7.8)

$$N_{cr} = \frac{6.193\pi^2 D}{a^2} \tag{d}$$

for the square plate under consideration.

7.6 PLATES WITH SMALL INITIAL CURVATURE

For a plate with an *initial curvature* subjected to the action of *lateral load only*, the governing equation (1.17) still holds, provided that at any point the magnitude of the *initial* deflection w_0 is *small* compared with the plate thickness t. The *total* deflection w is obtained by the superposition of w_0 and deflection w_1 due to the lateral load. Here w_1 is determined by solving Eq. (1.17) as in the case of flat plates.

However, as might be anticipated on physical grounds, the load-carrying capacity and deformation of a plate under *in-plane and lateral loading* are significantly affected by *any* initial curvature. In order to take into account the extent of this influence, Eq. (7.3) is modified as follows

$$\nabla^4 w_1 = \frac{1}{D}\left(p + N_x \frac{\partial^2 w}{\partial x^2} + N_y \frac{\partial^2 w}{\partial y^2} + 2N_{xy} \frac{\partial^2 w}{\partial x\, \partial y}\right) \tag{7.27}$$

where $w = w_0 + w_1$. This is the *governing differential equation* for deflection of thin plates with *small initial curvature*. It is noted that the left- and the right-hand sides of the above depend upon the *change* in the curvature and the *total* curvature of the plate, respectively.

Upon comparison of Eqs. (1.17) and (7.27) it is observed that the influence of an initial curvature on the deflection is identical with that of a fictitious *equivalent lateral load*

$$N_x \frac{\partial^2 w_0}{\partial x^2} + N_y \frac{\partial^2 w_0}{\partial y^2} + 2N_{xy} \frac{\partial^2 w_0}{\partial x\, \partial y}$$

One concludes therefore that a plate experiences bending under direct forces *only* if it has an initial curvature.

Consider, as an example, a simply supported plate for which the *unloaded shape* is described by

$$w_0 = \sum_m^\infty \sum_n^\infty a_{mn} \sin \frac{m\pi x}{a} \sin \frac{n\pi y}{b} \tag{a}$$

Assume that the plate edges at $x = 0$ and $x = a$ are subjected to uniform compressive forces $N_x = -N$ (Fig. 7.3). The differential equation (7.27), with Eq. (*a*) introduced is

$$\nabla^4 w_1 = \frac{1}{D}\left(N \frac{a_{mn}\pi^2}{a^2} \sin \frac{m\pi x}{a} \sin \frac{n\pi y}{b} - N \frac{\partial^2 w_1}{\partial x^2}\right) \tag{b}$$

When the series solution

$$w_1 = \sum_m^\infty \sum_n^\infty b_{mn} \sin \frac{m\pi x}{a} \sin \frac{n\pi y}{b} \tag{c}$$

is inserted into Eq. (*b*), it is found that

$$b_{mn} = \frac{a_{mn} N}{\pi^2 D/a^2 [m + (n^2 a^2/mb^2)]^2 - N} \tag{d}$$

The solution of Eq. (*b*) is determined by substituting the above into Eq. (*c*). The total deflection is then

$$w = w_0 + w_1 = \sum_m^\infty \sum_n^\infty \frac{a_{mn}}{1-\alpha} \sin \frac{m\pi x}{a} \sin \frac{n\pi y}{b} \tag{7.28}$$

where

$$\alpha = \frac{N}{\pi^2 D/a^2[m + (n^2a^2/mb^2)]^2} \tag{7.29}$$

For $m = n = 1$, the maximum deflection of the plate, which occurs at the center, is given by

$$w_{\max} = \frac{a_{11}}{1-\alpha} \tag{7.30}$$

in which now

$$\alpha = \frac{N}{\pi^2 D/a^2[1 + (a^2/b^2)]^2}$$

Note that the deflection expression derived above is analogous to the relation for initially curved columns.[1]

From Eq. (*d*), it is seen that b_{mn} (and thus w_1) increases with increasing N. When N reaches the critical value given by Eq. (7.8), the denominator of Eq. (*d*) vanishes and w_1 grows without limit. This means that the plate buckles, as described in the alternate manner in Sec. 7.3.

To write the equation for a plate subjected to uniform *tensile forces* (Fig. 7.2), it is required only to *change the sign* of N in the equations of the foregoing example. By following an approach similar to that described above, the deflection of an initially curved plate under simultaneous action of in-plane forces N_x, N_y, and N_{xy} may also be readily obtained.

7.7 BENDING TO A CYLINDRICAL SURFACE

Assume that a *long* rectangular plate of uniform thickness t is bent into a *cylindrical surface* with its generating line parallel to the y axis. For this case, $w = w(x)$ and the *governing equation for deflection*, Eq. (7.3), reduces to

$$D\frac{d^4w}{dx^4} - N_x\frac{d^2w}{dx^2} = p \tag{7.31}$$

When axial force N_x is zero, the foregoing agrees with Eq. (*c*) of Example 1.1. Observe that the calculation of the plate deflection simplifies to the solution of Eq. (7.31), which is of the same form as the differential equation for deflection of beams under the action of lateral and axial forces.

If plate edges are *not free* to move horizontally, a tension in the plate is

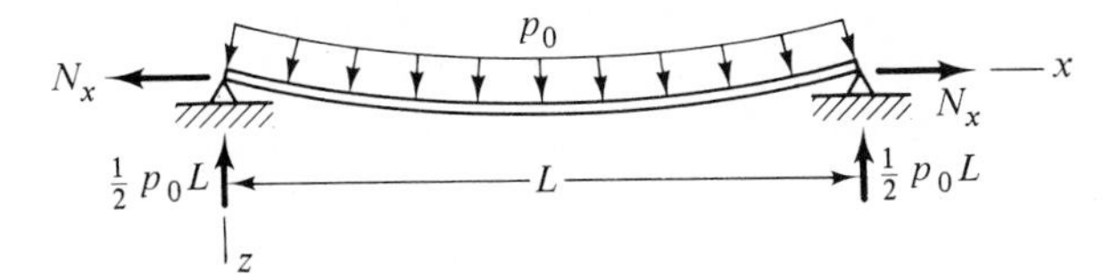

Figure 7.9

produced *depending upon* the magnitude of *lateral deflection* w. The problem then becomes complicated. The tensile forces in the plate carry part of the lateral loading through membrane action.

A typical case in which bending to a cylindrical surface occurs is illustrated below.

Example 7.6 A rectangular plate, the length of which is large in comparison with its width, is subjected to a uniform loading of intensity p_0. The longitudinal edges of the plate are free to rotate but otherwise immovable. Determine the lateral deflection and stresses in the plate.

SOLUTION A strip of unit width removed from a plate of this type will be in the same condition as a laterally and axially loaded beam or so-called *tie-rod* (Fig. 7.9). The value of the axial tensile forces $N_x = N$ is such that horizontal movement of the edges is prevented. The bending moment at any section x of the strip is described by

$$M(x) = \tfrac{1}{2} p_0 L x - \tfrac{1}{2} p_0 x^2 - N w$$

In terms of this moment Eq. (7.31) may be written

$$D \frac{d^2 w}{dx^2} = -M(x) \tag{7.32}$$

We observe that differentiating Eq. (7.32) twice results in Eq. (7.31).

Employing the notation

$$\lambda^2 = \frac{N L^2}{4D} \tag{7.33}$$

the solution of Eq. (7.32) appears as follows:

$$w = C_1 \sinh \frac{2\lambda x}{L} + C_2 \cosh \frac{2\lambda x}{L} + \frac{p_0 L^3 x}{8\lambda^2 D} - \frac{p_0 L^2 x^2}{8\lambda^2 D} - \frac{p_0 L^4}{16\lambda^4 D} \tag{a}$$

Since the deflections vanish at the ends, $w = 0$ at $x = 0$ and $x = L$, of the strip, and thus

$$C_1 = \frac{p_0 L^4}{16\lambda^4 D} \frac{1 - \cosh 2\lambda}{\sinh 2\lambda} \qquad C_2 = \frac{p_0 L^4}{16\lambda^2 D} \tag{b}$$

Employing the identities

$$\cosh 2\lambda = \cosh^2 \lambda + \sinh^2 \lambda \qquad \sinh 2\lambda = 2 \sinh \lambda \cosh \lambda$$

$$\cosh^2 \lambda = 1 + \sinh^2 \lambda$$

and substituting Eqs. (*b*) into Eq. (*a*), we obtain w in convenient form. It follows that the deflection curve of the strip is

$$w = \frac{p_0 L^4}{16\lambda^4 D}\left\{\frac{\cosh \lambda[1 - (2x/L)]}{\cosh \lambda} - 1\right\} + \frac{p_0 L^2 x}{8\lambda^2 D}(L - x) \tag{7.34}$$

The maximum displacement occurs at midspan ($x = L/2$):

$$w_{\text{max}} = \frac{5p_0 L^4}{384D} f_1(\lambda) \tag{7.35}$$

where

$$f_1(\lambda) = \frac{24}{5\lambda^4}\left(\text{sech} - 1 + \frac{\lambda^2}{2}\right)$$

From Eq. (7.34), we find that at $x = L/2$

$$M_{\text{max}} = -D\frac{d^2 w}{dx^2} = \frac{p_0 L^2}{8} f_2(\lambda) \tag{7.36}$$

where

$$f_2(\lambda) = \frac{2}{\lambda^2}(1 - \text{sech}\ \lambda)$$

Interestingly, if there were no tensile reactions at the ends of the strip, the maximum deflection and moment would be

$$w_{\text{max}} = \frac{5p_0 L^4}{384D} \qquad M_{\text{max}} = \tfrac{1}{8} p_0 L^2 \tag{c}$$

The effects of N upon displacement and moment are given by $f_1(\lambda)$ and $f_2(\lambda)$ which diminish rapidly with increasing λ.

It is observed that the displacement and moment depend upon the quantity λ, and hence the axial forces, as defined by Eq. (7.33). To determine N, we must consider the deformations. The extension of the strip produced by the axial tensile forces is, from Eq. (7.12*a*),

$$\frac{1}{2}\int_0^L \left(\frac{dw}{dx}\right)^2 dx$$

During bending the lateral contraction of the strip in the plane of the plate is assumed to be zero; $\varepsilon_y = 0$. Hooke's law,

$$\varepsilon_x = \frac{1}{Et}(N_x - \nu N_y) \qquad 0 = \frac{1}{Et}(N_y - \nu N_x)$$

yields $N_y = \nu N_x$ and $\varepsilon_x = N(1 - \nu^2)/Et$. It follows that

$$\frac{N(1 - \nu^2)L}{Et} = \frac{1}{2}\int_0^L \left(\frac{dw}{dx}\right)^2 dx \tag{d}$$

A good *approximation* for N is found by selecting a deflection curve of the form

$$w = \frac{a_1}{1 + \alpha} \sin \frac{\pi x}{L} \tag{e}$$

in which a_1 designates the midspan deflection produced by the *lateral load only*. The term $a_1/(1 + \alpha)$ is given by Eq. (7.30), wherein α is replaced by $-\alpha$, as the axial forces in the present case are tensile. The quantity α in Eq. (7.29), with $m = 1$, $n = 0$, $b = 1$, and $a = L$, becomes

$$\alpha = \frac{NL^2}{\pi^2 D} \tag{f}$$

Introducing Eq. (e) into Eq. (d), and integrating, we have

$$\frac{NL(1 - v^2)}{Et} = \frac{\pi^2 a_1^2}{4L(1 + \alpha)^2}$$

Finally, inserting Eqs. (f) and (1.11) into the above we find that

$$\alpha(1 + \alpha)^2 = \frac{3a_1^2}{t^2} \tag{7.37}$$

In *any* particular case, the above is solved for α, and the quantity λ, from Eqs. (7.33) and (f), is then

$$\lambda^2 = \frac{NL^2}{4D} = \frac{\pi^2 \alpha}{4} \tag{g}$$

For the present case, from Eqs. (c),

$$a_1 = \frac{5p_0 L^4}{384D}$$

and Eq. (7.37) becomes

$$\alpha(1 + \alpha)^2 = \frac{3}{t^2}\left(\frac{5p_0 L^4}{384D}\right)^2 \tag{h}$$

Consider, for example, a steel plate for which $E = 200$ GPa and $v = 0.3$, and of dimensions $L = 1.2$ m, and $t = 10$ mm, under a uniformly distributed load $p = 70$ kPa. We have

$$D = 200 \times 10^9 (0.01)^3/12(1 - 0.09) = 1.8315 \times 10^4.$$

Equation (h) then yields

$$\alpha(1 + \alpha)^2 = \frac{3}{(0.01)^2}\left[\frac{5 \times 70{,}000(1.2)^4}{384 x 1.83 x 10^4}\right]^2 = 319.94$$

from which

$$\alpha = 6.1895 \qquad \text{and} \qquad \lambda = \frac{\pi}{2}\sqrt{\alpha} = 3.9079$$

The in-plane tensile stress is now readily calculated as follows:

$$\sigma_{xp} = \frac{N}{t} = \frac{4\lambda^2 D}{L^2 t} = \frac{E\lambda^2}{3(1-\nu^2)}\left(\frac{t}{L}\right)^2$$

$$= \frac{200(10^9)(3.9079)^2}{3(1-0.09)}\left(\frac{0.01}{1.2}\right)^2 = 77.69\text{ MPa}$$

The maximum bending moment is given by Eq. (7.36) and the corresponding maximum bending stress is given by

$$\sigma_{xb} = \frac{6M_{\max}}{t^2} = \frac{3}{4}p_0\left(\frac{L}{t}\right)^2 f_2(\lambda)$$

$$= \frac{3}{4}(70{,}000)\left(\frac{1.2}{0.01}\right)^2 \frac{2[1-\operatorname{sech}(3.9079)]}{(3.9079)^2} = 95.03\text{ MPa}$$

The maximum stress in the plate is therefore

$$\sigma_{\max} = \sigma_{xp} + \sigma_{xb} = 77.69 + 95.03 = 172.72\text{ MPa}$$

Plates having other end conditions may be treated similarly.

PROBLEMS

Secs. 7.1 to 7.3

7.1 A steel ship bulkhead is a rectangular plate of length a and width b. The plate, assumed to be simply supported, is subject to uniform lateral pressure p_0 and uniform tensile forces N along the four edges. Derive a general expression for the deflection surface and determine the maximum deflection w and maximum stress σ_x for $a = b$, $m = 1$, and $n = 1$.

7.2 A simply supported rectangular plate carries uniform tensile force N along sides $x = 0$ and $x = a$ (Fig. 7.2) and hydrostatic surface pressure described by

$$p(x, y) = p_0\frac{x}{a}$$

Derive an expression for the deformed surface.

7.3 A structural component in the interior of a spacecraft consists of a square plate of sides a and thickness t. The plate may be approximated as simply supported on all edges and subjected to uniform biaxial compression N. Determine the buckling stress by using Eq. (7.3).

7.4 If there are body forces acting in the midplane of the plate, show that the governing differential equation for deflection becomes

$$\nabla^4 w = \frac{1}{D}\left(p + N_x\frac{\partial^2 w}{\partial x^2} + N_y\frac{\partial^2 w}{\partial y^2} + 2N_{xy}\frac{\partial^2 w}{\partial x\,\partial y} - F_x\frac{\partial w}{\partial x} - F_y\frac{\partial w}{\partial y}\right) \tag{P7.4}$$

Here F_x and F_y denote the x and y directed body forces per unit area of the midplane of the plate.

Secs. 7.4 to 7.7

7.5 Redo Prob. 7.3 employing the energy method.

7.6 Redo Example 7.2 employing the energy approach.

7.7 A simply supported square plate is subjected to uniform compression loads $N_x = -N$ applied to two opposite sides. Determine the buckling load, using the method of finite differences with $n = 3$.

7.8 Verify the result given by Eq. (d) of Example 7.5.

7.9 Apply the finite difference approach to obtain the buckling load for a clamped square plate under uniform biaxial pressure N per unit of boundary length. Use $n = 3$.

7.10 A simply supported rectangular plate with an initial deflection defined by

$$w_0 = a_0 \sin \frac{\pi x}{a} \sin \frac{\pi y}{b}$$

is subject to uniform biaxial tensile forces N. Determine the maximum displacement for $a = b$.

7.11 A long rectangular plate with edges clamped carries a uniform load of itensity p_0. An elemental strip of unit plate width, bent into a cylindrical surface, is similar to that shown in Fig. 7.9, except that now the ends are fixed and bending moments M_0 thus occur there. Demonstrate that the differential equation of the deflection curve is expressed by

$$D\frac{d^2w}{dx^2} - Nw = -\frac{p_0 Lx}{2} + \frac{p_0 x^2}{2} - M_0 \tag{P7.11a}$$

Obtain the following solution

$$w = -\frac{p_0 L^4}{16\lambda^3 D} \sinh \frac{2\lambda x}{L} + \frac{p_0 L^4}{16\lambda^3 D} \coth \lambda \cosh \frac{2\lambda x}{L} + \frac{p_0 L^3 x}{8\lambda^2 D} - \frac{p_0 L^2 x^2}{8\lambda^2 D} - \frac{p_0 L^4}{16\lambda^3 D} \coth \lambda \tag{P7.11b}$$

CHAPTER

EIGHT

LARGE DEFLECTIONS OF PLATES

8.1 INTRODUCTION

In previous sections of the text, relatively small plate deflection is assumed ($w < t$). In some applications of thin plates, however, the maximum deflection is equal to or larger than the plate thickness. Because of these *large* displacements ($x \geq t$) the midplane stretches, and hence the in-plane tensile stresses developed within the plate stiffen and add considerable load resistance to it, not predicted by the small-deflection bending theory. For such situations, an extended plate theory must be employed, accounting for the effects of large deflections. The *large-deflection theory* of plates assumes that the deflections are no longer small in comparison with the thickness but are nevertheless small compared with the remaining plate dimensions.

The large-displacement behavior for plates of simple form is illustrated in Sec. 8.2, principally to give some idea of the additional load-carrying action. The behavior described is also generally valid for plates of any other shape. Section 8.3 describes the differences between the small- and large-deflection theories. The general analytical solution of plate problems is formulated in Sec. 8.4. This is followed by the application of the energy method to the solution of problems involving large-plate deflections. The final section presents a numerical treatment for the bending of plates experiencing large deflections.

8.2 PLATE BEHAVIOR WHEN DEFLECTIONS ARE LARGE

In discussing large-deflection behavior, we must distinguish the cases of midplane deformation into "developable" or "nondevelopable" surfaces. A developable surface completely recovers its original flat shape and dimensions; developable behavior implies the absence of any deformation. Cylinders and cones have developable surfaces, while a sphere or a saddle is a nondevelopable surface. When a plate bends into a cylindrical geometry, for example, the tension

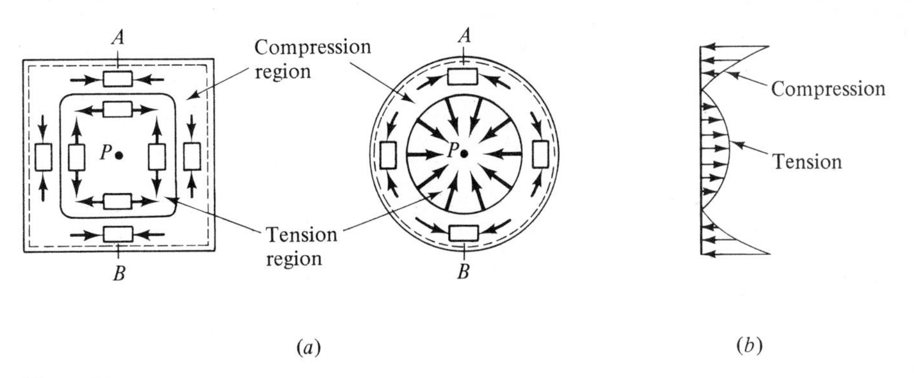

Figure 8.1

in the midplane can be produced only if the end supports are immovable (Sec. 7.7). Thus, as mentioned in Sec. 1.2, the limitation $w_{max} < t$ does *not* hold in cases where a simply supported plate bends into a developable surface, and the classical formulas are valid until *yielding impends* or $w_{max} \to a$, where a is the smaller span length of the plate.

For plates of ordinary proportions, the edge conditions have a pronounced effect upon the magnitude of the direct tensile stresses that may be developed within them. In the case of a simply supported plate undergoing a general large displacement, the plate edges will be *free* of stresses in the direction normal to the boundary, while all other points within the plate will not be stress-free. The tensile stresses, which increase with the distance from an edge, are caused by compressive stresses in a tangential direction near the plate edges. In practice, the latter stresses often produce wrinkling or buckling near the edges of a simply supported plate. On the other hand, normal and tangential stresses may both occur at the edges of a built-in plate. Clearly, only when the plate edges are clamped and fixed, are the direct tensile stresses utilized fully by carrying some of the lateral loading.

Consider the behavior of a square or circular simply supported plate under a concentrated center load P (Fig. 8.1*a*). While the plate is being subjected to an increasing load, it goes through the pure bending, small-displacement range to the large-displacement range. At this stage, the relationship between P and the deflection w is *no longer linear* (Sec. 8.3), owing to the *change in plate geometry*. At higher loads, a central tensile and outer compressive stress region develops. The distribution of stress at a section through points A and B of the plate is illustrated by the sketch shown in Fig. 8.1*b*.

8.3 COMPARISON OF SMALL- AND LARGE-DEFLECTION THEORIES

It is now demonstrated that all relationships derived thus far for plates bending into nondevelopable surfaces are true in general *only* if the maximum deflection is small in comparison with the thickness of the plate. For this purpose, we

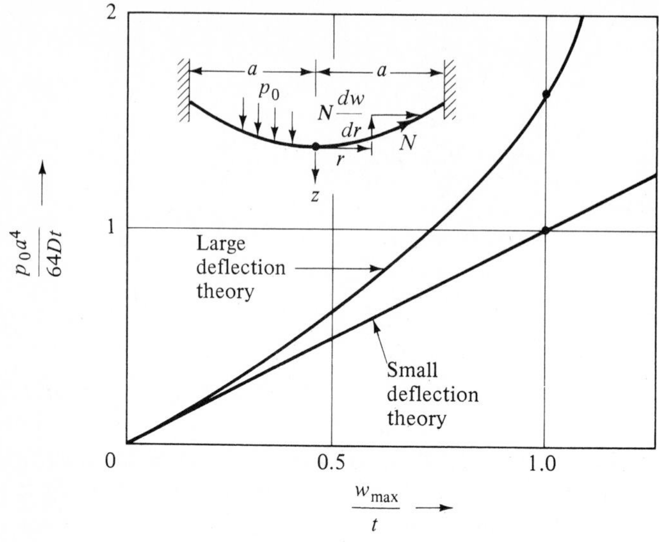

Figure 8.2

employ a simple *approximate approach* which is in good agreement with the exact theory for the circular plate. According to this method, the bending solution and the *membrane or very thin plate* solution (Sec. 10.1) are treated separately. The partial loads carried by the membrane and the bending actions in the plate are then added and equated to the actual plate loading.

As an example, consider the case of a clamped-edge circular plate subjected to a uniform load of intensity p_0 (Fig. 8.2). The *bending solution* for the maximum deflection occurring in the center is, from Eq. (2.14):

$$w_{max} = \frac{p_0 a^4}{64D} \qquad \text{or} \qquad p_0 = \frac{64D}{a^4} w_{max} \tag{a}$$

To derive the *membrane solution*, refer to Fig. 8.2, where N denotes the *constant* tensile force per unit length. The static equilibrium of vertical forces is expressed by $2\pi r N(dw/dr) = p_0 \pi r^2$, from which $dw/dr = p_0 r/2N$. Integration of the latter expression for the slope at $r = a$ leads to:

$$w_{max} = \frac{p_0 a^2}{4N} \tag{b}$$

Determination of the value of N in Eq. (b) requires consideration of the mid-plane deformations.

The radial elongation produced by the deflection w is found from Eq. (7.12a) as follows

$$\frac{1}{2}\int_0^a \left(\frac{dw}{dr}\right)^2 dr = \frac{1}{2}\int_0^a \left(\frac{p_0 r}{2N}\right)^2 dr = \frac{p_0^2 a^3}{24N^2}$$

The strain is therefore

$$\varepsilon = \frac{p_0^2 a^2}{24N^2} \tag{c}$$

This strain is the *same* in all directions. Hence, $(1 - \nu)N/t = E\varepsilon$ together with Eq. (*c*) yields

$$N^3 = \frac{E}{1 - \nu} \frac{p_0^2 a^2 t}{24} \tag{d}$$

Finally, elimination of N between Eqs. (*b*) and (*d*) results in the membrane solution:

$$p_0 = \frac{8}{3} \frac{E}{1 - \nu} \frac{t}{a} \left(\frac{w_{max}}{a} \right)^3 \tag{e}$$

The load actually carried by the plate equals the sum of the partial loads resisted by the bending and the membrane actions. Upon application of Eqs. (*a*) and (*e*) we have

$$p_0 = \frac{64D}{a^3} \left(\frac{w_{max}}{a} \right) + \frac{8}{3} \frac{E}{1 - \nu} \frac{t}{a} \left(\frac{w_{max}}{a} \right)^3 \tag{f}$$

An alternate form of Eq. (*f*), by taking $\nu = 0.3$, is

$$\frac{p_0 a^4}{64Dt} = \frac{w_{max}}{t} \left[1 + 0.65 \left(\frac{w_{max}}{t} \right)^2 \right] \tag{g}$$

Clearly, the first and the second bracketed terms represent the bending and the membrane solutions, respectively.

To illustrate the variation of load and deflection for a uniformly loaded circular plate with clamped edge, w_{max}/t and $p_0 a^4/64Dt$ are plotted in Fig. 8.2. We observe that the *small*-deflection theory is *satisfactory for* $w_{max} < t/2$ and that larger deflections produce greater error. When $w_{max} = t$, for example, there is a 65 percent error in the load according to the bending theory alone. The experimental data agrees well with the result given by large-deflection theory.

8.4 THE GOVERNING EQUATIONS FOR LARGE DEFLECTIONS

A modified form of Eq. (7.3) may be employed in the analysis of deformation in a plate under transverse loading resulting in relatively large elastic deflection. It is important to note that the in-plane forces N_x, N_y, and N_{xy} *do not, as before, depend only* upon the external loading acting in the xy plane. They are now effected also by the *stretching of the midplane* of the plate produced by large deflections *owing to bending*. For a thin plate element, the x and y equilibria of direct forces are expressed by Eqs. (7.1) and (7.2). To ascertain the values of N_x, N_y, and N_{xy} in these two equations, a third expression will be developed from the midplane strain displacement relations as follows.

When the stretching of the midplane occurs *during the bending* of the plate, any point in the *midplane* experiences x, y, and z directed displacements, $u_0 = u$, $v_0 = v$, and w, respectively (Sec. 1.3). Then resultant strain components are found by combining Eqs. (1.1) and (7.12). Thus, the following are the *general* midplane strain-displacement relations owing to bending and stretching of the plate

$$\varepsilon_x = \frac{\partial u}{\partial x} + \frac{1}{2}\left(\frac{\partial w}{\partial x}\right)^2$$
$$\varepsilon_y = \frac{\partial v}{\partial y} + \frac{1}{2}\left(\frac{\partial w}{\partial y}\right)^2 \qquad (8.1)$$
$$\gamma_{xy} = \frac{\partial v}{\partial x} + \frac{\partial u}{\partial y} + \frac{\partial w}{\partial x}\frac{\partial w}{\partial y}$$

As the strains are evidently not independent of one another, by differentiating ε_x twice with respect to y, ε_y twice with respect to x, and γ_{xy} with respect to x and y, the following familiar relationship is obtained:

$$\frac{\partial^2 \varepsilon_x}{\partial y^2} + \frac{\partial^2 \varepsilon_y}{\partial x^2} - \frac{\partial^2 \gamma_{xy}}{\partial x\,\partial y} = \left(\frac{\partial^2 w}{\partial x\,\partial y}\right)^2 - \frac{\partial^2 w}{\partial x^2}\frac{\partial^2 w}{\partial y^2} \qquad (a)$$

According to Hooke's law,

$$\varepsilon_x = \frac{1}{Et}(N_x - \nu N_y) \qquad \varepsilon_y = \frac{1}{Et}(N_y - \nu N_x) \qquad \gamma_{xy} = \frac{N_{xy}}{Gt} \qquad (b)$$

Upon introduction of Eqs. (b) into Eq. (a), we thus have the third equation in terms of N_x, N_y, and N_{xy}. This expression, as well as Eqs. (7.1) and (7.2), is identically satisfied by the *stress function* $\phi(x, y)$, related to the direct forces as follows:

$$N_x = t\frac{\partial^2 \phi}{\partial y^2} \qquad N_y = t\frac{\partial^2 \phi}{\partial x^2} \qquad N_{xy} = -t\frac{\partial^2 \phi}{\partial x\,\partial y} \qquad (c)$$

Substitution of the foregoing into Eqs. (b) yields

$$\varepsilon_x = \frac{1}{E}\left(\frac{\partial^2 \phi}{\partial y^2} - \nu\frac{\partial^2 \phi}{\partial x^2}\right)$$
$$\varepsilon_y = \frac{1}{E}\left(\frac{\partial^2 \phi}{\partial x^2} - \nu\frac{\partial^2 \phi}{\partial y^2}\right) \qquad (d)$$
$$\gamma_{xy} = -\frac{2(1+\nu)}{E}\frac{\partial^2 \phi}{\partial x\,\partial y}$$

Inserting Eqs. (d) into Eq. (a) results in

$$\frac{\partial^4 \phi}{\partial x^4} + 2\frac{\partial^4 \phi}{\partial x^2 y^2} + \frac{\partial^4 \phi}{\partial y^4} = E\left[\left(\frac{\partial^2 w}{\partial x\, \partial y}\right)^2 - \frac{\partial^2 w}{\partial x^2}\frac{\partial^2 w}{\partial y^2}\right] \tag{8.2}$$

and introduction of Eqs. (c) into Eq. (7.3) leads to

$$\frac{\partial^4 w}{\partial x^4} + 2\frac{\partial^4 w}{\partial x^2\, \partial y^2} + \frac{\partial^4 w}{\partial y^4} = \frac{t}{D}\left[\frac{p}{t} + \frac{\partial^2 \phi}{\partial y^2}\frac{\partial^2 w}{\partial x^2} + \frac{\partial^2 \phi}{\partial x^2}\frac{\partial^2 w}{\partial y^2} - 2\frac{\partial^2 \phi}{\partial x\, \partial y}\frac{\partial^2 w}{\partial x\, \partial y}\right] \tag{8.3}$$

Expressions (8.2) and (8.3) are the *governing differential equations* for *large deflections* of thin plates. Determination of ϕ and w requires the solution of these equations which must, of course, satisfy the boundary conditions. Once the stress function is known, the midplane stresses are obtained through the use of Eqs. (c). Knowing the deflection w, upon application of Eqs. (1.8), (1.14), and (1.13), as in the case of small deflection, the normal and shear stresses, respectively, are determined.

Equations (8.2) and (8.3) were introduced by *von Kármán* in 1910. Unfortunately, where realistic problems are concerned, solving these coupled, nonlinear, partial differential equations may be a formidable task. A number of approximate solutions have been determined for the uniformly loaded plates of simple regular shapes.[7,11] Only as a result of recent progress in the development of numerical approaches has the general problem of plates been treated satisfactorily.[17,25] We shall discuss the powerful finite element solution of large deflections of thin plates in Sec. 8.6.

In concluding this discussion, consider the bending of the plate into a cylindrical surface (Sec. 7.7). If the generating line of the cylindrical surface is parallel to the y axis, $w = w(x)$, and $\partial^2\phi/\partial x^2$ and $\partial^2\phi/\partial y^2$ are constants. Hence, Eq. (8.2) is satisfied identically, and Eq. (8.3) reduces to Eq. (7.31). This particular case of bending has already been treated in Examples 1.1 and 7.6.

8.5 DEFLECTIONS BY THE RITZ METHOD

In this section the discussion is limited to *very thin plates*, that is, those in which deflections are many times greater than their thickness and on which only in-plane or membrane forces do work. Plates of this type may thus be regarded as flexible *membranes* having no resistance to bending action.

The strain energy associated with the stretching of the midplane of a membrane is given by (Sec. 1.9):

$$U_m = \frac{1}{2}\iint (N_x \varepsilon_x + N_y \varepsilon_y + N_{xy}\gamma_{xy})\, dx\, dy \tag{8.4a}$$

or

$$U_m = \frac{Et}{2(1-\nu^2)}\iint \left[\varepsilon_x^2 + \varepsilon_y^2 + 2\nu\varepsilon_x\varepsilon_y + \tfrac{1}{2}(1-\nu)\gamma_{xy}^2\right] dx\, dy \tag{8.4b}$$

Upon substituting the values of the strains from Eqs. (8.1) into the above, the strain energy is obtained in the following form:

$$U_m = \frac{Et}{2(1-\nu^2)} \iint \left\{ \left(\frac{\partial u}{\partial x} \right)^2 + \frac{\partial u}{\partial x} \left(\frac{\partial w}{\partial x} \right)^2 + \left(\frac{\partial v}{\partial y} \right)^2 + \frac{\partial v}{\partial y} \left(\frac{\partial w}{\partial y} \right)^2 \right.$$
$$+ \frac{1}{4} \left[\left(\frac{\partial w}{\partial x} \right)^2 + \left(\frac{\partial w}{\partial y} \right)^2 \right]^2 + 2\nu \left[\frac{\partial u}{\partial x} \frac{\partial v}{\partial y} + \frac{1}{2} \frac{\partial v}{\partial y} \left(\frac{\partial w}{\partial x} \right)^2 + \frac{1}{2} \frac{\partial u}{\partial x} \left(\frac{\partial w}{\partial y} \right)^2 \right]$$
$$+ \frac{1-\nu}{2} \left[\left(\frac{\partial u}{\partial y} \right)^2 + 2 \frac{\partial u}{\partial y} \frac{\partial v}{\partial x} + \left(\frac{\partial v}{\partial x} \right)^2 \right.$$
$$\left.\left. + 2 \frac{\partial u}{\partial y} \frac{\partial w}{\partial x} \frac{\partial w}{\partial y} + 2 \frac{\partial v}{\partial x} \frac{\partial w}{\partial x} \frac{\partial w}{\partial y} \right] \right\} dx\, dy \tag{8.5}$$

The Ritz method provides a simple approach to the determination of membrane deflections, as illustrated in the following example.

Example 8.1 What are the deflection and stress at the center of a very thin square plate of sides $2b$ due to a uniform load p_0 per unit surface area (Fig. 8.3)? The displacements u, v, and w are taken to be zero at the edges.

SOLUTION Assuming that the origin of coordinates x, y is placed at the center of the plate, the displacement may be represented by

$$w = a_0 \cos \frac{\pi x}{2b} \cos \frac{\pi y}{2b}$$
$$u = c_0 \sin \frac{\pi x}{b} \cos \frac{\pi y}{2b} \tag{a}$$
$$v = c_0 \sin \frac{\pi y}{b} \cos \frac{\pi x}{2b}$$

where a_0 and c_0 are unknown coefficients. The expressions (a) satisfy the boundary conditions. It is also observed that the requirement of symmetry of deformation is fulfilled by u, v, and w.

Introducing Eqs. (a) into Eq. (8.5), integrating, we find that for $\nu = 0.25$:

$$U_m = \frac{Et}{7.5} \left[\frac{5\pi^4}{64} \frac{a_0^4}{b^2} - \frac{17\pi^2}{6} \frac{a_0^2}{b} + c_0^2 \left(\frac{35\pi^2}{4} + \frac{80}{9} \right) \right] \tag{b}$$

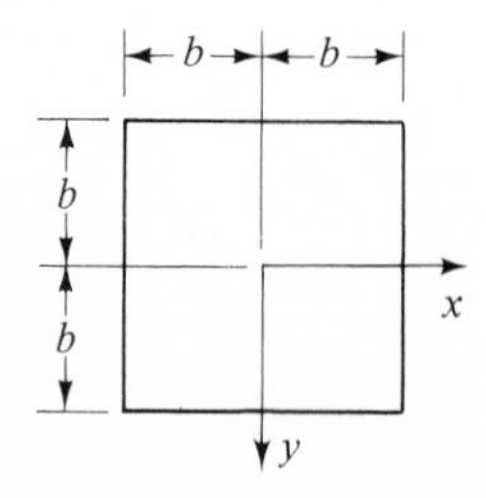

Figure 8.3

The work done by the uniformly distributed load is

$$W = \int_{-b}^{b} \int_{-b}^{b} p_0 a_0 \cos \frac{\pi x}{2b} \cos \frac{\pi y}{2b} \, dx \, dy \tag{c}$$

Potential energy $\Pi = U - W$. The conditions $\partial\Pi/\partial a_0 = 0$ and $\partial\Pi/\partial c_0$ yield two equations, the solution of which results in the following values for the coefficients:

$$a_0 = 0.802b \sqrt[3]{\frac{p_0 b}{Et}} \qquad c_0 = 0.147 \frac{a_0^2}{b} \tag{8.6}$$

The maximum lateral deflection occurs at the center and is given by $w_{max} = 0.802b \sqrt[3]{p_0 b/Et}$. On applying Eqs. ($a$), the tensile strain at $x = y = 0$ is found to be: $\varepsilon_x = \varepsilon_y = \pi c_0/b = 0.462a_0^2/b^2$. The associated tensile stress is

$$\sigma = \frac{E}{1 - \nu}\left(0.462 \frac{a_0^2}{b^2}\right) = 0.396 \sqrt[3]{\frac{p_0^2 E b^2}{t^2}} \tag{8.7}$$

The energy method can similarly be applied in the case of large deflections of plates having other boundary conditions and shapes.

8.6 THE FINITE ELEMENT SOLUTION

The finite element method (Chap. 5) is here applied to determining the large displacements and stresses in a plate of arbitrary shape under general loading. The behavior of plates experiencing large deflections is illustrated in Sec. 8.2. Clearly, we must now extend the formulations of Sec. 5.8 to include the effect of midplane plate strains and corresponding in-plane stresses.

To begin with, assume an initial system of in-plane forces applied to the plate and regard these as constants during bending (Sec. 7.4). On this basis, in-plane strains of an element will be represented, from Eqs. (8.1), as

$$\{\bar{\varepsilon}\}_e = \left\{ \frac{1}{2}\left(\frac{\partial w}{\partial x}\right)^2, \frac{1}{2}\left(\frac{\partial w}{\partial y}\right)^2, \frac{\partial w}{\partial x}\frac{\partial w}{\partial y} \right\} \tag{8.8}$$

The "generalized" bending-strain-displacement relations are given by Eqs. (5.17):

$$\{\varepsilon\}_e = \left\{ -\frac{\partial^2 w}{\partial x^2}, -\frac{\partial^2 w}{\partial y^2}, -2\frac{\partial^2 w}{\partial x \, \partial y} \right\} \tag{8.9}$$

The stress resultants are thus comprised of *constant* direct forces and moments:

$$\begin{Bmatrix} N_x \\ N_y \\ N_{xy} \end{Bmatrix}_e = t \begin{Bmatrix} \bar{\sigma}_x \\ \bar{\sigma}_y \\ \bar{\tau}_{xy} \end{Bmatrix}_e \qquad \begin{Bmatrix} M_x \\ M_y \\ M_{xy} \end{Bmatrix}_e = \int_{-t/2}^{t/2} \begin{Bmatrix} \sigma_x \\ \sigma_y \\ \tau_{xy} \end{Bmatrix}_e z \, dz \tag{8.10a}$$

or

$$\{N\}_e = t\{\bar{\sigma}\}_e \qquad \{M\}_e = \int_{-t/2}^{t/2} \{\sigma\}_e \, z \, dz \tag{8.10b}$$

The bending stresses $\{\sigma\}$ and the "generalized" bending strains $\{\varepsilon\}$ are related by Eq. (5.18). In-plane stresses $\{\bar{\sigma}\}$ and in-plane strains $\{\bar{\varepsilon}\}$ are connected by the Hooke's law

$$\{\bar{\sigma}\}_e = [\bar{D}]\{\bar{\varepsilon}\}_e \tag{8.11}$$

Here

$$[\bar{D}] = \frac{E}{1 - \nu^2}\begin{bmatrix} 1 & \nu & 0 \\ \nu & 1 & 0 \\ 0 & 0 & (1-\nu)/2 \end{bmatrix} \tag{8.12}$$

represent the *elasticity matrix for plane stress.*

Inasmuch as the direct and bending strains are taken to be *independent*, the expression (1.41) for the potential energy of plates is

$$\Pi = \frac{1}{2}\iint_A \{\varepsilon\}_e^T\{M\}_e \, dx \, dy + \frac{t}{2}\iint_A \begin{Bmatrix} \dfrac{\partial w}{\partial x} \\ \dfrac{\partial w}{\partial y} \end{Bmatrix}^T [\bar{\sigma}]_e \begin{Bmatrix} \dfrac{\partial w}{\partial x} \\ \dfrac{\partial w}{\partial y} \end{Bmatrix} dx \, dy - \iint_A (pw) \, dx \, dy \tag{8.13}$$

where

$$[\bar{\sigma}]_e = \begin{bmatrix} \bar{\sigma}_x & \bar{\tau}_{xy} \\ \bar{\tau}_{xy} & \bar{\sigma}_y \end{bmatrix}_e \tag{8.14}$$

We shall proceed with the discussion by considering the case of a specific element.

Rectangular finite element The bending properties of this element are developed in Sec. 5.10). We now treat its in-plane deformation properties. The derivatives (slopes) of w, are from Eq. (5.40),

$$\begin{Bmatrix} \dfrac{\partial w}{\partial x} \\ \dfrac{\partial w}{\partial y} \end{Bmatrix}_e = \begin{bmatrix} 0 & 1 & 0 & 2x & 0 & 0 & 3x^2 & 2xy & y & 0 & 3x^2y & y^3 \\ 0 & 0 & 1 & 0 & x & 2y & 0 & x^2 & 2xy & 3y^2 & x^3 & 3xy^2 \end{bmatrix}\{a_1, \ldots, a_{12}\} \tag{8.15a}$$

or

$$\{\theta\}_e = [S]\{a\} \tag{8.15b}$$

Applying Eq. (5.42), we have

$$\{\theta\}_e = [S][C]^{-1}\{\delta\}_e = [G]\{\delta\}_e \tag{8.16}$$

in which

$$[G] = [S][C]^{-1} \tag{8.17}$$

It is observed that $[G]$ is a matrix defined only in terms of the coordinates.

The potential-energy expression, upon introducing Eqs. (8.16) into Eq. (8.13), is therefore

$$\Pi = \frac{1}{2}\iint_A \{\varepsilon\}_e^T\{M\}_e \, dx\, dy + \frac{t}{2}\{\delta\}_e^T\left(\iint_A [G]^T[\bar{\sigma}][G]\, dx\, dy\right)\{\delta\}_e - \iint_A (pw)\, dx\, dy \tag{8.18}$$

As in Sec. 5.8, it can be demonstrated readily that the principle of potential energy now yields

$$\{Q\}_e = [k]_e\{\delta\}_e + [k_G]_e\{\delta\}_e = [k_T]_e\{\delta\}_e \tag{8.19}$$

These are modified *equations of equilibrium* of nodal forces. The new term

$$[k_G]_e = t[C^{-1}]^T\left(\iint_A [S]^T[\bar{\sigma}][S]\, dx\, dy\right)[C^{-1}] \tag{8.20}$$

is known as the *initial stress* or *geometric stress* matrix. Appropriate definitions of the small-displacement stiffness matrix $[k]_e$ and nodal force matrix $\{Q\}_e$ are given by Eqs. (5.23) and (5.24), respectively. The $[k_T]_e$ is termed the *total stiffness matrix* of the element. Equation (8.20) may be expressed in the following convenient form

$$[k_G]_e = \bar{\sigma}_x[k_{Gx}]_e + \bar{\sigma}_y[k_{Gy}]_e + \bar{\tau}_{xy}[k_{Gxy}]_e \tag{8.21}$$

wherein $[k_{Gx}]_e$, for instance, designates

$$[k_{Gx}]_e = t[C^{-1}]^T\left(\iint_A [S]^T\begin{bmatrix}1 & 0\\ 0 & 0\end{bmatrix}[S]\, dx\, dy\right)[C^{-1}] \tag{8.22}$$

Relationships for other components of the geometric matrix, $[k_{Gy}]_e$ and $[k_{Gxy}]_e$, may be written in a like manner.

The generalized procedure for solving a plate *large-deflection* problem is summarized as follows:

(1) Assume the initial direct stresses $\{\bar{\sigma}\}$ to be zero. Apply the procedure (steps 1 to 3) of Sec. 5.8 to obtain $[k]_e$, $\{Q\}_e$, and the hence small-deflection solution for the nodal displacements $\{\delta\}_e$.
(2) Compute the slopes at some representative point within the element (e.g., at the centroid) from Eq. (8.16): $\{\theta\}_e = [G]\{\delta\}_e$.
(3) Compute the in-plane strains $\{\bar{\varepsilon}\}_e$ from Eq. (8.8).
(4) Compute the in-plane stresses from Eq. (8.11): $\{\bar{\sigma}\}_e = [\bar{D}]\{\bar{\varepsilon}\}_e$.
(5) Compute the geometrical stiffness $[k_G]_e$ from Eq. (8.21) in terms of the given element properties.
(6) Compute the element total stiffness matrix from Eq. (8.19): $[k_T]_e = [k]_e + [k_G]_e$.

(7) Repeat steps (1) to (4), each time with a new $[k_T]_e$ found by applying steps (5) and (6) until satisfactory convergence of the in-plane stresses $\{\bar{\sigma}\}_e$ is attained.

The results determined by applying the finite element method to the large deflection of thin plates agrees well with analytical solutions.[18] It is interesting to note that some classical buckling problems of plates may also be treated by the approach described in this section.

PROBLEMS

Secs. 8.1 to 8.6

8.1 Verify that for axisymmetrically bent thin plates, the strain–large-displacement relations are

$$\varepsilon_r = \frac{du}{dr} + \frac{1}{2}\left(\frac{\partial w}{dr}\right)^2 \qquad \varepsilon_0 = \frac{u}{r} \tag{P8.1}$$

Here u and w are the r- and z-directed displacements, respectively.

8.2 Apply Hooke's law and Eqs. (P8.1) to derive the following stress-displacement relations for axisymmetrically loaded plates:

$$\begin{aligned}\sigma_r &= \frac{E}{1-\nu^2}\left[\frac{du}{dr} + \frac{1}{2}\left(\frac{dw}{dr}\right)^2 + \nu\frac{u}{r}\right] \\ \sigma_\theta &= \frac{E}{1-\nu^2}\left[\frac{u}{r} + \nu\frac{du}{dr} + \frac{\nu}{2}\left(\frac{dw}{dr}\right)^2\right]\end{aligned} \tag{P8.2}$$

8.3 Show that the system of equations (8.2) and (8.3), for axisymmetrically bent plates, assumes the form:

$$\begin{aligned}\nabla^4\phi &= -\frac{E}{2}L(w, w) \\ \nabla^4 w &= \frac{t}{D}L(w, \phi) + \frac{p}{D}\end{aligned} \tag{P8.3}$$

where

$$L(w, \phi) = \frac{1}{r}\frac{d\phi}{dr}\frac{d^2w}{dr^2} + \frac{1}{r}\frac{dw}{dr}\frac{d^2\phi}{dr^2}$$

and $L(w, w)$ is found upon replacement of w by ϕ in the above expression.

8.4 Verify that the strain energy, due to the stretching of the midplane, is expressed by

$$U_m = \frac{\pi E t}{1-\nu^2}\int\left[\left(\frac{du}{dr}\right)^2 + \frac{du}{dr}\left(\frac{dw}{dr}\right)^2 + \frac{u^2}{r^2} + \frac{2\nu u}{r}\frac{du}{dr} + \frac{\nu u}{r}\left(\frac{dw}{dr}\right)^2 + \frac{1}{4}\left(\frac{dw}{dr}\right)^4\right] r\,dr \tag{P8.4}$$

for a very thin axisymmetrically loaded circular plate.

CHAPTER

NINE

THERMAL STRESSES IN PLATES

9.1 INTRODUCTION

Solution for the deflection and stress in plates subjected to temperature variation requires reformulation of the stress-strain relationship. This is accomplished by superposition of the strain attributable to stress and that due to temperature. For homogeneous isotropic materials, *a change in temperature* $\Delta T = T - T_0$ produces uniform linear strain in every direction. Here T and T_0 are the final and the initial temperatures, respectively. In equation form, the *thermal strains* are expressed as

$$\varepsilon_t = \alpha(\Delta T) \tag{9.1}$$

where α, an experimentally determined material property, is termed the coefficient of thermal expansion. Over a moderate temperature change, α remains reasonably constant. In SI units, α is expressed in meters per meter per degree Celsius. Coefficients of thermal expansion for common materials are listed in Table 1.1. For isotropic materials, a change in temperature produces no shear strains, i.e., $\gamma_t = 0$. In this text, the modulus of elasticity E and coefficient of thermal expansion α are treated as constants over the temperature ranges involved.

Stresses owing to the restriction of thermally induced expansion or contraction of a body are termed *thermal stresses*. When a free plate is heated *uniformly*, there are produced normal strains but *no* thermal stresses. If, however, the plate experiences a *nonuniform* temperature field, or if the displacements are prevented from occurring freely because of the restrictions placed on the boundary even with a uniform temperature, or if the material displays *anisotropy* even with uniform heating, thermal stresses will occur.

9.2 STRESS, STRAIN, AND DISPLACEMENT RELATIONS

The total x and y strains, ε_x and ε_y, are obtained by adding to the thermal strains, the strains due to stress resulting from external forces:

$$\begin{aligned}\varepsilon_x &= \frac{1}{E}(\sigma_x - \nu\sigma_y) + \alpha(\Delta T)\\ \varepsilon_y &= \frac{1}{E}(\sigma_y - \nu\sigma_x) + \alpha(\Delta T)\\ \gamma_{xy} &= \frac{\tau_{xy}}{G}\end{aligned} \tag{9.2}$$

From Eqs. (9.2), the total stress components are given by

$$\begin{aligned}\sigma_x &= \frac{E}{1-\nu^2}[\varepsilon_x + \nu\varepsilon_y - (1+\nu)\alpha(\Delta T)]\\ \sigma_y &= \frac{E}{1-\nu^2}[\varepsilon_y + \nu\varepsilon_x - (1+\nu)\alpha(\Delta T)]\\ \tau_{xy} &= G\gamma_{xy}\end{aligned} \tag{9.3}$$

An increase in temperature ΔT is algebraically positive.

In deriving the strain-displacement relationships, a state of plane stress is assumed. It is further assumed that straight lines, initially normal to the midsurface, remain straight and normal to that surface after heating. Thus, the assumptions of Sec. 1.2, *with the exception* of (2), apply. Substituting Eqs. (*b*) of Sec. 1.3 into Eqs. (1.1*a*) to (1.1*c*), and denoting $u_0 = u$ and $v_0 = v$, we obtain the following expressions for the strains in terms of displacements:

$$\begin{aligned}\varepsilon_x &= \frac{\partial u}{\partial x} - z\frac{\partial^2 w}{\partial x^2}\\ \varepsilon_y &= \frac{\partial v}{\partial y} - z\frac{\partial^2 w}{\partial y^2}\\ \gamma_{xy} &= \left(\frac{\partial u}{\partial y} + \frac{\partial v}{\partial x}\right) - 2z\frac{\partial^2 w}{\partial x\, \partial y}\end{aligned} \tag{9.4}$$

In the foregoing, the first terms represent the strain components in the midplane of the plate, and $w(x, y)$ represents the transverse deflection.

9.3 STRESS RESULTANTS

The stresses distributed over the thickness of the plate result in *in-plane forces* and *moments* per unit length as shown in Figs. 7.1 and 1.5. The latter quantities are given by Eqs. (1.9). The in-plane force components are represented by

$$\begin{Bmatrix} N_x \\ N_y \\ N_{xy} \end{Bmatrix} = \int_{-t/2}^{t/2} \begin{Bmatrix} \sigma_x \\ \sigma_y \\ \tau_{xy} \end{Bmatrix} dz \tag{9.5}$$

Introducing Eqs. (9.3) and (9.4) into Eqs. (1.9) and (9.5), the *stress resultants* are obtained:

$$\begin{aligned}
N_x &= \frac{Et}{1-\nu^2}\left(\frac{\partial u}{\partial x} + \nu\frac{\partial v}{\partial y}\right) - \frac{N^*}{1-\nu} \\
N_y &= \frac{Et}{1-\nu^2}\left(\frac{\partial v}{\partial y} + \nu\frac{\partial u}{\partial x}\right) - \frac{N^*}{1-\nu} \\
N_{xy} &= \frac{E}{2(1+\nu)}\left(\frac{\partial u}{\partial y} + \frac{\partial v}{\partial x}\right) \\
M_x &= -D\left(\frac{\partial^2 w}{\partial x^2} + \nu\frac{\partial^2 w}{\partial y^2}\right) - \frac{M^*}{1-\nu} \\
M_y &= -D\left(\frac{\partial^2 w}{\partial y^2} + \nu\frac{\partial^2 w}{\partial x^2}\right) - \frac{M^*}{1-\nu} \\
M_{xy} &= -(1-\nu)D\frac{\partial^2 w}{\partial x\,\partial y}
\end{aligned} \tag{9.6}$$

Here the quantities

$$N^* = \alpha E\int_{-t/2}^{t/2}(\Delta T)\,dz \qquad M^* = \alpha E\int_{-t/2}^{t/2}(\Delta T)z\,dz \tag{9.7}$$

are termed the *thermal* stress resultants.

The components of stress may now be determined by substitution of Eqs. (9.4) into (9.3) and elimination of the displacement derivatives through the use of Eqs. (9.6):

$$\begin{aligned}
\sigma_x &= \frac{1}{t}\left(N_x + \frac{N^*}{1-\nu}\right) + \frac{12z}{t^3}\left(M_x + \frac{M^*}{1-\nu}\right) - \frac{E\alpha(\Delta T)}{1-\nu} \\
\sigma_y &= \frac{1}{t}\left(N_y + \frac{N^*}{1-\nu}\right) + \frac{12z}{t^3}\left(M_y + \frac{M^*}{1-\nu}\right) - \frac{E\alpha(\Delta T)}{1-\nu} \\
\tau_{xy} &= \frac{1}{t}N_{xy} + \frac{12z}{t^3}M_{xy}
\end{aligned} \tag{9.8}$$

Equations (9.8) permit the direct calculation of the stress components for a plate of any cross section subject to an arbitrary temperature distribution. To derive the stress equations in *polar coordinates*, one need only replace subscripts x by r and y by θ in these expressions.

The first and second terms of Eqs. (9.8) are due to the *in-plane forces* and *bending moments*, respectively. The third term represents in-plane stress in the case of uniform heating, and bending stress in the case of nonuniform heating. An example of the latter is a plate with its upper surface heated and lower surface cooled. At some location (x, y), there will develop compressive stress $-\alpha E(\Delta T)/(1 - \nu)$ in the upper half and tensile stress $+\alpha E(\Delta T)/(1 - \nu)$ in the lower half.

It is observed that Eqs. (9.8) are of the same form as those associated with beams under *compound* loading.

9.4 THE GOVERNING DIFFERENTIAL EQUATIONS

The *governing equation for deflection* w, derived using the procedure described in Sec. 1.6 and the modified moment resultants of Eqs. (9.6), are combined to yield

$$D\,\nabla^4 w = p - \frac{1}{1-\nu}\nabla^2 M^* \tag{9.9}$$

For the case of a nonuniformly heated free plate, i.e., $p = 0$, Eq. (9.9) reduces to

$$D\,\nabla^4 w = -\frac{1}{1-\nu}\nabla^2 M^* \tag{9.10}$$

It is usual to denote

$$p^* = -\frac{1}{1-\nu}\nabla^2 M^* \tag{9.11}$$

where p^* is termed the *equivalent* transverse load. We observe from Eqs. (9.9) and (9.10) that it is possible to superimpose the deflections owing to the temperature alone with those owing to transverse loads alone.

The two-dimensional equilibrium and compatibility equations (in the xy plane) are employed to obtain the forces N as shown in the outline that follows, similar to that employed in Sec. 8.4. The differential equations of equilibrium in the plane of the plate are given by Eqs. (7.1) and (7.2). These expressions are identically satisfied by the stress function $\phi(x, y)$, related to the force resultants as follows:

$$N_x = \frac{\partial^2 \phi}{\partial y^2} \qquad N_{xy} = -\frac{\partial^2 \phi}{\partial x\,\partial y} \qquad N_y = \frac{\partial^2 \phi}{\partial x^2} \tag{9.12}$$

For a plate of constant thickness and negligible weight, the equation of compatibility is

$$\frac{\partial^2}{\partial y^2}(N_x - \nu N_y + N^*) + \frac{\partial^2}{\partial x^2}(N_y - \nu N_x + N^*) = 2(1+\nu)\frac{\partial^2 N_{xy}}{\partial x\,\partial y} \tag{a}$$

Table 9.1

Edge	Clamped	Simply supported	Free
At	$w = 0$	$w = 0$	$\dfrac{\partial^2 w}{\partial x^2} = -\dfrac{M^*}{(1-\nu)D}$
$x = a$	$\dfrac{\partial w}{\partial x} = 0$	$\dfrac{\partial^2 w}{\partial x^2} = -\dfrac{M^*}{(1-\nu)D}$	$D\left[\dfrac{\partial^3 w}{\partial x^3} + (2-\nu)\dfrac{\partial^3 w}{\partial x\, \partial y^2}\right] + \dfrac{1}{1-\nu}\dfrac{\partial M^*}{\partial x} = 0$

Substitution of Eqs. (9.12) into the above yields

$$\nabla^4 \phi + \nabla^2 M^* = 0 \tag{9.13}$$

What has been accomplished is the formulation of a plane stress problem in *thermoeleasticity* in such a way as to require the solutions of a single partial differential equation (which must, of course, fulfill the boundary conditions).

In summary, the solution of a thermoelastic-plate problem requires the solution of Eq. (9.13) for the midplane forces and Eq. (9.9) for the deflection w and corresponding moments. These *two solutions may be obtained independently* of one another and the stresses owing to each added if the total stresses are required. The second problem, the determination of the transverse deflection, is the *direct concern* of plate theory. The first problem is a concern of the theory of elasticity, and is *not* treated in this text.

The boundary conditions at edge $x = a$ of a rectangular plate (Fig. 1.7), [Eqs. (1.25) through (1.27)], are now expressed as shown in Table 9.1. We observe from the table that the boundary conditions for a simply supported and a free-edge thermoelastic plate are nonhomogeneous. Other kinds of boundary conditions may also be obtained by employing the stress resultants derived in this chapter and the procedure presented in Sec. 1.7.

In the sections to follow, several methods are described, useful in the solution of *thermal bending problems* of elastic thin plates. A number of references are available for those seeking a more thorough treatment.[26,27] *Design data* and other useful information dealing with thermal stresses are presented in the papers of Goodier[28] and others.[7]

9.5 SIMPLY SUPPORTED RECTANGULAR PLATE SUBJECT TO AN ARBITRARY TEMPERATURE DISTRIBUTION

This section deals with the deflection and stress owing to nonuniform heating of a simply supported rectangular plate (Fig. 3.1). The approach introduced may be extended to all polygonal simply supported plates.

The boundary conditions are represented by (Table 9.1)

$$\begin{aligned} w &= 0 \qquad M_x = 0 \qquad (x = 0,\ x = a) \\ w &= 0 \qquad M_y = 0 \qquad (y = 0,\ y = b) \end{aligned} \tag{a}$$

From Eqs. (9.6) and (*a*), the following expression applies to the boundary:

$$D\,\nabla^2 w = -\frac{M^*}{1-\nu} \tag{b}$$

The governing equation, Eq. (9.10), is equivalent to

$$D\,\nabla^2 w + \frac{M^*}{1-\nu} = f(x, y) \qquad \nabla^2 f = 0 \tag{c,d}$$

Equations (*b*) and (*c*) result in $f = 0$, which is an appropriate solution of Eq. (*d*).

The problem at hand, as in Sec. 7.5, is thus represented by a second-order differential equation

$$D\,\nabla^2 w = -\frac{M^*}{1-\nu} \tag{9.14}$$

which must of course satisfy the boundary conditions

$$w = 0 \qquad (x = 0, a;\ y = 0, b) \tag{9.15}$$

The solution is obtained by the application of Fourier series for load (moment) and deflection:

$$M^* = \sum_{m=1}^{\infty}\sum_{n=1}^{\infty} p_{mn} \sin\frac{m\pi x}{a} \sin\frac{n\pi y}{b} \tag{9.16a}$$

$$w = \sum_{m=1}^{\infty}\sum_{n=1}^{\infty} a_{mn} \sin\frac{m\pi x}{a} \sin\frac{n\pi y}{b} \tag{9.16b}$$

The coefficients p_{mn} are, from Eq. (3.3):

$$p_{mn} = \frac{4}{ab}\int_0^b\int_0^a M^* \sin\frac{m\pi x}{a} \sin\frac{n\pi y}{b}\,dx\,dy \tag{9.17}$$

Substitution of Eq. (9.16*a*), (9.17), and (9.16*b*) into Eq. (9.14) leads to

$$a_{mn} = \frac{1}{(1-\nu)\pi^2 D}\frac{p_{mn}}{(m/a)^2 + (n/b)^2} \tag{9.18}$$

The deflection w corresponding to the thermal loading $M^*(x, y)$ has thus been determined.

9.6 SIMPLY SUPPORTED RECTANGULAR PLATE WITH TEMPERATURE DISTRIBUTION VARYING OVER THE THICKNESS

The solution of a simply supported plate subjected to nonuniform heating such that the temperature varies through the thickness only, $T(z)$, can readily be obtained from the results of Sec. 9.5. In this case, the thermal loading M^* is constant and Eq. (9.17), after integration, leads to

$$p_{mn} = \frac{4M^*}{\pi^2 mn}[1 - (-1)^m][1 - (-1)^n] \tag{a}$$

Substitution of Eqs. (*a*) and (9.18) into Eq. (9.16*b*) yields the following expression for deflection:

$$w = \frac{16M^*}{(1 - \nu)D\pi^4} \sum_m^\infty \sum_n^\infty \frac{\sin(m\pi x/a)\sin(n\pi y/b)}{mn[(m/a)^2 + (n/b)^2]} \qquad (m, n = 1, 3, \ldots) \tag{9.19}$$

The bending moments and stresses in the plate may now be calculated from Eqs. (9.6) and (9.8). As already noted in Example 3.2, while the expression for deflection (9.19) converges very rapidly, the relationship for moments does not. An alternate solution[26] of Eqs. (9.14) and (9.15), more suitable to the computation of moments, may be obtained by the use of simple series for w and M^* rather than the double series as before (Sec. 3.4).

9.7 ANALOGY BETWEEN THERMAL AND ISOTHERMAL PLATE PROBLEMS

We now demonstrate that an analogy exists between the thermal and isothermal plate-bending problems, serving as a basis of a convenient procedure to determine the deflection. The analogy is *complete only for the determination of deflection*. The *thermal stresses are ascertained by adding* $-\alpha E(\Delta T)/1 - \nu)$ to the stress components σ_x and σ_y of the isothermal solution.

Plates with clamped edges The problem of the bending of built-in plates as a result of nonuniform thermal load requires the solution of Eq. (9.10) together with the specified boundary conditions given in Table 9.1. Note that the boundary conditions for a clamped edge do not involve explicitly the temperature. Thus, it is observed from a comparison of Eqs. (9.9) and (9.10), that the solution sought is identical with that for the same shaped clamped plate subject to the equivalent transverse load p^*. The thermal problem is therefore reduced to an isothermal one, and the results and techniques of the latter case are valid for the problem under consideration. Table 9.2 provides a list of some examples.

Table 9.2 Plates with clamped edges

Geometry	Loading (p^*)	Solution
Rectangular	Uniform	Secs. 3.7 and 3.12
Circular	Uniform	Sec. 2.4
(solid)	Radial	Sec. 9.8
Annular	Radial	Sec. 9.8
	Uniform	Table 2.3

In the case of a plate of *arbitrary shape* undergoing thermal variation *through the thickness only*, we have $\nabla^2 M^* = 0$ and $p^* = 0$. According to the analogy, $w = 0$, and the corresponding stresses, from Eqs. (9.8), are

$$\sigma_x = \frac{1}{t}\left(N_x + \frac{N^*}{1-\nu}\right) - \frac{\alpha E(\Delta T)}{1-\nu}$$

$$\sigma_y = \frac{1}{t}\left(N_y + \frac{N^*}{1-\nu}\right) - \frac{\alpha E(\Delta T)}{1-\nu} \qquad (a)$$

$$\tau_{xy} = \frac{N_{xy}}{t}$$

The first and the second terms represent the *plane-* and the *bending*-stress components, respectively.

Plates with simply supported or free edges An analogy also exists between heated and unheated plates with other than clamped supports. In this case a modification of the edge conditions is required inasmuch as they contain the temperature. At a simply supported edge of the analogous isothermal plate, $w = 0$ as before, but a bending moment $M^*/(1 - \nu)$ must be assumed to apply. In a like manner, at a free edge of the analogous unheated plate, a force equal to $(\partial M^*/\partial x)/(1 - \nu)$ must be applied. It is thus observed that a thermal solution can always be determined by *superposition of various isothermal solutions.*

Consider, for example, the bending caused by a nonuniform temperature distribution of a simply supported plate. The deflection of the plate is determined by adding the deflection of an unheated, simply supported plate subject to the surface load p^*, to the deflection of an unheated plate carrying no transverse load but subject to the moment $M^*/(1 - \nu)$ acting at its edges.

The foregoing analogy is also useful in the experimental analysis of elastic heated plates. This is because it may be easier to test a plate at constant temperature with given transverse and edge loadings and then to impose upon it arbitrary temperature distributions.[29]

Example 9.1 An aircraft window, which can be represented approximately as a simply supported circular plate, is subjected to uniform temperature T_1 and uniform temperature T_2 at the lower and the upper surfaces, respectively (Fig. 9.1*a*). Determine the deflection and bending stress if the plate is free of

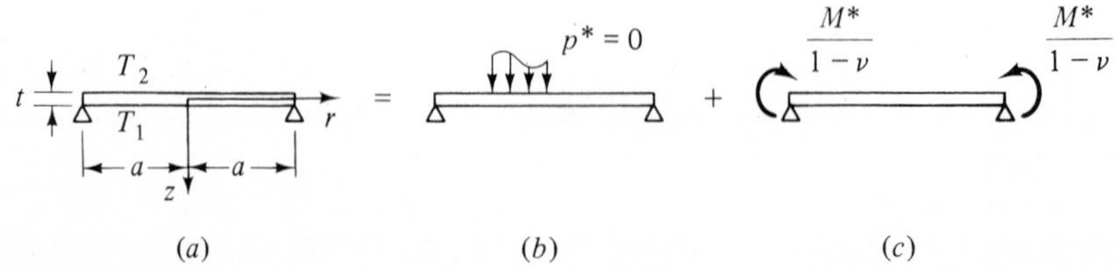

Figure 9.1

stress at 0°C. Assume that the temperature through the thickness varies linearly and that $T_1 > T_2$.

SOLUTION The plate of Fig. 9.1*a* is replaced by the plates shown in Figs. 9.1*b* and 9.1*c*. The temperature difference between the faces is $T_1 - T_2$ and that between either face and the midsurface is $\Delta T = \frac{1}{2}(T_1 + T_2)$. Since for the present case M^* is a function of z only, Eqs. (9.11) give an equivalent loading $p^* = 0$. The thermal stress resultant is, from Eq. (9.7),

$$M^* = \alpha E \int_{-t/2}^{t/2} \left[\tfrac{1}{2}(T_1 + T_2) + \tfrac{1}{2}(T_1 - T_2)\frac{z}{t/2} \right] z\, dz = \frac{\alpha E t^2}{12}(T_1 - T_2) \qquad (9.20)$$

For the plate of Fig. 9.1*b*, the bending stress at the faces is

$$\sigma_r = \sigma_\theta = -\frac{\alpha E(\Delta T)}{1 - \nu} = -\frac{\alpha E}{2(1 - \nu)}(T_1 - T_2) \qquad (b)$$

Also, there is no deflection of this plate ($w = 0$).

In the case of the plate in Fig. 9.1*c*, introducing Eq. (9.20) into Eq. (2.27) and setting $b = 0$, $M_1 = 0$, and $M_2 = M^*/(1 - \nu)$, we have

$$w = \frac{\alpha(a^2 - r^2)}{2t}(T_1 - T_2)$$

$$M_r = M_\theta = \frac{M^*}{1 - \nu} = \frac{\alpha E t^2}{12(1 - \nu)}(T_1 - T_2) \qquad (9.21a,b)$$

The bending stresses at the faces of the plate, upon introduction of Eqs. (9.20) and (9.21*b*) into (9.8), are as follows:

$$\sigma_r = \sigma_\theta = \frac{\alpha E}{2(1 - \nu)}(T_1 - T_2) \qquad (c)$$

The resultant stress in the original plate is obtained by addition of the stresses given by Eqs. (*b*) and (*c*):

$$\sigma_r = \sigma_\theta = 0$$

This is the result expected. Equation (9.21*a*) leads to the relationship

$$w_{\max} = \frac{\alpha a^2}{2t}(T_1 - T_2) \qquad (9.22)$$

for the maximum deflection of the original plate.

9.8 AXISYMMETRICALLY HEATED CIRCULAR PLATES

Consider the bending of an axisymmetrically heated circular plate having simply supported or clamped edge conditions and subjected to temperatures varying with the r and z coordinates, $T(r, z)$, such that the equivalent transverse load

$p^* = p^*(r)$. The expressions for moments, Eqs. (2.9), for the situation described become

$$M_r = -D\left(\frac{d^2w}{dr^2} + \frac{\nu}{r}\frac{dw}{dr}\right) - \frac{M^*}{1-\nu}$$
$$M_\theta = -D\left(\frac{1}{r}\frac{dw}{dr} + \nu\frac{d^2w}{dr^2}\right) - \frac{M^*}{1-\nu} \tag{9.23}$$

The plate deflection must satisfy the differential equation (2.10*b*)

$$\frac{1}{r}\frac{d}{dr}\left\{r\frac{d}{dr}\left[\frac{1}{r}\frac{d}{dr}\left(r\frac{dw}{dr}\right)\right]\right\} = \frac{p^*}{D} \tag{9.24}$$

with the *boundary conditions*

$$w = 0 \qquad \frac{dw}{dr} = 0$$
$$w = 0 \qquad D\left(\frac{d^2w}{dr^2} + \frac{\nu}{r}\frac{dw}{dr}\right) + \frac{M^*}{1-\nu} = 0 \tag{9.25}$$

Equations (9.25) refer to *clamped* and *simply supported edges*, respectively. They are derived by employing a procedure identical with that described in Sec. 1.7 together with Eqs. (9.23).

The general solution of Eq. (9.24), referring to Eq. (2.13), may be expressed

$$w = c_1 \ln\frac{r}{b} + c_2 r^2 \ln\frac{r}{b} + c_3(r^2 - b^2) + c_4 + w_p \tag{9.26}$$

where the c's are determined from the boundary conditions. Note that inner radius $b = 0$ for solid plates. The outer radius of the plate is designated a (Fig. 2.5). The particular solution, denoted by w_p in Eq. (9.26), can be determined by four successive integrations of Eq. (9.24):

$$w_p = \int_b^r \frac{1}{r_4}\int_b^{r_4} r_3 \int_b^{r_3} \frac{1}{r_2}\int_b^{r_2} \frac{r_1 p^*}{D}\, dr_1\, dr_2\, dr_3\, dr_4 \tag{9.27}$$

In the case of an *annular plate* under a temperature distribution such that $p^* =$ constant, Eq. (9.27) appears as follows (Prob. 9.9):

$$w_p = \frac{p^*}{64D}\left[r^4 - 5b^4 - 4b^2(b^2 + 2r^2)\ln\frac{r}{b} + 4b^2r^2\right] \tag{9.28}$$

When $b = 0$, Eq. (9.28) yields the particular solution of a solid plate.

For the sake of simplicity in the representation of the results, let

$$w_p^{(n)} = \frac{d^n w_p}{dr^n} \qquad (r = a) \tag{9.29}$$

The values of constants are listed[26] in terms of the foregoing notation, (Prob. 9.6) in Table 9.3. In this table, M_a^* represents the value of the thermal

Table 9.3

Geometry	Constants in Eq. (9.26)
A. Solid plate (clamped at edge $r = a$)	$c_1 = c_2 = 0$ $c_3 = -\frac{1}{2a} w_p^{(1)}, \quad c_4 = \frac{a}{2} w_p^{(1)} - w_p^{(0)}$
B. Solid plate (supported at edge $r = a$)	$c_1 = c_2 = 0$ $c_3 = -\frac{1}{2(1+\nu)}\left[\frac{M_a^*}{(1-\nu)D} + w_p^{(2)} + \frac{\nu}{a} w_p^{(1)}\right]$ $c_4 = \frac{a^2}{2(1+\nu)}\left[\frac{M_a^*}{(1-\nu)D} + w_p^{(2)} + \frac{\nu}{a} w_p^{(1)}\right] - w_p^{(0)}$
C. Annular plate (clamped at inner $r = b$ and outer $r = a$ edges)	$c_1 = \frac{b^2 a w_p^{(1)}[a^2 - b^2 - 2a^2 \ln(a/b)] + 4a^2 b^2 w_p^{(0)} \ln(a/b)}{(a^2 - b^2)^2 - 4a^2 b^2 \ln^2(a/b)}$ $c_2 = \frac{-a w_p^{(1)}[a^2 - b^2 - 2b^2 \ln(a/b)] + 2(a^2 - b^2) w_p^{(0)}}{(a^2 - b^2)^2 - 4a^2 b^2 \ln^2(a/b)}$ $c_3 = \frac{a w_p^{(1)}[(a^2 - b^2)\ln(a/b)] + w_p^{(0)}[b^2 - a^2 - 2a^2 \ln(a/b)]}{(a^2 - b^2)^2 - 4a^2 b^2 \ln^2(a/b)}$ $c_4 = 0$

stress resultant at $r = a$. Given a temperature distribution $T(r, z)$, the deflection and moment in a solid or annular plate with simply supported or built-in edges can thus be obtained from Eqs. (9.26) and (9.23). Plates with other edge conditions can similarly be treated.

Example 9.2 Redo the problem discussed in Example 9.1 using the relationships developed in this section.

SOLUTION As $p^* = 0$, we have, from Eqs. (9.29), $w_p^{(n)} = 0$ for all n. Referring to row B of Table 9.3,

$$c_1 = c_2 = 0 \qquad -c_3 = \frac{c_4}{a^2} = \frac{M_a^*}{2(1-\nu^2)D} = \frac{6M^*}{Et^3}$$

where M_a^* is replaced by M^*, because the thermal loading does not vary with radius. Equation (9.26) then yields

$$w = \frac{6M^*}{Et^3}(a^2 - r^2)$$

This expression with Eq. (9.20) results in a solution of w, identical to that found in Example 9.1. Upon inserting Eqs. (9.21*a*) and (9.20) into Eqs. (9.23), we find that $M_r = M_\theta = 0$. Applying Eqs. (9.8), we again obtain $\sigma_r = \sigma_\theta = 0$.

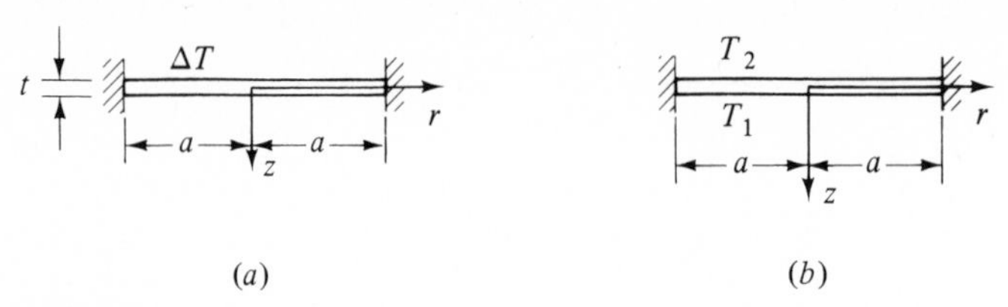

Figure 9.2

Example 9.3 Determine the deflections and stresses in a circular plate with clamped edge for the following cases: (*a*) the plate is stress-free at temperature T_0 and uniformly heated to a temperature T_1 (Fig. 9.2*a*); (*b*) the plate experiences a steady-state, linear temperature variation between its faces (Fig. 9.2*b*). Assume that the plate is free of stress at 0°C and that $T_1 > T_2$.

SOLUTION (*a*) The uniform temperature differential is $\Delta T = T_1 - T_0$. We have $p^* = 0$ and thus $w_p^{(n)} = 0$ for all n. Row A of Table 9.3 then gives

$$c_1 = c_2 = c_3 = c_4 = 0$$

Hence, Eq. (9.26) yields

$$w = 0 \tag{a}$$

The thermal stress resultant is

$$M^* = \alpha E(\Delta T) \int_{-t/2}^{t/2} z\,dz = 0$$

Expressions (9.23) then lead to $M_r = M_\theta = 0$. The magnitude of the bending stresses at the plate surfaces, from Eqs. (9.8), is therefore

$$\sigma_r = \sigma_\theta = \frac{E\alpha(\Delta T)}{1 - \nu} \tag{9.30}$$

(*b*) We now have $\Delta T = (T_1 - T_2)/2$ (Fig. 9.2*b*). The displacement $w = 0$ as before. The thermal stress resultant M^* is defined by Eq. (9.20) and the bending moments are found from Eqs. (9.23):

$$M_r = M_\theta = -\frac{\alpha E t^2}{12(1 - \nu)}(T_1 - T_2)$$

By means of Eqs. (9.8), we have

$$\sigma_r = \sigma_\theta = \frac{E\alpha(\Delta T)}{1 - \nu} = \frac{E\alpha}{2(1 - \nu)}(T_1 - T_2) \tag{9.31}$$

for the magnitude of the bending stress at the upper and the lower faces of the plate.

It is noted that inasmuch as $p^* = 0$, solutions (9.30) and (9.31) may readily be found without any calculation by the use of analogy between heated and isothermal clamped plates.

PROBLEMS

Secs. 9.1 to 9.7

9.1 An aluminum airplane wing panel is assumed to be stress-free at 20°C. After a time at cruising speed, the temperatures on the heated (upper) and the cooled (lower) surfaces are 54 and 24°C, respectively. Compute the thermal stress resultants for a linear temperature transition. Assume that the panel edges are clamped. Use $E = 70$ GPa, $\alpha = 23.2 \times 10^{-6}$ per °C, $\nu = 0.3$, and $t = 6$ mm.

9.2 Derive expressions for the bending moments in a simply supported square plate of sides a with temperature distribution linearly varying over the thickness only.

9.3 Determine the maximum deflection w and the maximum moment M_x in a simply supported square plate of sides a, subject to a temperature field Az^3, where A is a constant. Retain only the first term of the series solutions.

9.4 Consider a built-in edge, square plate of sides a, under an equivalent transverse thermal loading $p^*(x, y)$ (Fig. 3.13). Employ the analogy between isothermal and thermal problems and apply the Ritz method, taking $m = n = 1$ to obtain an expression for the deflection surface w.

9.5 A square plate is simply supported at $x = 0$ and $x = a$. The remaining edges at $y = \pm a/2$ are clamped (Fig. 3.9a). The plate is subjected to a temperature distribution Ay^2z^3, where A is a constant. Derive an approach to the evaluation of the center displacement w. Use the analogy with the isothermal solutions given in Secs. 3.6 and 3.7.

Sec. 9.8

9.6 Verify the results given in Cases A and B of Table 9.3.

9.7 Determine the deflection w and the stress σ_r at the center of a simply supported circular plate experiencing a temperature field Azr^3, where A is a constant.

9.8 Redo Prob. 9.7 for a plate clamped on all edges.

9.9 Determine the displacement w at $r = 2a$ of a hollow plate having inner ($r = a$) and outer ($r = 3a$) edges built in. Assume a temperature distribution Bz^5r^2, where B is a constant.

CHAPTER

TEN

MEMBRANE STRESSES IN SHELLS

10.1 GENERAL BEHAVIOR AND COMMON THEORIES OF SHELLS

Until now, our concern was with the analysis of thin flat plates. We now extend the discussion to *curved surface structures* termed *thin shells*. Examples of shells include pressure vessels, airplane wings, pipes, the exterior of rockets, missiles, automobile tires, incandescent lamps, caps, roof domes, factory or car sheds, and a variety of containers. Each of these has walls that are curved. Inasmuch as a *curved plate* can be viewed as a portion of a shell, the general equations for thin shells are also applicable to curved plates. We shall limit our treatment (except Secs. 11.11 and 11.12) to shells of constant thickness, small in comparison with the other two dimensions. As in the treatment of plates, the plane bisecting the shell thickness is called the *midsurface*. To describe the shape of a shell, we need only know the geometry of the midsurface and the thickness of the shell at each point. Shells of technical significance are often defined as *thin* when the ratio of thickness t to radius curvature r is equal to or less than 1/20. For thin shells of practical importance, this ratio may be 1/1000 or smaller.

The analysis of shell structures often embraces two distinct, commonly applied theories. The first of these, the *membrane theory*, usually applies to a rather large part of the entire shell. A membrane, either flat or curved, is identified as a body of the same shape as a plate or shell, but incapable of conveying moments or shear forces. In other words, a membrane is a two-dimensional analog of a flexible string with the exception that it can resist compression. The second, the *bending theory* or *general theory*, includes the effects of bending. Thus it permits the treatment of discontinuities in the stress distribution taking place in a limited region in the vicinity of a load or structural discontinuity. However, information relative to shell *membrane stresses* is usually of much greater practical significance than the knowledge of the *bending stresses*. The former are also far simpler to calculate. For thin shells having no abrupt changes in thickness, slope, or curvature, the meridional stresses are uniform throughout the wall thickness. The bending theory generally comprises a membrane solution, corrected in those areas in which discontinuity effects are pronounced. The goal is thus not the improvement of the membrane solution, but rather the analysis of stresses and strains owing to the edge forces or concentrated loadings, which cannot be accomplished by the membrane theory only.

It is important to note that membrane forces are independent of bending and are completely defined by the conditions of static equilibrium. As no material properties are used in the derivation of these forces, the *membrane theory applies to all shells* made of any material (e.g., metal, fabric, reinforced concrete, sandwich shell, soap film, gridwork shell, and plywood). Various relationships developed for bending theory in Chaps. 11 to 13 are, however, restricted to homogeneous, elastic, isotropic shells.

The basic kinematic assumptions associated with the deformation of a thin shell as used in small-deflection analysis are as follows:

(1) The ratio of shell thickness to radius of curvature of the midsurface is small in comparison with unity.
(2) Deflections are small compared with shell thickness.
(3) Plane sections through a shell taken normal to the midsurface, remain plane and become normal to the *deformed* midsurface after the shell is subjected to bending. This hypothesis implies that strains γ_{xz} and γ_{yz} are negligible. Normal strain, ε_z, owing to transverse loading may also be omitted.
(4) The z-directed normal stress, σ_z, is neglible.

In this chapter, we consider only shells and loadings for which bending stresses are negligibly small. Applications are also presented of the governing equations of the membrane theory to specific practical cases.

10.2 LOAD RESISTANCE ACTION OF A SHELL

The common deformational behavior of beams, thin plates, and thin shells is illustrated by the unified set of assumptions (Secs. 1.2 and 10.1). The load-carrying mechanisms of these members do not resemble one another, however. That the load-resisting action of a shell differs from that of other structural forms is underscored by noting the extraordinary capacity of an eggshell or an

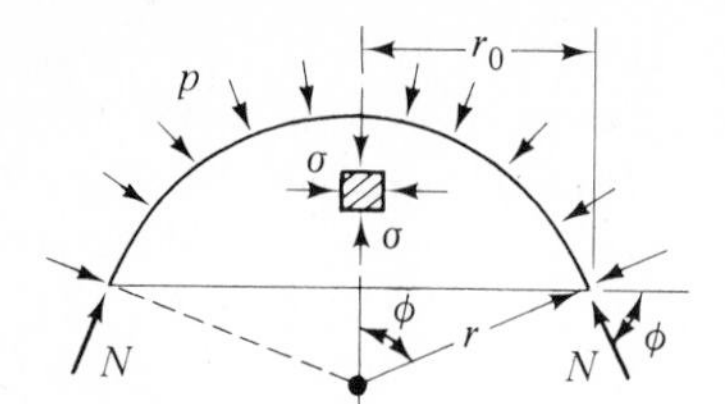

Figure 10.1

electric light bulb to withstand normal forces, this despite their thinness and fragility. (A hen's egg has a radius along the axis of revolution $r = 20$ mm and a thickness $t = 0.4$ mm; thus $t/r = 1/50$.) The above behavior contrasts markedly with similar materials in plate or beam configurations under lateral loading. A shell, being curved, can develop in-plane forces (thrusts) to form the primary resistance action in addition to those forces and moments existing in a plate or beam.

To describe the phenomenon, consider a part of a spherical shell of radius r and thickness t, subjected to a uniform pressure of intensity p (Fig. 10.1). The condition that the sum of vertical forces be zero is expressed: $2\pi r_0 N \sin\phi - p\pi r_0^2 = 0$ or

$$N = \frac{pr_0}{2\sin\phi} = \frac{pr}{2}$$

in which N is the in-plane force per unit of circumference. This relationship is valid anywhere in the shell, as N is seen not to vary with ϕ. We note that, in contrast to the case of plates, load is sustained by the midsurface.

It is next demonstrated that the bending stresses play an insignificant role in the load-resisting action. Based upon the symmetry of the shell and the loading, the state of stress at a point, represented by an infinitesimal element, is as shown in Fig. 10.1. The compressive *direct stress* has the form

$$\sigma = -\frac{pr}{2t} \tag{a}$$

The stress normal to the midsurface is negligible and thus the direct strain, from Hooke's law, is

$$\varepsilon = \frac{1}{E}(\sigma - \nu\sigma) = -(1-\nu)\frac{pr}{2Et} \tag{b}$$

Associated with this strain, the reduced circumference is $2\pi r' = 2\pi(r + r\varepsilon)$, or $r' = r(1+\varepsilon)$. The *variation* in curvature χ (chi) is thus

$$\chi = \frac{1}{r'} - \frac{1}{r} = \frac{1}{r}\left(\frac{1}{1+\varepsilon} - 1\right) = -\frac{\varepsilon}{r}\left(\frac{1}{r+\varepsilon}\right) = -\frac{\varepsilon}{r}(1 - \varepsilon + \varepsilon^2 - \cdots)$$

Neglecting higher-order terms owing to their small magnitude, and introducing Eq. (*b*), the above expression becomes

$$\chi = -\frac{\varepsilon}{r} = \frac{(1-\nu)p}{2Et} \tag{c}$$

A relationship for the shell-bending moment is derived from the plate formulas. For the spherical shell under consideration we have $\chi = \chi_x = \chi_y$. Thus, Eqs. (1.10) and (*c*) lead to

$$M = -D(\chi_x + \nu\chi_y) = -D(1-\nu^2)\frac{p}{2Et} = -\frac{pt^2}{24}$$

Hence, the bending stress is given by

$$\sigma_b = \frac{6M}{t^2} = -\frac{p}{4} \tag{d}$$

The ratio of the direct stress to the bending stress is

$$\frac{\sigma}{\sigma_b} = \frac{2r}{t} \tag{e}$$

It is observed that the direct or membrane stress is very much larger than the bending stress as $(t/2r) \ll 1$. We are led to conclude therefore that the *applied load is resisted predominantly by the in-plane stressing* of the shell.

The foregoing discussion relates to the simplest shell configuration. However, the conclusions drawn with respect to the basic action apply to any geometry and loading at locations away from the edges or points of application of concentrated load. Should there be asymmetries in load or shape, shearing stresses as well as the membrane and bending stresses will be present. Treated in greater detail in the sections which follow is the state of membrane stress in shells of various geometry.

It is noted that thin shells may be vulnerable to local *buckling* under compression stresses. The problem of shell *instability* is treated in Chap. 13. For reference purposes, we now introduce the following relation[30] useful in the prediction of the critical stress σ_{cr} for *local buckling* of a thin shell:

$$\sigma_{cr} = k\frac{Et}{a} \tag{f}$$

Here E is the modulus of elasticity, and the value of constant k can be determined by rational analysis, such as described in Secs. 13.7 and 13.8, and incorporated with an empirical factor in order to relate theoretical values to actual test data. The value $k \approx 0.25$ is often used.

Consider, for example, a concrete or masonry shell for which $E = 20$ GPa and $t/a = 1/500$. We have

$$\sigma_{cr} = 0.25 \times 20 \times 10^9\left(\tfrac{1}{500}\right) = 10 \text{ MPa}$$

This, compared with the ultimate strength of the concrete of 21 MPa, demonstrates the importance of buckling analysis in predicting allowable load.

Shell structures should thus be checked for the possibility of buckling in compressed areas as well as for yielding or fracture in those sections subjected to tensile forces (Example 10.1*b*).

10.3 GEOMETRY OF SHELLS OF REVOLUTION

Consider a particular type of shell described by a *surface of revolution* (Fig. 10.2). Examples include the sphere, cylinder, and cone. The midsurface of a shell of revolution is generated by rotation of a so-called *meridian curve* about an axis lying in the plane of the curve. Figure 10.2 shows that a point on the shell is conveniently located by coordinates θ, ϕ, r_0, and that elemental surface $ABCD$ is defined by two *meridians* and two *parallel circles* or *parallels*.

The planes associated with the principal radii of curvature r_1 and r_2 at any point on the midsurface of the shell are the *meridian plane* and the *parallel plane* at the point in question, respectively. The radii of curvature r_1 and r_2 are thus related to sides CD and AC. The principal radius r_2 generates the shell surface in the direction perpendicular to the direction of the tangent to the meridian curve. The two radii r_0 and r_2 are related inasmuch as $r_0 = r_2 \sin \phi$ (Fig. 10.2). It follows that lengths of the curvilinear shell element are

$$L_{AC} = r_0\, d\theta = r_2 \sin \phi\, d\theta \qquad L_{CD} = r_1\, d\phi \tag{a}$$

It is assumed in the above description that the principal radii of curvature of the shell r_1 and r_2 are *known* constants. In the case of a radius of curvature which varies from point to point, the radii are computed applying the equation which defines the shell shape, along with various relationships of differential geometry of a surface,[31] as illustrated in Example 10.4.

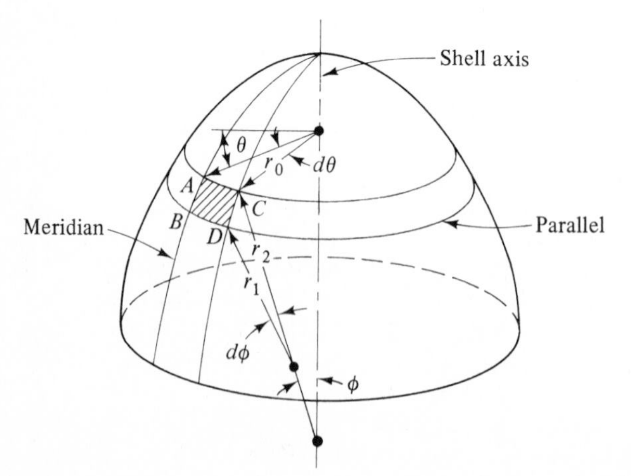

Figure 10.2

10.4 SYMMETRICALLY LOADED SHELLS OF REVOLUTION

In axisymmetrical problems involving shells of revolution, no shear forces exist and there are only two unknown membrane forces per unit length, N_θ and N_ϕ. The governing equations for these forces are derived from two equilibrium conditions. Figures 10.3*a* and 10.3*b* show two different views of the element *ABCD* cut from the shell of Fig. 10.2. Prescribed by the condition of symmetry, the membrane forces and the loading display no variation with θ. The externally applied forces per unit surface area are represented by the components p_y and p_z in the y and z directions, respectively.

Description of the z equilibrium requires that the z components of the loading as well as of the forces acting on each edge of the element be considered. The z-directed distributed load carried on the surface area of the element is

$$p_z r_0 r_1 \, d\theta \, d\phi$$

The force acting on the top edge of the element equals $N_\phi r_0 \, d\theta$. Neglecting higher terms, the force on the bottom edge is also $N_\phi r_0 \, d\theta$. The z-directed component at each edge is then $N_\phi r_0 \, d\theta \sin (d\phi/2)$. This force is nearly equal to $N_\phi r_0 \, d\theta \, d\phi/2$, yielding the following magnitude of the resultant for both edges:

$$N_\phi r_0 \, d\theta \, d\phi$$

As the cross-sectional area along each of the two sides of the element is $r_1 \, d\phi$, the force on these areas is $N_\theta r_1 \, d\phi$. The resultant in the direction of the radius of the parallel plane for both such forces is $N_\theta r_1 \, d\phi \cdot d\theta$, producing the following component in the z direction

$$N_\theta r_1 \, d\phi \, d\theta \sin \phi$$

For the forces considered above, from $\sum F_z = 0$, we have

$$N_\phi r_0 + N_\theta r_1 \sin \phi + p_z r_0 r_1 = 0 \tag{a}$$

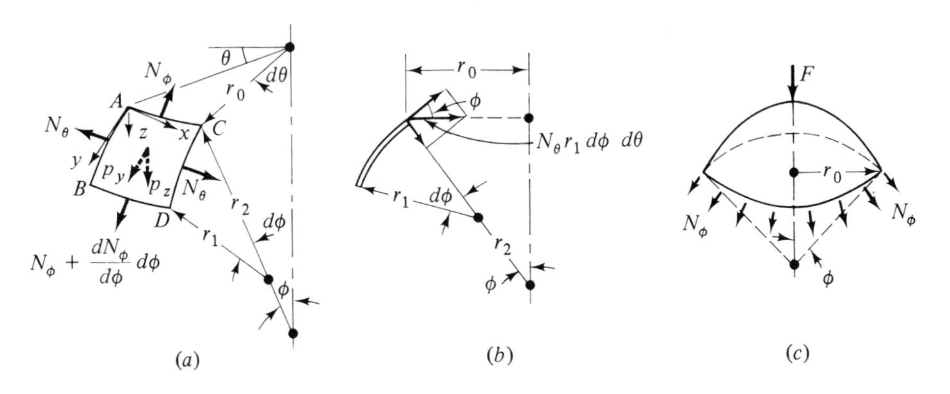

Figure 10.3

This expression may be converted to simpler form by dividing by $r_0 r_1$ and replacing r_0 by $r_2 \sin \phi$. By so doing, one of the basic relations for the axisymmetrically loaded shell is found as follows

$$\frac{N_\phi}{r_1} + \frac{N_\theta}{r_2} = -p_z \tag{10.1a}$$

The equilibrium of forces in the direction of the meridional tangent, that is, in the y direction is expressed

$$\frac{d}{d\phi}(N_\phi r_0)\, d\phi\, d\theta - N_\theta r_1\, d\phi\, d\theta \cos \phi + p_y r_1\, d\phi\, r_0\, d\theta = 0 \tag{b}$$

The first term represents the sum of normal forces acting on edges AC and BD, while the third term is the loading component. The second term of Eq. (b) is the component in the y direction of the radial resultant force $N_\theta r_1\, d\phi\, d\theta$ acting on faces AB and CD. Dividing Eq. (b) by $d\theta\, d\phi$ the equation of equilibrium of the y directed forces is now

$$\frac{d}{d\phi}(N_\phi r_0) - N_\theta r_1 \cos \phi = -p_y r_1 r_0 \tag{10.2}$$

It is noted that an equation of equilibrium which can be used instead of Eq. (10.2) follows readily by isolating part of the shell intercepted by angle ϕ (Fig. 10.3c). Here force F represents the *resultant of all external loading* applied to this free body. Recall that from conditions of symmetry, forces N_ϕ are constant around the edge. Equilibrium of the vertical forces is therefore described by $2\pi r_0 N_\phi \sin + F = 0$ and it follows that

$$N_\phi = -\frac{F}{2\pi r_0 \sin \phi} \tag{10.1b}$$

We verify below that Eq. (10.1b) is an alternative form of Eq. (10.2). Substitution of N_θ from Eq. (10.1a) into (10.2) and multiplication of the resulting expression by $\sin \phi$ leads to

$$\frac{d}{d\phi}(r_0 N_\phi) \sin \phi + r_0 N_\phi \cos \phi = -r_1 r_2 p_z \cos \phi \sin \phi - r_1 r_2 p_y \sin^2 \phi$$

Clearly, the left-hand side of the foregoing equation may be written

$$\frac{d}{d\phi}(r_0 N_\phi \sin \phi) = \frac{d}{d\phi}(r_2 N_\phi \sin^2 \phi)$$

and force N_ϕ determined through integration:

$$N_\phi = -\frac{1}{r_2 \sin^2 \phi}\left[\int r_1 r_2 (p_z \cos \phi + p_y \sin \phi) \sin \phi\, d\phi + c\right] \tag{c}$$

Here constant c represents the effects of the loads which may be applied to a shell element (Fig. 10.3c). Thus, introduction of $2\pi c = F$, $p_z = p_y = 0$, and $r_0 = r_2 \sin \phi$ into Eq. (c), results in the value of N_ϕ, defined by Eq. (10.1b).

Equations (10.1) are sufficient to determine the so-called *hoop force* N_θ and the *meridional force* N_ϕ from which the stresses are readily determined. Negative algebraic results indicate compressive stresses.

Because of their freedom of motion in the z direction, for the axisymmetrically loaded shells of revolution considered, strains are produced such as to assure consistency with the field of stress and compatibility with one another.[32] The action cited demonstrates the basic difference between the problem of a shell membrane and one of *plane stress*. In the latter case, a compatibility equation is required. However, it is clear that when a shell is subject to the action of concentrated surface loadings or is constrained at its boundaries, membrane theory cannot everywhere fulfill the conditions of deformation. The complete solution is obtained only by application of bending theory.

10.5 SOME TYPICAL CASES OF SHELLS OF REVOLUTION

The membrane stresses in any particular axisymmetrically loaded shell in the form of a surface of revolution may be determined from the governing expressions of equilibrium developed in the preceding section. Treated in the following paragraphs are several common structural members.

Spherical shell For spherical shells one can set the *mean* radius $a = r_1 = r_2$. Then Eqs. (10.1) appear in the form

$$N_\phi + N_\theta = -p_z a$$
$$N_\phi = -\frac{F}{2\pi a \sin^2 \phi} \tag{10.3}$$

The simplest case is that of a spherical shell subjected to *constant internal gas pressure p*, like a balloon. We now have $p = -p_z$, $\phi = 90°$, and $F = -\pi a^2 p$. Inasmuch as any section through the center results in the identical free body, $N_\phi = N_\theta = N$. The stress, from Eqs. (10.3), is therefore

$$\sigma = \frac{N}{t} = \frac{pa}{2t} \tag{10.4}$$

where t is the thickness of the shell. The expansion of the sphere, applying Hook's law, is then

$$\delta_s = \frac{a}{Et}(N - \nu N) = \frac{pa^2}{2Et}(1 - \nu) \tag{10.5}$$

Conical shell (Fig. 10.4) In this typical case, angle ϕ is a constant ($r_1 = \infty$) and can no longer serve as a coordinate on the meridian. Instead we introduce

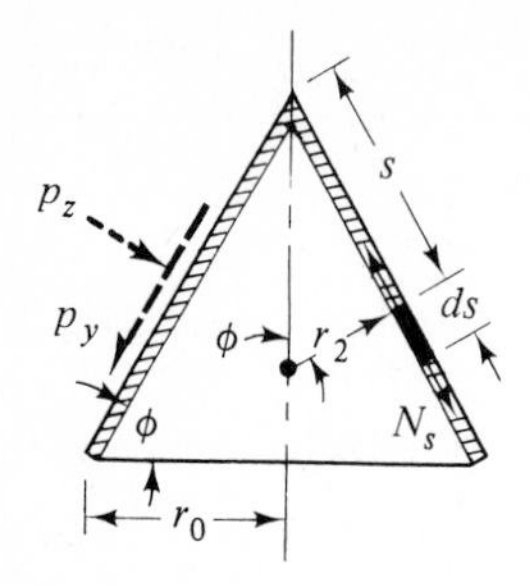

Figure 10.4

coordinate s, the distance of a point of the midsurface, *usually measured from the vertex*, along the generator. Accordingly, the length of a meridional element $ds = r_1\, d\phi$. Hence

$$\frac{d}{d\phi} = r_1 \frac{d}{ds} \tag{a}$$

Also,

$$r_0 = s \cos \phi \qquad r_2 = s \cot \phi \qquad N_\phi = N_s \tag{b}$$

These relationships, when introduced into Eqs. (10.2) and (10.1a), lead to

$$\begin{aligned} \frac{d}{ds}(N_s s) - N_\theta &= -p_y s \\ N_\theta = -p_z s \cot \phi &= -\frac{p_z r_0}{\sin \phi} \end{aligned} \tag{10.6a, b}$$

where r_0 is the mean radius at the base. Clearly, load components p_y and p_z are in the s and radial directions, respectively. The sum of Eqs. (10.6) yields

$$\frac{d}{ds}(N_s s) = -(p_y + p_z \cot \phi)s$$

The meridional force, upon integration of the above expression, is

$$N_s = -\frac{1}{s}\int (p_y + p_z \cot \phi)s\, ds \tag{10.7}$$

An alternate form of Eq. (10.6a) may be obtained from Eqs. (10.1b) and (b). The membrane forces are then

$$\begin{aligned} N_s &= -\frac{F}{2\pi r_0 \sin \phi} \\ N_\theta &= -\frac{p_z r_0}{\sin \phi} \end{aligned} \tag{10.8}$$

It is observed that given an external load distribution, hoop and meridional stresses can be computed *independently*.

Circular cylindrical shell To obtain the stress resultants in a circular cylindrical shell, one can begin with the cone equations, setting $\phi = \pi/2$, $p_z = p_r$ and mean radius $a = r_0 =$ constant. Hence Eqs. (10.8) become

$$N_s = N_x = -\frac{F}{2\pi a} \qquad N_\theta = -p_r a \tag{10.9}$$

in which x is measured in the axial direction.

For a closed-end cylindrical vessel under *constant internal pressure*, $p = -p_r$ and $F = -\pi a^2 p$. Equations (10.9) then yield the following axial and hoop stresses:

$$\sigma_x = \frac{pa}{2t} \qquad \sigma_\theta = \frac{pa}{t} \tag{10.10}$$

From Hooke's law, the extension of the radius of the cylinder under the action of the stresses given above is

$$\delta_c = \frac{a}{E}(\sigma_\theta - \nu\sigma_x) = \frac{pa^2}{2Et}(2 - \nu) \tag{10.11}$$

Solutions of various other cases of practical significance may be obtained by employing a procedure similar to that described in the foregoing paragraphs, as demonstrated in Examples 10.1 through 10.6.

Example 10.1 Consider a simply supported *covered market dome* of radius a and thickness t, carrying only its own weight p per unit area. (a) Determine the stresses, for a dome of half-spherical geometry (Fig. 10.5a). (b) Assume that the hemispherical dome is constructed of 70-mm-thick concrete of unit weight 23 kN/m³ and span $2a = 56$ m. Apply the maximum principal stress

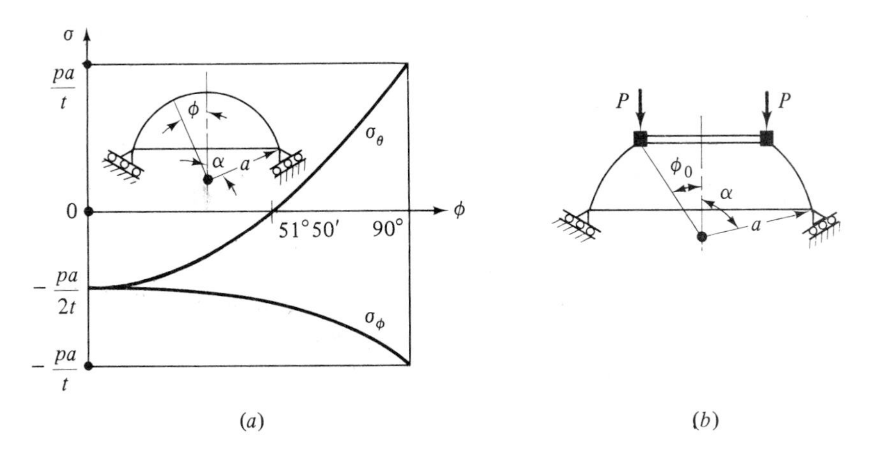

Figure 10.5

theory to evaluate the shell's ability to resist failure by fracture. The ultimate compressive strength or crushing strength of concrete $\sigma_{u''} = 21$ MPa, and $E = 20$ GPa. Also check the possibility of local buckling. (*c*) Determine the stresses in a dome which is a truncated half sphere (Fig. 10.5*b*).

SOLUTION The *components of the dome weight* are

$$p_x = 0 \qquad p_y = p \sin \phi \qquad p_z = p \cos \phi \tag{10.12}$$

(*a*) Referring to Fig. 10.5*a*, the weight of that part of the dome subtended by ϕ is

$$F = \int_0^{\phi} p \cdot 2\pi a \sin \phi \cdot a \, d\phi = 2\pi a^2 p(1 - \cos \phi) \tag{c}$$

Introduction into Eqs. (10.3) of p_z and F given by Eqs. (10.12) and (*c*), and division of the results by t yields the membrane stresses:

$$\begin{aligned} \sigma_\phi &= -\frac{ap(1 - \cos \phi)}{t \sin^2 \phi} = -\frac{ap}{t(1 + \cos \phi)} \\ \sigma_\theta &= -\frac{ap}{t}\left(\cos \phi - \frac{1}{1 + \cos \phi}\right) \end{aligned} \tag{10.13}$$

These stresses are plotted for a *hemisphere* in Fig. 10.5*a*. Clearly, σ_ϕ is always compressive; its value increases with ϕ from $-pa/2t$ at the crown to $-pa/t$ at the edge. The sign of σ_θ on the other hand, changes with the value of ϕ. The second of the above equations yields $\sigma_\theta = 0$ for $\phi = 51°50'$. When ϕ is smaller than this value, σ_θ is compressive. For $\phi > 51°50'$, σ_θ is tensile, as shown in the figure.

(*b*) The maximum compressive stress in the dome is $\sigma_\phi = pa/t = (0.023t)a/t = 0.023 \times 28 = 0.644$ MPa. Note that no failure occurs as $|\sigma_\phi| < |\sigma_{u''}|$. Clearly, even for large domes *the stress level due to dead weight is far from the limit stress of the material, at least in compression*. Note also that concrete is weak in tension and a different conclusion may emerge from consideration of failure owing to direct tensile forces. If the tensile strength of the material is smaller than 0.644 MPa, an assessment of tensile reinforcement will be required to assure satisfactory design.

Upon application of Eq. (*f*) of Sec. 10.2, the stress level at which local buckling occurs in the dome is found to be

$$\sigma_{cr} = 0.25(20 \times 10^9)(\tfrac{7}{2800}) = 12.5 \text{ MPa}$$

It is observed that there is no possibility of local buckling, as $\sigma_\phi < \sigma_{cr}$.

(*c*) Most domes are not closed at the upper portion and have a lantern, a small tower for lighting, and ventilation. In this case, a reinforcing ring is used to support the upper structure as shown in Fig. 10.5*b*. Let $2\phi_0$ be the angle corresponding to the opening and P be the vertical load per unit length acting on the reinforcement ring. The resultant of the total load on that portion of the dome subtended by the angle ϕ, Eq. (c), is then

$$F = 2\pi \int_{\phi_0}^{\phi} a^2 p \sin\phi \, d\phi + P \cdot 2\pi a \sin\phi_0$$

$$= 2\pi a^2 p(\cos\phi_0 - \cos\phi) + 2\pi P \sin\phi_0$$

Using Eqs. (10.3), we obtain

$$\begin{aligned} \sigma_\phi &= -\frac{ap}{t}\frac{\cos\phi_0 - \cos\phi}{\sin^2\phi} - \frac{P\sin\phi_0}{t\sin^2\phi} \\ \sigma_\theta &= \frac{ap}{t}\left(\frac{\cos\phi_0 - \cos\phi}{\sin^2\phi} - \cos\phi\right) + \frac{P\sin\phi_0}{t\sin^2\phi} \end{aligned} \tag{10.14}$$

for the hoop and the meridional stresses.

Note that the circumferential strain in the dome under the action of membrane stresses may be computed from Hooke's law

$$\varepsilon_\theta = \frac{1}{E}(\sigma_\theta - \nu\sigma_\phi)$$

This strain contributes to a change $\varepsilon_\theta \cdot a \sin\alpha$ in edge radius. The simple support shown in the figure, free to move as the shell deforms under loading, ensures that no bending is produced in the neighborhood of the edge.

Example 10.2 Consider a planetarium dome that may be approximated as an edge-supported *truncated cone*. Derive expressions for the hoop and meridional forces for two conditions of loading: (a) the shell carries its own weight p per unit area (Fig. 10.6a); (b) the shell carries a snow load assumed to be uniformly distributed over the plan (Fig. 10.6b).

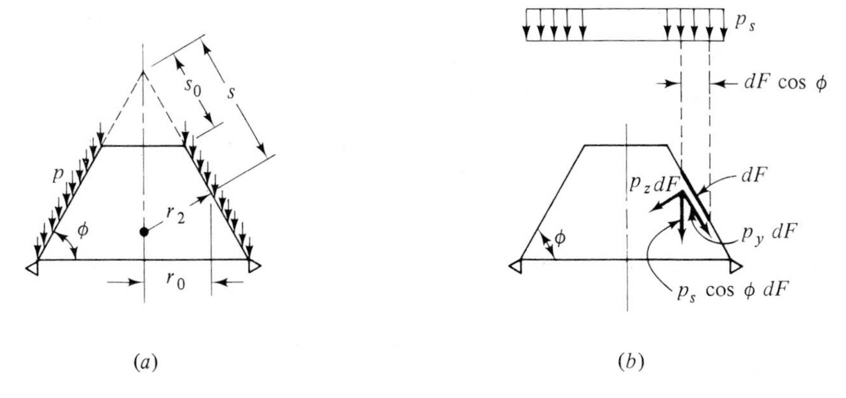

Figure 10.6

SOLUTION (*a*) Referring to Fig. 10.6*a*,

$$r_2 = s \cot \phi \qquad r_0 = s \cos \phi \qquad p_z = p \cos \phi \qquad (d)$$

The weight of that part of the cone defined by $s - s_0$ is determined from

$$F = \int_{s_0}^{s} p \cdot 2\pi r_2 \sin \phi \cdot ds = 2\pi p \int_{s_0}^{s} s \cot \phi \sin \phi \, ds$$

or

$$F = \pi p \cos \phi (s^2 - s_0^2) + c$$

As no force acts at the top edge, $c = 0$. Now the s- and θ-directed stress resultants can readily be obtained. Substituting F and Eqs. (*d*) into Eqs. (10.8),

$$N_s = -\frac{p}{2s} \frac{s^2 - s_0^2}{\sin \theta}$$
$$N_\theta = -ps \frac{\cos^2 \phi}{\sin \phi} \qquad (10.15)$$

(*b*) To analyze the components of the *snow load* p_s, we use a sketch of the forces acting on a midsurface element dF (Fig. 10.6*b*). Referring to this figure, one has

$$p_x = 0 \qquad p_y = p_s \sin \phi \cos \phi \qquad p_z = p_2 \cos^2 \phi \qquad (10.16)$$

From Eqs. (10.8), (*d*), and (10.16), we have

$$N_\theta = -p_s s \frac{\cos^3 \phi}{\sin \phi} \qquad (10.17a)$$

Similarly, Eqs. (10.7), (*d*), and (10.16) yield

$$N_s = -\frac{1}{s} \int (p_s \sin \phi \cos \phi + p_s \cos^2 \phi \cot \phi) s \, ds + \frac{c}{s}$$
$$= -\frac{p_s s}{2} \cot \phi + \frac{c}{s}$$

The condition that $N_s = 0$ at $s = s_0$ leads to $c = \frac{1}{2} p_s s_0^2 \cot \phi$. Hence

$$N_s = -\frac{P_s}{2s}(s^2 - s_0^2) \cot \phi \qquad (10.17b)$$

Upon dividing Eqs. (10.17) by t, the membrane stresses are obtained.

It is interesting to note that the *three typical loads* (weight per unit surface area, snow load per plan area, and wind load per surface area) are of the *same order*. For ordinary structures these might approximate 1500 to 2000 Pa.

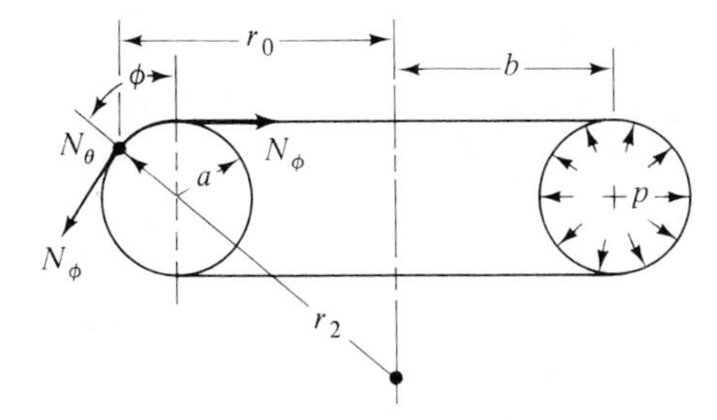

Figure 10.7

Example 10.3 A shell in the shape of a *torus* or *doughnut* of circular cross section is subjected to internal pressure p (Fig. 10.7). Determine membrane stresses σ_ϕ and σ_θ.

SOLUTION Consider the portion of the shell defined by ϕ. The vertical equilibrium of forces leads to

$$2\pi r_0 \cdot N_\phi \sin\phi = \pi p(r_0^2 - b^2)$$

or

$$N_\phi = \frac{p(r_0^2 - b^2)}{2r_0 \sin\phi} = \frac{pa(r_0 + b)}{2r_0}$$

Introducing N_ϕ into Eq. (10.1a), setting $p_z = -p$ and $r_1 = a$, we obtain

$$N_\theta = \frac{pr_2(r_0 - b)}{2r_0} = \frac{pa}{2}$$

The stresses are then

$$\sigma_\phi = \frac{pa(r_0 + b)}{2r_0 t} \qquad \sigma_\theta = \frac{pa}{2t} \tag{10.18}$$

It is noted that σ_θ is constant throughout the shell from the condition of symmetry.

Example 10.4 Figure 10.8 represents the end enclosure of a cylindrical vessel in the form of a half *ellipsoid* of semiaxes a and b. Determine the membrane stresses resulting from an internal steam pressure p.

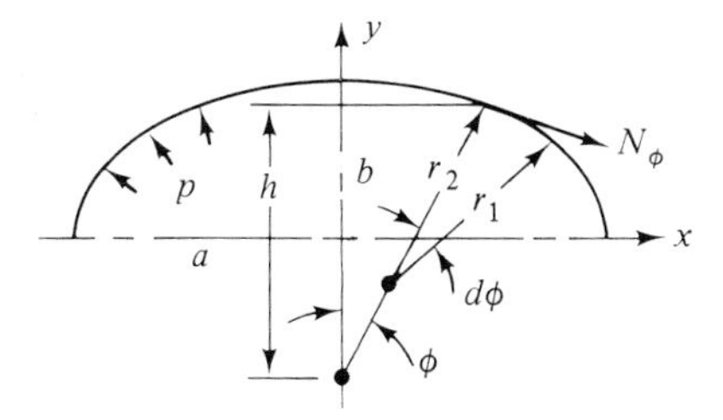

Figure 10.8

SOLUTION Expressions for the principal radii of curvature r_1 and r_2 will be required. The equation of the ellipse $b^2x^2 + a^2y^2 = a^2b^2$ leads to $y = \pm b\sqrt{a^2 - x^2}/a$. The magnitude of the derivatives of this expression are

$$y' = \frac{bx}{a\sqrt{a^2 - x^2}} = \frac{b^2x}{a^2y} \qquad y'' = \frac{b^4}{a^2y^3} \tag{e}$$

Referring to the figure,

$$\tan\phi = y' = \frac{x}{h} \qquad r_2 = \sqrt{h^2 + x^2} \tag{f}$$

From the first of Eqs. (*e*) and (*f*), $h = a\sqrt{a^2 - x^2}/b$, which, when substituted into the second of Eqs. (*f*), yields r_2. Introduction of Eqs. (*e*) into the familiar expression for the curvature, $[1 + (y')^2]^{3/2}/y''$, gives the radius r_1. Thus,

$$r_1 = \frac{(a^4y^2 + b^4x^2)^{3/2}}{a^4b^4} \qquad r_2 = \frac{(a^4y^2 + b^4x^2)^{1/2}}{b^2} \tag{10.19}$$

The load resultant is represented by $F = \pi p r_2^2 \sin^2\phi$. The membrane forces can then be determined from Eqs. (10.1) in terms of the principal curvatures. It follows that

$$\sigma_\phi = \frac{pr_2}{2t} \qquad \sigma_\theta = \frac{p}{t}\left(r_2 - \frac{r_2^2}{2r_1}\right) \tag{10.20}$$

At *the crown* (top of the shell) $r_1 = r_2 = a^2/b$ and Eqs. (10.20) reduce to

$$\sigma_\phi = \sigma_\theta = \frac{pa^2}{2bt}$$

At *the equator* (base of the shell) $r_1 = b^2/a$ and $r_2 = a$ and Eqs. (10.20) appear as

$$\sigma_\phi = \frac{pa}{2t} \qquad \sigma_\theta = \frac{pa}{t}\left(1 - \frac{a^2}{2b^2}\right)$$

It is observed that the hoop stress σ_θ becomes compressive for $a^2 > 2b^2$. Clearly, the meridian stresses σ_ϕ are always tensile. A ratio $a/b = 1$, the case of a sphere, yields the lowest stress.

Example 10.5 Analyze the membrane stresses in a thin metal *container* of *conical shape*, supported from the top. Consider two specific cases: (*a*) the shell is subjected to an internal pressure p; (*b*) the shell is filled with a liquid of specific weight γ (Fig. 10.9).

SOLUTION (*a*) For this case, $p_z = -p$, $\phi = (\pi/2) + \alpha$ and $F = -p\pi r_0^2$. Expressions (10.8), after dividing by the thickness t, then become

$$\sigma_s = \frac{pr_0}{2t\cos\alpha} \qquad \sigma_\theta = \frac{pr_0}{t\cos\alpha} \tag{10.21}$$

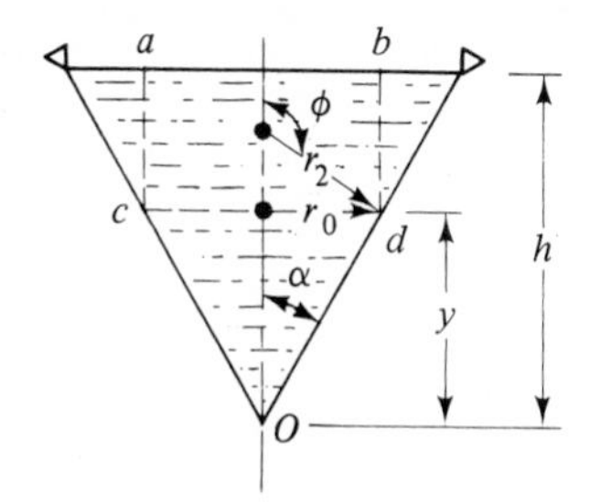

Figure 10.9

(*b*) According to the familiar laws of hydrostatics, the pressure at any point in the shell equals *the weight of a column of unit cross-sectional area of the liquid* at that point. At any arbitrary level y, the *pressure* is therefore:

$$p = -p_z = \gamma(h - y) \tag{g}$$

Employing $r_0 = y \tan \alpha$, the second of Eqs. (10.8), after division by t, becomes

$$\sigma_\theta = \frac{\gamma(h - y)y}{t} \frac{\tan \alpha}{\cos \alpha} \tag{10.22a}$$

Differentiating with respect to y and equating to zero reveals that the maximum value of the above stress occurs at $y = h/2$ and is given by

$$\sigma_{\theta,\,\max} = \frac{\gamma h^2}{4t} \frac{\tan \alpha}{\cos \alpha}$$

The load is equal to the weight of the liquid of volume *acOdb*. That is,

$$F = -\pi\gamma y^2(h - y + \tfrac{1}{3}y) \tan^2 \alpha$$

Introducing this value into the first of Eqs. (10.8), and dividing the resulting expression by t leads to

$$\sigma_s = \frac{\gamma(h - 2y/3)}{2t} \frac{\tan \alpha}{\cos \alpha} \tag{10.22b}$$

The maximum value of this stress, $\sigma_{s,\,\max} = 3h^2\gamma \tan \alpha/16t \cos \alpha$, occurs at $y = 3h/4$.

Example 10.6 Determine the membrane forces in a *spherical storage tank* filled with liquid of specific weight γ, and supported on a cylindrical pipe (Fig. 10.10).

SOLUTION The loading is expressed

$$p = -p_z = \gamma a(1 - \cos \phi)$$

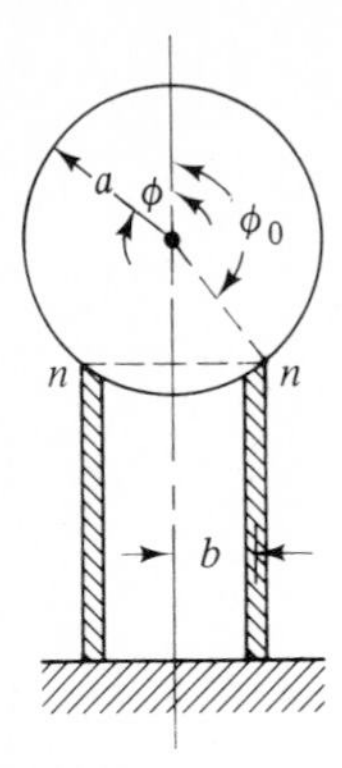

Figure 10.10

Owing to this pressure, the resultant force F for the portion intercepted by ϕ is:

$$F = -2\pi a^2 \int_0^\phi \gamma a(1 - \cos\phi)\sin\phi\cos\phi\,d\phi$$

$$= -2\pi a^3\gamma[\tfrac{1}{6} - \tfrac{1}{2}\cos^2\phi(1 - \tfrac{2}{3}\cos\phi)]$$

Inserting the above into Eqs. (10.3),

$$N_\phi = \frac{\gamma a^2}{6\sin^2\phi}[1 - \cos^2\phi(3 - 2\cos\phi)] = \frac{\gamma a^2}{6}\left(1 - \frac{2\cos^2\phi}{1 + \cos\phi}\right)$$

$$N_\theta = \frac{\gamma a^2}{6}\left(5 - 6\cos\phi + \frac{2\cos^2\phi}{1 + \cos\phi}\right) \tag{10.23}$$

Equations (10.23) are valid for $\phi > \phi_0$.

In determining F for $\phi > \phi_0$, the sum of the vertical support reactions $4\gamma\pi a^3/3$, must also be taken into account in addition to the internal pressure loading. That is

$$F = -\tfrac{4}{3}\pi a^3\gamma - 2\pi a^3\gamma[\tfrac{1}{6} - \tfrac{1}{2}\cos^2\phi(1 - \tfrac{2}{3}\cos\phi)]$$

Equations (10.3) now yield

$$N_\phi = \frac{\gamma a^2}{6}\left(5 + \frac{2\cos^2\phi}{1 - \cos\phi}\right)$$

$$N_\theta = \frac{\gamma a^2}{6}\left(1 - 6\cos\phi - \frac{2\cos^2\phi}{1 - \cos\phi}\right) \tag{10.24}$$

From Eqs. (10.23) and (10.24) it is observed that both forces N_ϕ and N_θ change values abruptly at the support ($\phi = \phi_0$). A discontinuity in N_θ means a discontinuity of the deformation of the parallel circles on the immediate sides of the *nn*. Thus, the deformation associated with the membrane solution is *not compatible* with the continuity of the structure at support *nn*.

10.6 AXIALLY SYMMETRIC DEFORMATION

We now discuss the displacements in symmetrically loaded shells of revolution by considering an element AB of length $r_1\, d\phi$ of the meridian in an unstrained shell. Let the displacements in the direction of the tangent to the meridian and in the direction normal to the midsurface be denoted by v and w, respectively (Fig. 10.11). After straining, AB is displaced to position $A'B'$. In the analysis which follows, the small deformation approximation is employed, and higher-order infinitesimal terms are neglected. The deformation experienced by an element of infinitesimal length $r_1\, d\phi$ may be regarded as composed of an increase in length $(dv/d\phi)\, d\phi$, owing to the tangential displacements, and a decrease in length $w\, d\phi$ produced by the radial displacement w. The meridional strain ε_ϕ, the total deformation per unit length of the element AB, is thus

$$\varepsilon_\phi = \frac{1}{r_1}\frac{dv}{d\phi} - \frac{w}{r_1} \tag{a}$$

The deformation of an element of a parallel circle may be treated in a like manner. It can be shown that the increase in radius r_0 of the circle, produced by the displacements v and w is $v \cos \phi - w \sin \phi$. Inasmuch as the circumference of the parallel circle expands in direct proportion to its radius

$$\varepsilon_\theta = \frac{1}{r_0}(v \cos \phi - w \sin \phi) \tag{b}$$

Recalling that $r_0 = r_2 \sin \phi$, the hoop strain is written as follows:

$$\varepsilon_\theta = \frac{1}{r_2}(v \cot \phi - w) \tag{c}$$

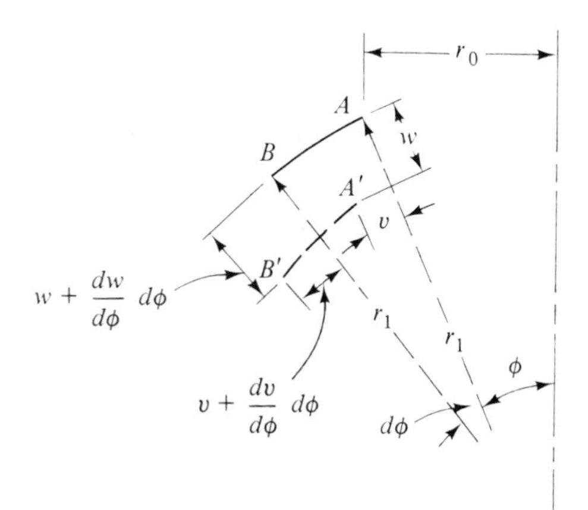

Figure 10.11

Elimination of w from Eqs. (a) and (c) leads to the following differential equation for v:

$$\frac{dv}{d\phi} - v \cot \phi = r_1 \varepsilon_\phi - r_2 \varepsilon_\theta \tag{d}$$

The strains are related to the membrane stresses by Hooke's law

$$\varepsilon_\phi = \frac{1}{E}(\sigma_\phi - \nu\sigma_\theta) \qquad \varepsilon_\theta = \frac{1}{E}(\sigma_\theta - \nu\sigma_\phi) \tag{e}$$

Introduction of the above into Eq. (d) gives

$$\frac{dv}{d\phi} - v \cot \phi = \frac{1}{E}[\sigma_\phi(r_1 + \nu r_2) - \sigma_\theta(r_2 + \nu r_1)] \tag{10.25}$$

We observe that the symmetric deformations of a shell of revolution may be determined by integrating Eq. (10.25) when the membrane stresses are known. Next, we let

$$\frac{dv}{d\phi} - v \cot \phi = f(\phi)$$

This equation has the solution

$$v = \left[\int \frac{f(\phi)}{\sin \phi}\, d\phi + c\right] \sin \phi \tag{f}$$

The constant of integration c is determined from a boundary condition. Once v has been found, one can readily obtain w from Eq. (c).

Example 10.7 Determine the displacements of the spherical roof dome supporting its own weight (Fig. 10.5a).

SOLUTION For the half sphere under consideration $r_1 = r_2 = a$ and stresses σ_ϕ and σ_θ are given by Eqs. (10.13). Equation (10.25) is therefore

$$\frac{dv}{d\phi} - v \cot \phi = \frac{a^2 p(1+\nu)}{Et}\left(\cos \phi - \frac{2}{1 + \cos \phi}\right) = f(\phi)$$

Inserting this expression into Eq. (f), we obtain

$$v = \frac{a^2 p(1+\nu)}{Et}\left[\sin \phi \ln (1 + \cos \phi) - \frac{\sin \phi}{1 + \cos \phi}\right] + c \sin \phi \tag{g}$$

It is necessary to choose c such that $v = 0$ at $\phi = \alpha$ (Fig. 10.5a). It follows that

$$c = \frac{a^2 p(1+\nu)}{Et}\left[\frac{1}{1 + \cos \alpha} - \ln (1 + \cos \alpha)\right] \tag{h}$$

Upon substituting this value of c into Eq. (g), deflection v is obtained and Eq. (c) then yields w. It is noted that if the *support* displacement w is to be determined, one need not employ Eq. (g), as $v = 0$ there; the second of Eqs. (e) and Eq. (c) directly give the solution.

10.7 ASYMMETRICALLY LOADED SHELLS OF REVOLUTION

In the bending of a shell of revolution under *unsymmetrical* loading, not only do normal forces N_ϕ and N_θ act on the sides of an element, but shearing forces $N_{\theta\phi}$ and $N_{\phi\theta}$ as well (Fig. 10.12). The moment equilibrium requires that $N_{\theta\phi} = N_{\phi\theta}$ *as is always the case for a thin shell* (Sec. 11.2). The surface load, referred to the unit area of the midsurface, has components p_x, p_y, and p_z.

The x-directed forces are as follows. The force

$$\frac{\partial N_\theta}{\partial \theta} r_1 \, d\theta \, d\phi \tag{a}$$

is owing to the variation of N_θ. Horizontal components of the forces $N_{\theta\phi} \cdot r_1 \, d\phi$ acting on the faces AB and CD of the element make an angle $d\theta$, and thus have the following resultant in the x direction:

$$N_{\theta\phi} \cdot r_1 \, d\phi \cdot \cos\phi \cdot d\theta \tag{b}$$

The difference of the shearing forces acting on faces AC and BD of the element is expressed

$$N_{\theta\phi} \frac{dr_0}{d\phi} d\phi \, d\theta + \frac{\partial N_{\theta\phi}}{\partial \phi} r_0 \, d\phi \, d\theta = \frac{\partial}{\partial \phi}(r_0 N_{\theta\phi}) \, d\theta \, d\phi \tag{c}$$

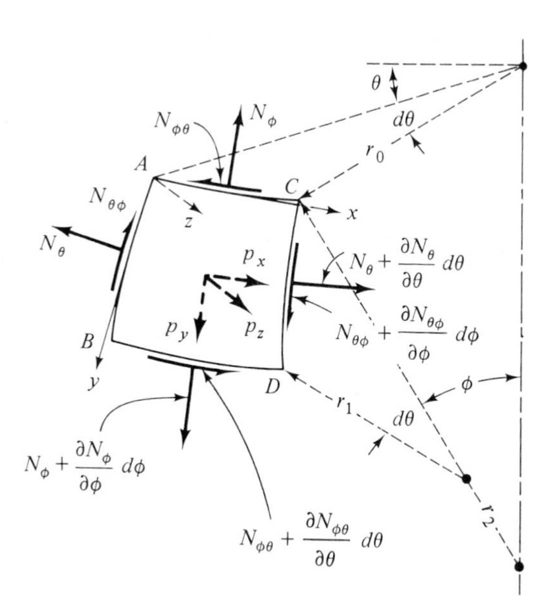

Figure 10.12

The component of the external force is

$$p_x r_0 r_1 \, d\theta \, d\phi \tag{d}$$

The x equilibrium condition thus reads:

$$\frac{\partial}{\partial \phi}(r_0 N_{\theta\phi}) + \frac{\partial N_\theta}{\partial \theta} r_1 + N_{\theta\phi} r_1 \cos \phi + p_x r_0 r_1 = 0 \tag{10.26a}$$

To the expression governing the y equilibrium of the symmetrically loaded case (Sec. 10.4), we must add the force

$$\frac{\partial N_{\theta\phi}}{\partial \theta} r_1 \, d\theta \, d\phi$$

produced by the difference in the shearing forces acting on the faces AB and CD of the element. Inasmuch as the projection of the shearing forces on the z axis vanishes, Eq. (10.1*a*) remains valid for the present case as well. The equilibrium of y- and z-directed forces is therefore satisfied by the expressions:

$$\frac{\partial}{\partial \phi}(N_\phi r_0) + \frac{\partial N_{\theta\phi}}{\partial \theta} r_1 - N_\theta r_1 \cos \phi + p_y r_1 r_0 = 0 \tag{10.26b}$$

$$\frac{N_\phi}{r_1} + \frac{N_\theta}{r_2} = -p_z \tag{10.26c}$$

Equations (10.26) permit determination of the membrane forces in a shell of revolution with nonsymmetrical loading that may, in general, vary with θ and ϕ. Such a case is discussed in the next section.

We note that the governing equations of equilibrium for the spherical, conical, and cylindrical shells may readily be deduced from Eqs. (10.26) upon following a procedure identical with that described in Sec. 10.5.

10.8 SHELLS OF REVOLUTION UNDER WIND LOADING

It is usual to represent dynamic loading such as wind and earthquake effects by statically equivalent or pseudostatic loading adequate for purposes of design. The *wind load* on shells is composed of pressure on the wind side and suction of the leeward side. Only the *load component acting perpendicular to the midsurface* p_z is considered important. Components p_x and p_y are due to friction forces and are of negligible magnitude. Assuming for the sake of simplicity that the wind acts in the direction of the meridian plane $\theta = 0$, the components of wind pressure are as follows:

$$p_x = 0 \qquad p_y = 0 \qquad p_z = p \sin \phi \cos \theta \tag{10.27}$$

In the above, p represents the *static* wind pressure intensity. For purposes of illustration, Fig. 10.13 shows the distribution of the static-design wind load on a spherical dome. This distribution should be regarded as a rough approximation.[31,33]

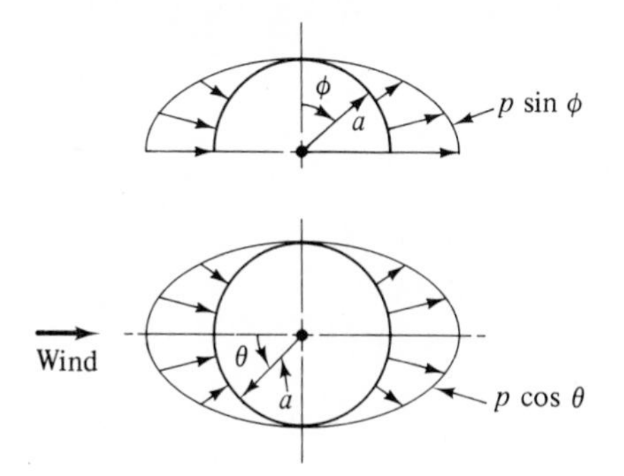

Figure 10.13

Proceeding with the solution, we substitute Eqs. (10.27) into (10.26) to obtain

$$\frac{\partial}{\partial \phi}(r_0 N_\phi) + \frac{\partial N_{\theta\phi}}{\partial \theta} r_1 - N_\theta r_1 \cos \phi = 0$$

$$\frac{\partial}{\partial \phi}(r_0 N_{\theta\phi}) + \frac{\partial N_\theta}{\partial \theta} r_1 + N_{\theta\phi} r_1 \cos \phi = 0 \tag{a}$$

$$N_\phi r_0 + N_\theta r_1 \sin \phi = -p r_0 r_1 \sin \phi \cos \theta$$

The third expression, when substituted into the first and second, eliminates N_θ. The *equations of equilibrium* for shells of revolution *under* the action of *wind pressure* are then

$$\frac{\partial N_\phi}{\partial \phi} + \left(\frac{1}{r_0}\frac{dr_0}{d\phi} + \cot \phi\right) N_\phi + \frac{r_1}{r_0}\frac{\partial N_{\theta\phi}}{\partial \theta} = -p r_1 \cos \phi \cos \theta$$

$$\frac{\partial N_{\theta\phi}}{\partial \phi} + \left(\frac{1}{r_0}\frac{dr_0}{d\phi} + \frac{r_1}{r_2}\cot \phi\right) N_{\theta\phi} - \frac{1}{\sin \phi}\frac{\partial N_\phi}{\partial \theta} = -p r_1 \sin \theta \tag{10.28a–c}$$

$$N_\theta = -p r_0 \cos \theta - \frac{N_\phi r_0}{r_1 \sin \phi}$$

Illustrated in the solution of the following problem, is the determination of membrane stresses.

Example 10.8 Consider a *mushroom-like shelter*, a shell having the shape of a circular cone, supported by a column at the vertex (Fig. 10.14). Find the

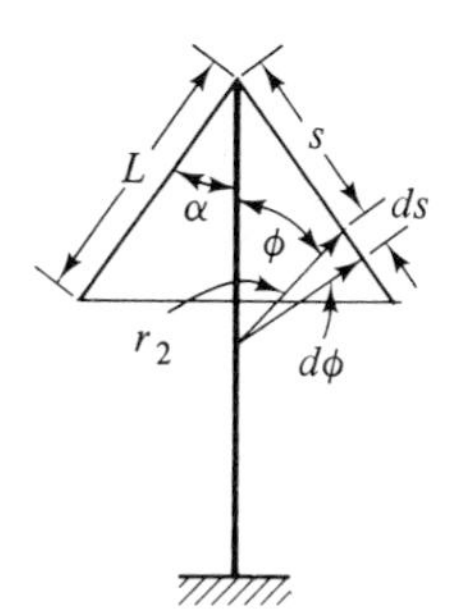

Figure 10.14

hoop, meridian, and shear stresses if the shell is submitted to a wind pressure described by Eqs. (10.27).

SOLUTION Referring to Fig. 10.14,

$$r_1 = \infty \qquad r_2 = s \tan \alpha \qquad ds = r_1\, d\phi \qquad r_0 = s \sin \alpha \tag{b}$$

from which

$$\frac{d}{d\phi} = r_1 \frac{d}{ds} \qquad \frac{dr_0}{ds} = \sin \alpha \tag{c}$$

In addition, we replace N_ϕ by N_s and $N_{\theta\phi}$ by $N_{\theta s}$. Upon introducing these together with Eqs. (10.27) into Eq. (10.28b), we obtain, after integration

$$N_{\theta s} = -\frac{1}{s^2}\left(\frac{ps^3}{3} + c\right)\sin \theta \tag{d}$$

The condition that the edge $y = L$ of the shell is free is fulfilled by setting $c = -pL^3/3$ in the above expression. We thus have

$$\tau_{\theta s} = \frac{N_{\theta s}}{t} = \frac{p}{3t}\frac{L^3 - s^3}{s^2} \sin \theta \tag{10.29a}$$

Equation (10.28a), after substituting Eqs. (b), (c), and (10.29a), becomes

$$\frac{\partial N_s}{\partial s} + \frac{N_s}{s} = -\left(\frac{p}{3}\frac{L^3 - s^3}{s^3 \sin \alpha} + p \sin \alpha\right)\cos \theta$$

Integration of the above leads to

$$\sigma_s = \frac{N_s}{t} = \frac{p \cos \theta}{t \sin \alpha}\left(\frac{L^3 - s^3}{3s^2} - \frac{L^2 - s^2}{2s}\cos^2 \alpha\right) \tag{10.29b}$$

Expression (10.28c) then results in the hoop stress:

$$\sigma_\theta = \frac{N_\theta}{t} = -\frac{ps \sin \alpha \cos \theta}{t} \tag{10.29c}$$

It is seen from Eqs. (10.29a) and (10.29b) that meridian and shear stresses grow without limit at the top ($s = 0$), as expected at a point support. It can be shown that the vertical resultant of forces N_s and $N_{\theta\phi}$ transmitted in a parallel circle approaches the total load of the shell when $s = s_1$. Here s_1 is a particular finite length. To avoid infinite stresses, the conical shell must be assumed to be fastened to the column along a circle corresponding to s_1.

10.9 CYLINDRICAL SHELLS OF GENERAL SHAPE

A cylindrical shell is formed by moving a straight line, the *generator*, parallel to its initial direction along a closed path. Depicted in Fig. 10.15a is a cylindical shell of *arbitrary* cross section. A shell element is bordered by two adjacent

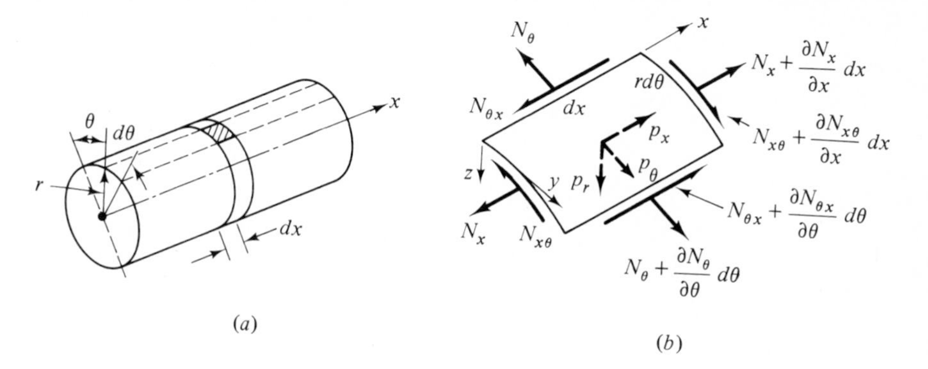

Figure 10.15

generators and two planes normal to the axial axis x, spaced dx apart. The element so described is located by coordinates x and θ.

Assume a *nonuniform* loading to act on the shell. Then, a free body diagram of a membrane element contains the forces shown in Fig. 10.15*b*. The x and θ components of the externally applied forces per unit area are labeled p_x and p_θ and are indicated to act in the directions of increasing x and θ. The radial or normal component of the loading p_r, acts in the positive, inward direction. The equilibrium of forces in the x, θ, and r directions is represented by

$$\frac{\partial N_x}{\partial x} dx \cdot r\, d\theta + \frac{\partial N_{x\theta}}{\partial \phi} d\theta \cdot dx + p_x \cdot dx \cdot r\, d\theta = 0$$

$$\frac{\partial N_\theta}{\partial \theta} d\theta \cdot dx + \frac{\partial N_{x\theta}}{\partial x} dx \cdot r\, d\theta + p_\theta \cdot dx \cdot r\, d\theta = 0$$

$$N_\theta\, dx \cdot d\theta + p_r \cdot dx \cdot r\, d\theta = 0$$

Dividing by the differential quantities, we obtain the equations of equilibrium of a cylindrical shell. It follows that

$$N_\theta = -p_r r$$

$$\frac{\partial N_{x\theta}}{\partial x} + \frac{1}{r}\frac{\partial N_\theta}{\partial \theta} = -p_\theta \qquad (10.30a\text{–}c)$$

$$\frac{\partial N_x}{\partial x} + \frac{1}{r}\frac{\partial N_{x\theta}}{\partial \theta} = -p_x$$

As already mentioned in Sec. 10.7 the above expressions could be obtained directly from Eqs. (10.26). It is observed that they are simple in structure and may be solved one by one.

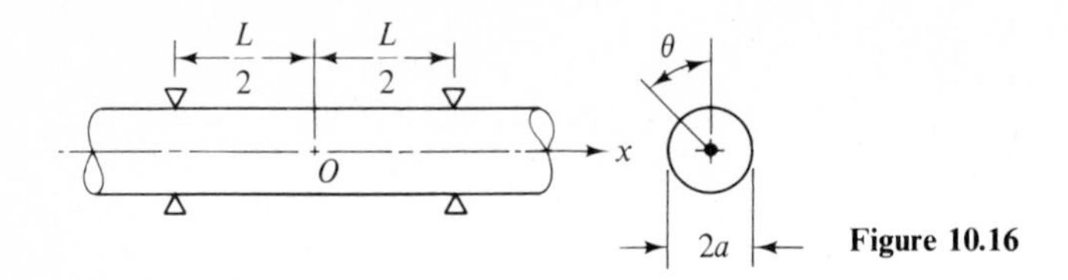

Figure 10.16

For a prescribed loading, N_θ is readily found from Eq. (10.30*a*). Subsequently, $N_{x\theta}$ and N_x are determined by integrating Eqs. (10.30*b*) and (10.30*c*). In so doing, we have

$$
\begin{aligned}
N_\theta &= -p_r r \\
N_{x\theta} &= -\int \left(p_\theta + \frac{1}{r}\frac{\partial N_\theta}{\partial \theta} \right) dx + f_1(\theta) \\
N_x &= -\int \left(p_x + \frac{1}{r}\frac{\partial N_{x\theta}}{\partial \theta} \right) dx + f_2(\theta)
\end{aligned}
\tag{10.31}
$$

where $f_1(\theta)$ and $f_2(\theta)$ are arbitrary functions of integration to be evaluated on the basis of the edge conditions. These functions arise because of the integration of partial derivatives.

Example 10.9 A long, horizontal, cylindrical *conduit* is supported as shown in Fig. 10.16 and filled to capacity with a liquid of specific weight γ. Determine the membrane forces under two assumptions: (*a*) there is free spanning, with expansion joints at both ends; (*b*) both ends are rigidly fixed.

SOLUTION At an arbitrary level defined by angle θ the pressure is $\gamma a\ (1 - \cos\,\theta)$ and the external forces are thus

$$p_r = -\gamma a(1 - \cos\,\theta) \qquad p_\theta = p_x = 0 \tag{a}$$

where the minus sign indicates the outward direction. Substituting the above into Eqs. (10.31), we have

$$
\begin{aligned}
N_\theta &= \gamma a^2(1 - \cos\,\theta) \\
N_{x\theta} &= -\int \gamma a \sin\,\theta\, dx + f_1(\theta) = -\gamma a x \sin\,\theta + f_1(\theta) \\
N_x &= \int \gamma x \cos\,\theta\, dx - \frac{1}{a}\int \frac{df_1}{d\theta}\, dx + f_2(\theta) \\
&= \frac{\gamma x^2}{2} \cos\,\theta - \frac{x}{a}\frac{df_1}{d\theta} + f_2(\theta)
\end{aligned}
\tag{b}
$$

(*a*) The conditions for the simply supported edges are represented by

$$N_x = 0 \qquad \left(x = \pm\frac{L}{2} \right) \tag{c}$$

Introduction of Eqs. (*b*) into (*c*) yields

$$0 = \frac{\gamma L^2}{8} \cos\theta - \frac{L}{2a}\frac{df_1}{d\theta} + f_2(\theta)$$

$$0 = \frac{\gamma L^2}{8} \cos\theta + \frac{L}{2a}\frac{df_1}{d\theta} + f_2(\theta)$$

Adding and subtracting the above expressions provide, respectively

$$f_2(\theta) = -\frac{\gamma L^2}{8} \cos\theta \tag{d}$$

$$\frac{df_1}{d\theta} = 0 \qquad \text{or} \qquad f_1(\theta) = 0 + c \tag{e}$$

From the second of Eqs. (*b*), we observe that c in Eq. (*e*) represents the value of the uniform load $N_{x\theta}$ at $x = 0$. This load is not present because the pipe is free of torque, thus $c = 0$. The solution, then, from Eqs. (*b*) and (*d*), is written as follows:

$$N_\theta = \gamma a^2(1 - \cos\theta)$$
$$N_{x\theta} = -\gamma a x \sin\theta \tag{10.32a–c}$$
$$N_x = -\frac{\gamma}{8}(L^2 - 4x^2)\cos\theta$$

Note that the shear $N_{x\theta}$ and the normal N_x forces, respectively, represent identical spanwise distributions with the shear force and the bending moment of a *uniformly loaded beam.*

Through the application of Hooke's law,

$$u = \frac{1}{Et}\int_{-L/2}^{L/2} (N_x - \nu N_\theta)\, dx \tag{f}$$

Upon introducing the membrane forces, Eqs. (10.32), the above expression provides the axial deformation.

(*b*) In this case, no change occurs in the length of the generator:

$$\int_{-L/2}^{L/2} (N_x - \nu N_\theta)\, dx = 0 \tag{g}$$

where

$$N_x = -\frac{\gamma}{8}(L^2 - 4x^2)\cos\theta + f_2(\theta)$$
$$N_\theta = \gamma a^2(1 - \cos\theta) \tag{h}$$

Introduction of Eqs. (*h*) into Eq. (*g*) yields

$$f_2(\theta) = \nu\gamma a^2(1 - \cos\theta) - \frac{\gamma L^2}{24}\cos\theta$$

Hence,

$$N_x = \nu\gamma a^2(1 - \cos\theta) + \frac{\gamma}{2}\left(x^2 - \frac{L^2}{12}\right)\cos\theta \tag{10.33}$$

One finds that the circumferential strain, $\varepsilon_\theta = (N_\theta - \nu N_x)/Et$, is not zero. Because clamped edges inhibit any such deformations at the ends, some bending of the pipe occurs near the supports. The membrane solution, Eqs. (10.32*a*), (10.32*b*), and (10.33), thus will agree very well with measurement at distances approximately a from the supports. The detail of the distribution of support-reaction forces is obtained by application of the bending theory.

10.10 BREAKDOWN OF ELASTIC ACTION IN SHELLS

From a design point of view, it is clear that for practical purposes, it is necessary to obtain the proper dimensions of a member which can sustain a prescribed loading without suffering failure by breakdown of elastic action. As already alluded to (Sec. 1.10), the dimensions that should be assigned to a loaded element depend upon the failure theory held concerning the cause of the yielding. This will be illustrated in the examples to follow for two particular shell structures.

Example 10.10 A circular cylindrical vessel with closed ends is subjected to internal pressure p (Fig. 10.17). The tube is fabricated of a material having a yield point stress σ_{yp}. Determine the required wall thickness t remote from the ends, according to the various theories of failure. Sketch the results.

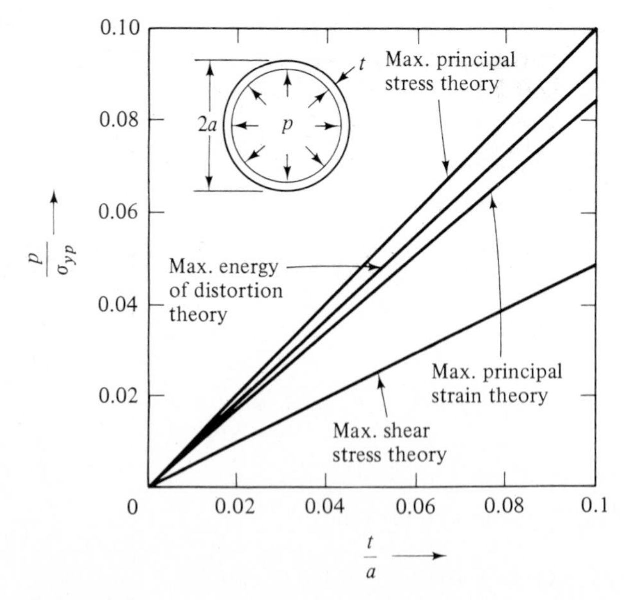

Figure 10.17

SOLUTION The hoop and the axial shell stresses are given by Eqs. (10.10):

$$\sigma_1 = \sigma_\theta = \frac{pa}{t} \qquad \sigma_2 = \sigma_x = \frac{pa}{2t} \tag{a}$$

Maximum principal stress theory. From Eqs. (1.44) and (*a*),

$$t = \frac{pa}{\sigma_{yp}} \tag{b}$$

Maximum shear stress theory. As σ_1 and σ_2 are of the same sign, we obtain, from Eqs. (1.45*b*) and (*a*):

$$t = \frac{pa}{2\sigma_{yp}} \tag{c}$$

Maximum principal strain theory. On the basis of Eqs. (1.46) and (*a*):

$$t = \frac{pa}{\sigma_{yp}}\left(1 - \frac{\nu}{2}\right) \tag{d}$$

Maximum energy of distortion theory. Formulas (1.47) and (*a*) yield

$$t = \frac{\sqrt{3}}{2}\frac{pa}{\sigma_{yp}} \tag{e}$$

The relationship between internal pressure and wall thickness as given by Eqs. (*b*) to (*e*) is represented in Fig. 10.17 for $\nu = 1/3$. It is observed that the maximum shear stress theory is the most conservative; the maximum principal stress theory leads to the least conservative result.

Example 10.11 A conical tank is constructed of steel of yield point stress σ_{yp} in tension and $\frac{3}{2}\sigma_{yp}$ in compression (Fig. 10.9). The tank is filled with a liquid of specific weight γ and is edge-supported. Determine the proper wall thickness t using a safety factor N. Apply (*a*) the maximum principal stress theory and (*b*) the maximum energy of distortion theory.

SOLUTION Expressions for the hoop and longitudinal stresses, from Eqs. (10.22), are

$$\sigma_1 = \sigma_\theta = \gamma(a - y)y\frac{\tan\alpha}{t\cos\alpha} \qquad \sigma_2 = \sigma_s = \gamma(a - \tfrac{2}{3}y)y\frac{\tan\alpha}{2t\cos\alpha} \tag{f}$$

The largest values of the principal stresses are given by:

$$\begin{aligned} \sigma_{1,\max} &= \frac{\gamma h^2 \tan\alpha}{4t\cos\alpha} \qquad \left(y = \frac{a}{2}\right) \\ \sigma_{2,\max} &= \frac{3\gamma h^2 \tan\alpha}{16t\cos\alpha} \qquad \left(y = \frac{3a}{4}\right) \end{aligned} \tag{g}$$

(*a*) *Maximum principal stress theory* Equations (10.43) together with Eqs. (*g*) yield

$$\frac{\sigma_{yp}}{N} = \frac{\gamma a^2}{4t}\frac{\tan\alpha}{\cos\alpha} \qquad \frac{3\sigma_{yp}}{2N} = \frac{3\gamma a^2}{16t}\frac{\tan\alpha}{\cos\alpha}$$

or

$$t = 0.250\frac{\gamma a^2 N}{\sigma_{yp}}\frac{\tan\alpha}{\cos\alpha} \qquad t = 0.125\frac{\gamma a^2 N}{\sigma_{yp}}\frac{\tan\alpha}{\cos\alpha} \tag{h}$$

(*b*) *Maximum energy of distortion theory* Expressions (*g*) demonstrate that the principal stresses assume their maximum values at different locations. The location at which the combined principal stresses is critical is ascertained as follows. First, Eqs. (*f*) are substituted into Eqs. (1.47):

$$\left[\gamma(a-y)y\frac{\tan\alpha}{t\cos\alpha}\right]^2 + \left[\left(a-\tfrac{2}{3}y\right)y\frac{\tan\alpha}{2t\cos\alpha}\right]^2 - \left[\gamma(a-y)y\frac{\tan\alpha}{t\cos\alpha}\right]\left[\left(a-\tfrac{2}{3}y\right)y\frac{\tan\alpha}{2t\cos\alpha}\right] = \sigma_{yp}^2 \tag{i}$$

Next, the derivative of the foregoing with respect to y is set equal to zero to yield

$$y = 0.52a \tag{j}$$

Substitution of Eq. (*j*) into (*i*) results in

$$t = 0.225\frac{\gamma a^2 N}{\sigma_{yp}}\frac{\tan\alpha}{\cos\alpha} \tag{k}$$

In this problem, the required wall thickness is the value given by the first of Eqs. (*h*). Hence, the thickness based upon the maximum principal stress theory is 10 percent greater than that predicted by the maximum energy of distortion theory.

PROBLEMS

Secs. 10.1 to 10.6

10.1 A spherical roof dome is subjected to a snow load p_s (Fig. 10.5*a*). Develop the expressions

$$\sigma_\phi = -\frac{p_s a}{2t} \qquad \sigma_\theta = -\frac{p_s a}{2t}\cos 2\phi \tag{P10.1}$$

for the meridional and hoop stresses. Assume that the dome is constructed of 8-cm-thick masonry having a span $2a = 70$ m, and a compressive ultimate strength $\sigma_{u''} = 22$ MPa. Determine the factor of safety according to the maximum principal stress theory, if there is snow accumulation over the dome such that $p_s = 2500$ Pa.

10.2 An observation dome of a pressurized aircraft is of ellipsoidal shape (Fig. 10.8). It is constructed of 6-mm-thick plastic material. Determine the limiting value of the pressure differential the shell can resist given a maximum stress of 14 MPa. The lengths of the semiaxes are $a = 0.15$ m and $b = 0.12$ m.

10.3 A conical aluminum container of thickness 3 mm, apex angle $\alpha = 45°$, and height $h = 3$ m, is filled with water (Fig. 10.9). Taking $\nu = 0.3$, compute the locations measured vertically above the apex, for which (*a*) the hoop strain is zero and (*b*) the hoop strain is maximum.

10.4 An edge-supported hemispherical container is filled with a liquid of specific weight γ (Fig. P10.4). (*a*) Demonstrate that the membrane stresses are given by

$$\sigma_\phi = \frac{(a + y)a\gamma}{4t} \qquad \sigma_\theta = \frac{a\gamma}{4t}(4t^2 y - a - y) \tag{P10.4}$$

(*b*) Calculate the radial (w) and the circumferential (v) deformations of the shell.

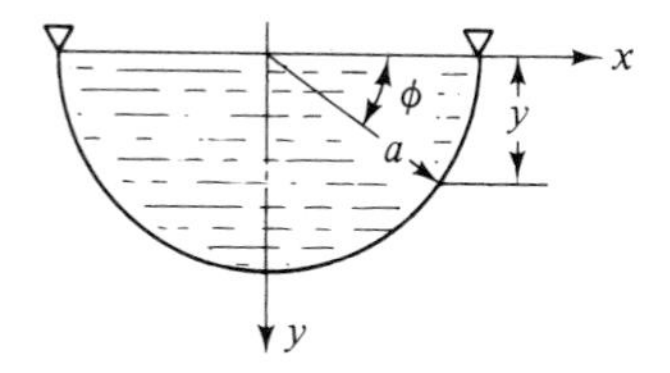

Figure P10.4

10.5 A supported truncated conical shell carries an upper edge load p_e (Fig. P10.5). Derive the following expressions for the membrane stresses

$$\sigma_s = -p_e \frac{s_0}{st} \frac{1}{\sin \phi} \qquad \sigma_\theta = \tau_{s\theta} = 0 \tag{P10.5}$$

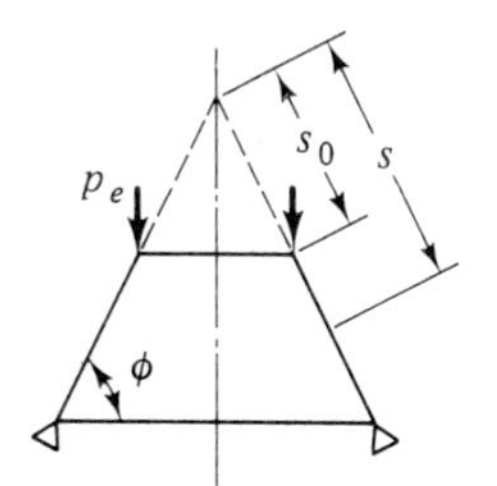

Figure P10.5

10.6 The *compound tank* shown in Fig. P10.6 consists of a conical shell with a spherical bottom. Show that the hoop and the meridional stresses in the conical part are given by

$$\sigma_\theta = \frac{\gamma}{t} y(h_1 - y) \frac{\tan \alpha}{\cos \alpha}$$

$$\sigma_s = \frac{\gamma}{6t}\left[(3h_1 - 2y)y \frac{\sin \alpha}{\cos^2 \alpha} + \frac{a^2(2a - h_2)\cos^2 \alpha}{y \sin \alpha} + \frac{a^3 \cos^2 \alpha}{y}\right] \tag{P10.6}$$

The solution for the bottom part is governed by Eqs. (P10.4). Note that at the juncture of the two parts, a ring must be provided to resist the difference in the horizontal components of the meridional forces N_s in the cone and the N_ϕ in the sphere.

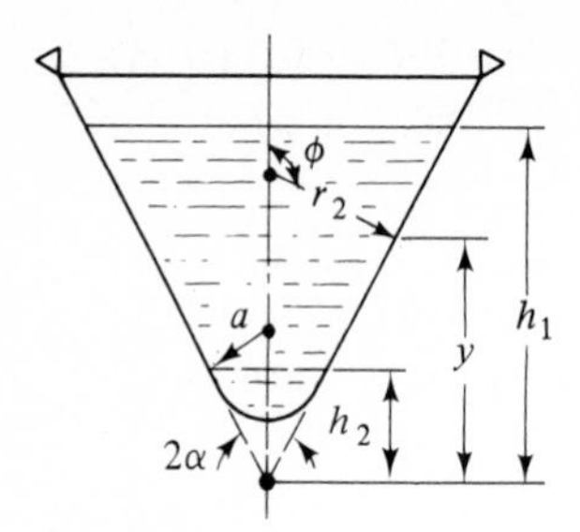

Figure P10.6

10.7 The mushroom-like shelter shown in Fig. 10.14 is assumed to support only its own weight per unit surface area p. Derive the following expressions for the membrane forces

$$N_\theta = -ps \sin \alpha \tan \alpha \qquad N_s = -\frac{p}{2}\frac{L^2 - s^2}{s \cos \alpha} \tag{P10.7}$$

10.8 Verify that, in the spherical tank described in Example 10.6, the maximum shear stress is represented by

$$\tau_{\max} = \frac{\gamma a^2}{6t}\left(1 + \frac{1}{1 + \cos \phi}\right)(1 - \cos \phi) \tag{P10.8}$$

10.9 Figure P10.9 shows a compound tank comprised of a cylindrical shell and a spherical bottom. The tank is filled to a level h with a liquid of specific weight γ. Derive the following expressions for the membrane stresses in the bottom part

$$\sigma_\theta = \frac{\gamma a^2}{6t}\left(\frac{3h}{a} + 5 - 6 \cos \phi + \frac{2 \cos^2 \phi}{1 + \cos \phi}\right)$$
$$\sigma_\phi = -\frac{\gamma a^2}{6t}\left(\frac{3h}{a} + 1 - \frac{2 \cos^2 \phi}{1 + \cos \phi}\right) \tag{P10.9}$$

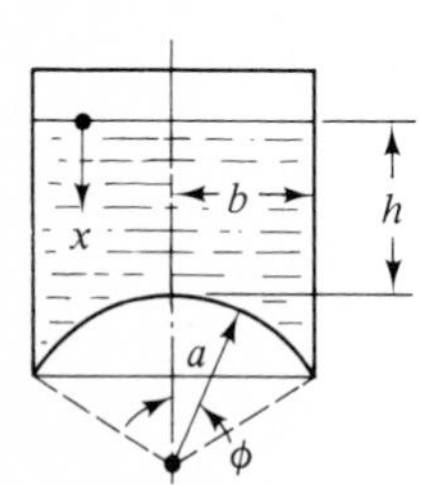

Figure P10.9

10.10 Determine the radial (w) and the circumferential (v) deformations in the spherical tank described in Example 10.6.

Secs. 10.7 to 10.10

10.11 Consider the tank of Prob. 10.9. Verify that the membrane stresses in the cylindrical part are given by

$$\sigma_\theta = \gamma b x \qquad \sigma_x = c \tag{P10.11}$$

Here the constant value c of the axial stress may be produced by the weight of a roof, for example.

10.12 A pipeline in the form of an *open semicircular channel* is filled with a liquid of specific weight γ. Refer to Fig. 10.16 for notation and assume that both ends are fixed. Derive the following expressions for the membrane forces

$$N_\theta = \gamma a^2 \sin\theta \qquad N_{x\theta} = \gamma a x \cos\theta$$
$$N_x = \frac{\gamma}{2}\left(\frac{L^2}{12} - x^2\right)\sin\theta + \nu N_\theta \tag{P10.12}$$

Note that θ is now measured from the horizontal axis.

10.13 The roof of an aircraft hangar, supporting its own weight p, may be approximated as a semicircular cross-sectional cylinder or so-called *barrel vault* (Fig. P10.13). At both ends of the shell, it is assumed that the conditions of simple support prevail. Develop the following expressions for the membrane stresses

$$\sigma_\theta = -\frac{pa}{t}\cos\theta \qquad \tau_{x\theta} = -\frac{2px}{t}\sin\theta$$
$$\sigma_x = \frac{p}{4at}(L^2 - 4x^2)\cos\theta \tag{P10.13}$$

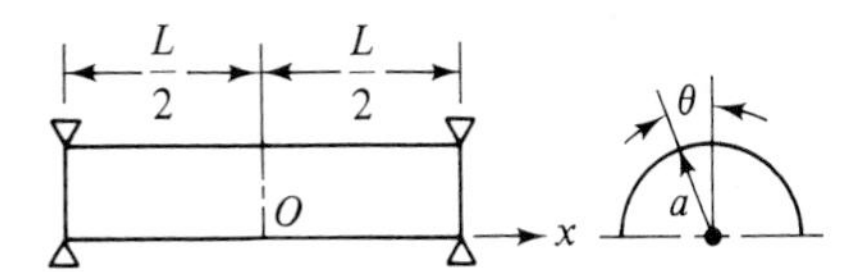

Figure P10.13

10.14 A horizontal circular pipe of radius a, length L, and thickness t, is filled with a gas at a constant pressure p. Determine the membrane stresses in the pipe if the ends are assumed to be built in.

10.15 A simply supported, horizontal, circular cylinder of radius a (Fig. 10.16), thickness t, and length L carries its own weight p. Derive the following expressions for the membrane stresses

$$\sigma_\theta = -\frac{pa}{t}\cos\theta \qquad \tau_{x\theta} = -\frac{2px}{t}\sin\theta$$
$$\sigma_x = -\frac{p}{4at}(L^2 - 4x^2)\cos\theta \tag{P10.15}$$

10.16 Redo Prob. 10.15 for the case of rigidly built-in cylinder ends. Verify that now the membrane stresses are represented by

$$\sigma_\theta = -\frac{pa}{t}\cos\theta \qquad \tau_{x\theta} = -\frac{2px}{t}\sin\theta$$
$$\sigma_x = \frac{p}{t}\left(\frac{x^2}{a} - \frac{L^2}{12a} - \nu a\right)\cos\theta \tag{P10.16}$$

10.17 Consider a thin-walled circular pipe of mean radius a, loaded as a cantilever (Fig. P10.17). Derive the following expressions for the membrane stresses

$$\sigma_\theta = 0 \qquad \tau_{x\theta} = -\frac{P}{\pi a t}\sin\theta$$
$$\sigma_x = \frac{Px}{\pi a^2 t}\cos\theta \tag{P10.17}$$

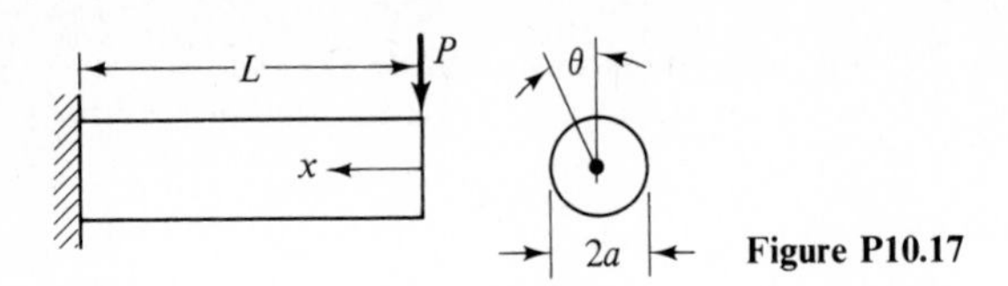

Figure P10.17

10.18 A circular toroidal shell is subjected to internal pressure $p = 1.5$ MPa (Fig. 10.7). Assume that the yield stress $\sigma_{yp} = 200$ MPa. The dimensions of the torus are $a = 0.06$ m, $b = 0.4$ m, and $t = 1$ mm. What is the factor of safety, assuming failure to occur in accordance with the energy of distortion theory.

CHAPTER

ELEVEN

BENDING STRESSES IN SHELLS

11.1 INTRODUCTION

It was observed in Chap. 10 that membrane theory cannot, in all instances, provide solutions compatible with the actual conditions of deformation. This theory also fails to predict the state of stress at the boundaries and in certain other areas of the shell. These shortcomings are avoided by application of bending theory, considering membrane forces, shear forces, and moments to act on the shell structure.

To develop the governing differential equations for the *midsurface* displacements u, v, and w which define the geometry or kinematics of deformation of a shell, one proceeds as in the case of plates. We shall begin by deriving the basic relationships between the stress resultants and the deformations for shells of general shape. In Secs. 11.4 and 11.5 are developed the relationships for the stresses and strain energy under an arbitrary loading.

The complete bending theory is mathematically intricate, and the first solutions involving shell-bending stresses date back to only 1920. With the exception of Sec. 11.11, we shall, in this chapter, limit consideration to the most significant practical case involving rotationally symmetrical loading. Stress analysis of cylindrical shells under general loads is postponed until Chap. 13, after applications to various common structural members are presented in Chap. 12.

11.2 SHELL STRESS RESULTANTS

In deriving an expression for the stress resultants, that is, the resultant forces and moments representing the internal stresses, consider an infinitestimal element (Fig. 11.1a). This element is defined by two pairs of planes, normal to the midsurface of the shell. The origin of a cartesian coordinate system is located at a

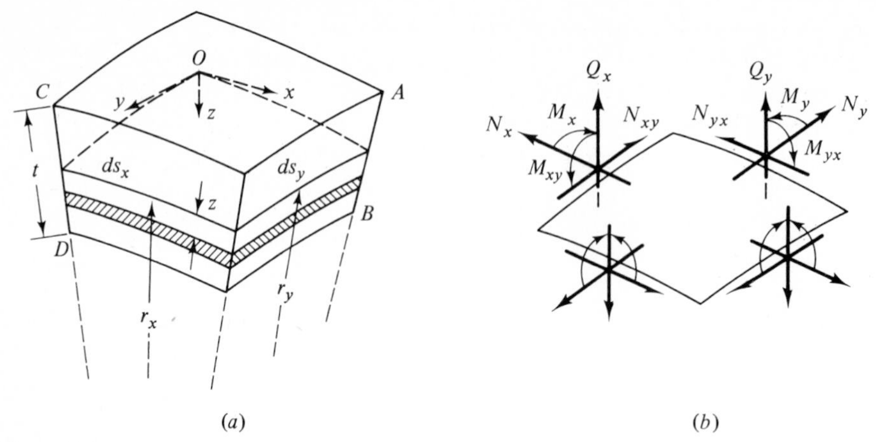

Figure 11.1

corner of the element, as shown, with the x and y axes tangent to the lines of principal curvature, and z perpendicular to the midsurface.

Because of shell curvature, the arc lengths of an element located a distance z from the midsurface are not simply ds_x and ds_y, the lengths measured on the midsurface, but rather

$$\frac{ds_x(r_x - z)}{r_x} = \left(1 - \frac{z}{r_x}\right) ds_x \qquad \frac{ds_y(r_y - z)}{r_y} = \left(1 - \frac{z}{r_y}\right) ds_y$$

where r_x and r_y are the radii of principal curvatures, in the xz and yz planes, respectively.

The stresses acting on the plane faces of the element are σ_x, σ_y, τ_{xy}, τ_{xz}, and τ_{yz}. Letting N_x represent the resultant normal force acting on plane face yz per unit length, and using the true arc length given above, we have

$$N_x \, ds_y = \int_{-t/2}^{t/2} \sigma_x \left(1 - \frac{z}{r_y}\right) ds_y \, dz$$

Cancelling arbitrary length ds_y, this becomes

$$N_x = \int_{-t/2}^{t/2} \sigma_x \left(1 - \frac{z}{r_y}\right) dz = \int_{-t/2}^{t/2} \sigma_x(1 - z\kappa_y) \, dz$$

Expressions for the remaining stress resultants per unit length are derived in a similar manner. The complete set shown in Fig. 11.1b is thus

$$\begin{Bmatrix} N_x \\ N_y \\ N_{xy} \\ N_{yx} \\ Q_x \\ Q_y \end{Bmatrix} = \int_{-t/2}^{t/2} \begin{Bmatrix} \sigma_x(1 - z\kappa_y) \\ \sigma_y(1 - z\kappa_x) \\ \tau_{xy}(1 - z\kappa_y) \\ \tau_{yx}(1 - z\kappa_x) \\ \tau_{xz}(1 - z\kappa_y) \\ \tau_{yz}(1 - z\kappa_x) \end{Bmatrix} dz \qquad \begin{Bmatrix} M_x \\ M_y \\ M_{xy} \\ M_{yx} \end{Bmatrix} = \int_{-t/2}^{t/2} \begin{Bmatrix} \sigma_x(1 - z\kappa_y) \\ \sigma_y(1 - z\kappa_x) \\ \tau_{xy}(1 - z\kappa_y) \\ \tau_{yx}(1 - z\kappa_x) \end{Bmatrix} z \, dz \qquad (11.1)$$

The sign convention is the same as in the treatment of plates.

It may be concluded that even though $\tau_{xy} = \tau_{yx}$, shearing forces N_{xy} and N_{yx} are not generally equal, nor are twisting moments M_{xy} and M_{yx}. This is because in general $r_x \neq r_y$. For *thin* shells, however, for which we are concerned, t is small relative to r_x and r_y, and consequently, z/r_x and z/r_y *may be neglected in comparison with unity*. On this basis, $N_{xy} = N_{yx}$ and $M_{xy} = M_{yx}$. The stress resultants are thus described by the *same* expressions as apply to thin plates.

11.3 FORCE, MOMENT, AND DISPLACEMENT RELATIONS

To relate the stress resultants to the shell deformations, σ_x, σ_y, and τ_{xy} must be evaluated in terms of strains. According to our assumption, the z-directed stress is neglected, $\sigma_z = 0$. Hooke's law is then written

$$\sigma_x = \frac{E}{1-\nu^2}(\varepsilon_x + \nu\varepsilon_y)$$

$$\sigma_y = \frac{E}{1-\nu^2}(\varepsilon_y + \nu\varepsilon_x) \qquad (a)$$

$$\tau_{xy} = \gamma_{xy} G$$

Let us first determine the strains appearing in the above expressions. Consider the deformed shell element of Fig. 11.2, noting that, by assumption (3) of Sec. 10.1, sides mn and $m'n'$ are straight lines. The figure shows the midsurface stretched and side mn rotated with respect to its original configuration. The unit elongation ε_x of a fiber of length l_f, located in the xz plane a distance z from the midsurface, is given by

$$\varepsilon_x = \frac{\Delta l_f}{l_f} \qquad (b)$$

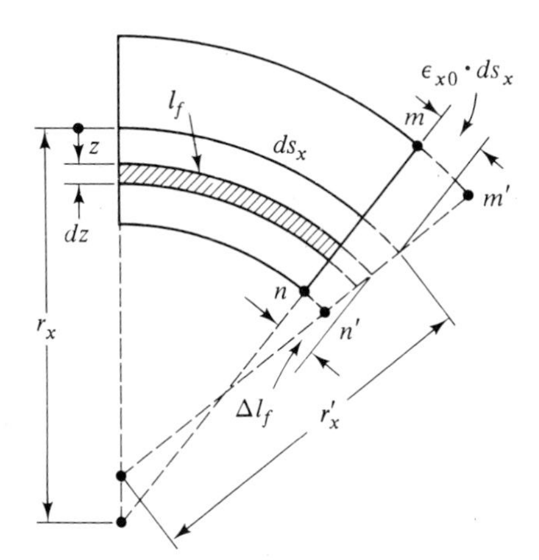

Figure 11.2

Here Δl_f is the elongation experienced by l_f. Referring to Fig. 11.2

$$l_f = ds_x(1 - z\kappa_x) \qquad \Delta l_f = ds_x(1 + \varepsilon_{x0})\left(1 - \frac{z}{r'_x}\right) - l_f \tag{c}$$

where ε_{x0} represents the x-directed *midsurface* unit deformation; r'_x, the radius of curvature after deformation; and ds_x, the length of the midsurface fiber. Substituting Eqs. (c) into Eq. (b), we have

$$\varepsilon_x = \frac{\varepsilon_{x0}}{1 - (z/r_x)} - \frac{z}{1 - (z/r_x)}\left[\frac{1}{(1 - \varepsilon_{x0})r'_x} - \frac{1}{r_x}\right]$$

in which r_x is the curvature prior to deformation. Because for the case under analysis, $t \ll r_x$, z/r_x may be omitted. In addition, it can be demonstrated that the *influence of* ε_{x0} *upon curvature* is negligible. Introducing the foregoing considerations, the above expression becomes

$$\varepsilon_x = \varepsilon_{x0} - z\left(\frac{1}{r'_x} - \frac{1}{r_x}\right) = \varepsilon_{x0} - z\chi_x \tag{d}$$

where χ represents the *change of curvature* of the midsurface. The unit elongation at any distance normal to the midsurface is thus related to the midsurface stretch and the change in curvature associated with deformation. For the y direction, a similar expression is obtained:

$$\varepsilon_y = \varepsilon_{y0} - z\left(\frac{1}{r'_y} - \frac{1}{r_y}\right) = \varepsilon_{y0} - z\chi_y \tag{e}$$

The nomenclature parallels that used in connection with ε_x.

The distribution of shear strain γ_{xy} is next evaluated. Let γ_{xy0} denote the shearing strain of the midsurface. Owing to the rotation of edge AB relative to Oz about the x axis (Fig. 11.1a) and γ_{xy0}, and referring to the third of Eqs. (1.3a) for plates, we have

$$\gamma_{xy} = \gamma_{xy0} - 2z\chi_{xy} \tag{f}$$

Here χ_{xy} designates the *twist* of the midsurface. Clearly, it represents the effect of the rotation of the shell elements about a normal to the midsurface.

Upon substitution of Eqs. (d), (e), and (f) into (a), we obtain

$$\begin{aligned}
\sigma_x &= \frac{E}{1 - \nu^2}[\varepsilon_{x0} + \nu\varepsilon_{y0} - z(\chi_x + \nu\chi_y)] \\
\sigma_y &= \frac{E}{1 - \nu^2}[\varepsilon_{y0} + \nu\varepsilon_{x0} - z(\chi_y + \nu\chi_x)] \\
\tau_{xy} &= (\gamma_{xy0} - 2z\chi_{xy})G
\end{aligned} \tag{11.2}$$

Finally, when Eq. (11.2) is introduced into Eq. (11.1), neglecting terms z/r_x and z/r_y as before, the stress resultants become

$$N_x = \frac{Et}{1-\nu^2}(\varepsilon_{x0} + \nu\varepsilon_{y0}) \qquad N_y = \frac{Et}{1-\nu^2}(\varepsilon_{y0} + \nu\varepsilon_{x0})$$

$$M_x = -D(\chi_x + \nu\chi_y) \qquad M_y = -D(\chi_y + \nu\chi_x) \tag{11.3}$$

$$N_{xy} = N_{yx} = \frac{\gamma_{xy0}Et}{2(1+\nu)} \qquad M_{xy} = M_{yx} = -D(1-\nu)\chi_{xy}$$

Here $D = Et^3/12(1-\nu^2)$ defines the *flexural rigidity* of the shell, the same as for a plate. Equations (11.3) are the *constitutive equations* for shells.

Should the actual conditions be such as to permit bending to be neglected, the analysis of stress is vastly simplified as M_x, M_y, and $M_{xy} = M_{yx}$ now vanish. What remains are the membrane forces N_x, N_y and $N_{xy} = N_{yx}$.

11.4 COMPOUND STRESSES IN A SHELL

We are now in a position to express the *compound stresses* in a shell produced by the forces and moments. For this purpose, we substitute the strains and deformations of Eqs. (11.3) into Eqs. (11.2) with the result that

$$\sigma_x = \frac{N_x}{t} + \frac{12M_x z}{t^3} \qquad \sigma_y = \frac{N_y}{t} + \frac{12M_y z}{t^3}$$

$$\tau_{xy} = \frac{N_{xy}}{t} + \frac{12M_{xy} z}{t^3} \tag{11.4}$$

The first terms above clearly describe membrane stress, and the second terms, bending stress. We observe that distribution of the stress components σ_x, σ_y, and τ_{xy} within the shell is *linear*.

It can be verified, as for plates or beams, that the vertical shearing stresses are governed by a *parabolic* distribution:

$$\tau_{xz} = \frac{3Q_x}{2t}\left(1 - \frac{4z^2}{t^2}\right) \qquad \tau_{yz} = \frac{3Q_y}{2t}\left(1 - \frac{4z^2}{t^2}\right) \tag{11.5}$$

Their values, as in case of plates, are small in comparison with the other plane-stress components.

The fundamental stress relationships are thus identical for beams, plates, and shells. In Chap. 10, methods for determining membrane forces were discussed for shells of various shapes. Cases involving bending moments are treated in the sections which follow, and in Chaps. 12 and 13. Knowing the stress resultants, one can readily compute the stress at any point within a shell through the application of Eqs. (11.4) and (11.5).

11.5 STRAIN ENERGY IN THE BENDING AND STRETCHING OF SHELLS

Equations (1.32) through (1.34), upon application of the appropriate stresses and strains, lead to a strain-energy expression for the shells. As in the bending of plates, we assume that the transverse shearing strains (γ_{xz}, γ_{yz}) and the normal stress (σ_z) vanish. The *components* of strain energy of a deformed shell are the *bending*-strain energy U_b and the *membrane*-strain energy U_m. That is,

$$U = U_b + U_m \tag{11.6}$$

The bending-strain energy, upon replacing the curvatures κ_x, κ_y, κ_{xy} in Eq. (1.34) by the changes in curvature χ_x, χ_y, χ_{xy}, is found to be

$$U_b = \tfrac{1}{2}D \iint_A [(\chi_x + \chi_y)^2 - 2(1 - \nu)(\chi_x \chi_y - \chi_{xy}^2)]\, dx\, dy \tag{11.7}$$

where A represents the surface area of the shell.

The membrane energy is associated with midsurface stretching produced by the in-plane forces and is given by

$$U_m = \tfrac{1}{2} \iint_A (N_x \varepsilon_{x0} + N_y \varepsilon_{y0} + N_{xy} \gamma_{xy0})\, dx\, dy \tag{11.8a}$$

Introduction of Eqs. (11.3) into the above expression leads to the following form involving the strains and elastic constants:

$$U_m = \frac{Et}{2(1 - \nu^2)} \iint_A [(\varepsilon_{x0} + \varepsilon_{y0})^2 - 2(1 - \nu)(\varepsilon_{x0}\varepsilon_{y0} - \tfrac{1}{4}\gamma_{xy0}^2)]\, dx\, dy \tag{11.8b}$$

Expressions (11.7) and (11.8) permit the energy to be evaluated readily for a number of commonly encountered shells of regular shape and regular loading. The strain energy plays an important role in treating the bending and buckling problems of shells (Chap. 13).

11.6 AXISYMMETRICALLY LOADED CIRCULAR CYLINDRICAL SHELLS

Pipes, tanks, boilers, and various other vessels under internal pressure exemplify the axisymmetrically loaded cylindrical shell. Owing to symmetry, an element cut from a cylinder of radius a will have acting on it only the stress resultants shown in Fig. 11.3: N_θ, M_θ, N_x, and Q_x. Furthermore, the circumferential force and moment, N_θ and M_θ, do not vary with θ. The circumferential displacement v thus vanishes and we need consider only the x and z displacements, u and w.

Subject to the foregoing simplifications, only three of the six equilibrium

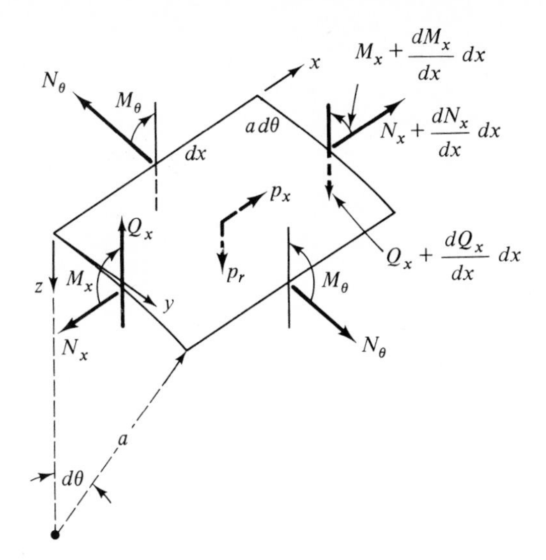

Figure 11.3

equations of the shell element remain to be satisfied. Suppose also that the external loading is as shown in Fig. 11.3. Equilibrium in the x (axial) and z (radial) directions now requires that

$$\frac{dN_x}{dx} dx \cdot a\,d\theta + p_x \cdot a\,d\theta \cdot dx = 0$$

$$\frac{dQ_x}{dx} dx \cdot a\,d\theta + N_\theta\,dx \cdot d\theta + p_r \cdot dx \cdot a\,d\theta = 0 \qquad (a)$$

Equilibrium of moments about the y axis is governed by

$$\frac{dM_x}{dx} dx \cdot a\,d\theta - Q_x \cdot a\,d\theta \cdot dx = 0 \qquad (b)$$

Equations (a) and (b) are, through cancellation of like terms, rewritten

$$\frac{dN_x}{dx} + p_x = 0$$

$$\frac{dQ_x}{dx} + \frac{1}{a} N_\theta + p_r = 0 \qquad (11.9a,b,c)$$

$$\frac{dM_x}{dx} - Q_x = 0$$

It is interesting to note that Eq. (11.9c) is a statement of the basic beam relationship: the shearing force is the first derivative of the bending moment. From Eq. (11.9a) the axial force N_x is

$$N_x = -\int p_x\,dx + c \qquad (c)$$

where c is a constant of integration. Clearly, the unknown quantities Q_x, N_θ, and M_x cannot be determined from Eqs. (11.9*b*) and (11.9*c*) alone and it is therefore necessary to examine the midsurface displacements.

Because $v = 0$, the strain-displacement relations are, from symmetry,

$$\varepsilon_x = \frac{du}{dx} \qquad \varepsilon_\theta = \frac{(a - w)\,d\theta - a\,d\theta}{a\,d\theta} = -\frac{w}{a} \tag{d}$$

Applying Hooke's law, we have

$$N_x = \frac{Et}{1 - \nu^2}(\varepsilon_x + \nu\varepsilon_\theta) = \frac{Et}{1 - \nu^2}\left(\frac{du}{dx} - \nu\frac{w}{a}\right)$$

from which

$$\frac{du}{dx} = \frac{1 - \nu^2}{Et}N_x + \nu\frac{w}{a} \tag{11.10a}$$

Then, from Hooke's law and Eqs. (*d*), the circumferential force is found to be

$$N_\theta = \frac{Et}{1 - \nu^2}(\varepsilon_\theta + \nu\varepsilon_x) = -\frac{Et}{1 - \nu^2}\left(\frac{w}{a} - \nu\frac{du}{dx}\right) \tag{11.11}$$

The bending moment displacement relations are the same as for a plane bent into a cylindrical surface. That is, because $d^2w/dy^2 = 0$,

$$M_x = -D\frac{d^2w}{dx^2} \qquad M_\theta = \nu M_x \tag{11.12}$$

where D is the fluxural rigidity of the shell, given by Eq. (1.11). Employing Eqs. (11.9*b*) and (11.9*c*) and eliminating Q_x, the following is obtained:

$$\frac{d^2M_x}{dx^2} + \frac{1}{a}N_\theta + p_r = 0$$

Finally, when the above expression is combined with Eqs. (11.10*a*), (11.11) and (11.12), we have

$$\frac{d^2}{dx^2}\left(D\frac{d^2w}{dx^2}\right) + \frac{Et}{a^2}w - \nu N_x - p_r = 0 \tag{11.13a}$$

For a shell of constant thickness, Eq. (11.13*a*) becomes

$$D\frac{d^4w}{dx^4} + \frac{Et}{a^2}w - \nu\frac{N_x}{a} - p_r = 0 \tag{11.13b}$$

A more convenient form of this expression is

$$\frac{d^4w}{dx^4} + 4\beta^4w - \frac{\nu N_x}{aD} = \frac{p_r}{D} \tag{11.10b}$$

Here

$$\beta^4 = \frac{Et}{4a^2D} = \frac{3(1 - \nu^2)}{a^2t^2} \tag{11.14}$$

and where *geometric parameter* β has the dimension of L^{-1}, the reciprocal of length. Equations (11.10*b*) or (11.13) and (11.10*a*) represent the *governing displacement conditions* for a symmetrically loaded circular cylindrical shell. When an *axial load does not exist*, $N_x = 0$, and Eqs. (11.10) simplify to

$$\frac{du}{dx} = \nu \frac{w}{a}$$
$$\frac{d^4 w}{dx^4} + 4\beta^4 w = \frac{p_r}{D} \qquad (11.15a, b)$$

The first of these, upon integration, directly yields u. Expression (11.15*b*) is an ordinary differential equation with constant coefficients. It also represents[1] the equation of a beam of flexural rigidity D, resting on an elastic foundation and subject to loading p_r.

The homogeneous solution of Eq. (11.15*b*) is given by

$$w_h = c_1 e^{m_1 x} + c_2 e^{m_2 x} + c_3 e^{m_3 x} + c_4 e^{m_4 x}$$

Here c_1, c_2, c_3, c_4 are constants and m_1, m_2, m_3, m_4 are the roots of the expression

$$m^4 + 4\beta^4 = 0$$

This equation may be written, by addition and subtraction of $4m^2\beta^2$, as $(m^2 + 2\beta^2)^2 - 4m^2\beta^2 = 0$. Hence, $m^2 + 2\beta^2 = \pm 2m\beta$. We thus have

$$m = \pm\beta(1 \pm i)$$

It follows that

$$w_h = e^{-\beta x}(c_1 e^{i\beta x} + c_2 e^{-i\beta x}) + e^{\beta x}(c_3 e^{i\beta x} + c_4 e^{-i\beta x})$$

Let $f(x)$ represent the particular solution w_p. It is noted that *the results of membrane theory can always be considered as the particular solutions of the equations of bending theory* (Sec. 12.4).

The general solution of Eq. (11.15*b*) may therefore be written

$$w = e^{-\beta x}(C_1 \cos \beta x + C_2 \sin \beta x) + e^{\beta x}(C_3 \cos \beta x + C_4 \sin \beta x) + f(x) \qquad (11.16)$$

where C_1, C_2, C_3, C_4 are arbitrary constants of integration, determined on the basis of the appropriate boundary conditions.

Section 11.7 serves to illustrate application of the theory.

11.7 A TYPICAL CASE OF THE AXISYMMETRICALLY LOADED CYLINDRICAL SHELL

This section deals with the bending problem of a cylinder with length very large compared with its diameter, the so-called *infinite* cylinder, subjected to a load P uniformly distributed along a circular section (Fig. 11.4). Inasmuch as there is no

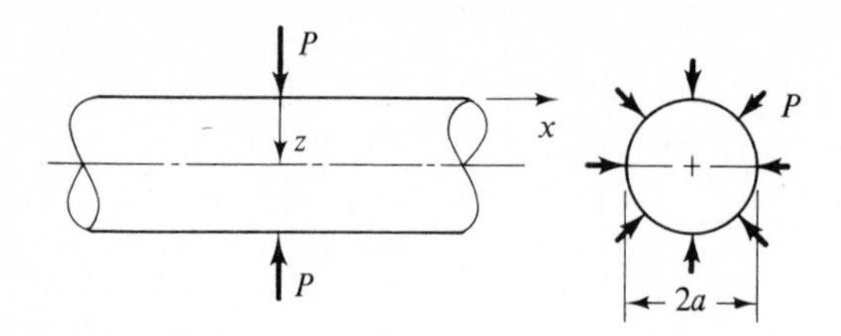

Figure 11.4

pressure p_r distributed over the surface of the shell and $N_x = 0$, we set $f(x) = 0$ in Eq. (11.16). The solution may be written as follows:

$$w = e^{-\beta x}(C_1 \cos \beta x + C_2 \sin \beta x) + e^{\beta x}(C_3 \cos \beta x + C_4 \sin \beta x) \quad (11.17)$$

Owing to shell symmetry, boundary conditions for the right half are deduced from the fact that as $x \to \infty$, the deflection and all derivatives of w with respect to x must vanish. These conditions are fulfilled if $C_3 = C_4 = 0$ in Eq. (11.17). We thus have

$$w = e^{-\beta x}(C_1 \cos \beta x + C_2 \sin \beta x) \quad (11.18)$$

Inasmuch as $N_x = 0$, Eqs. (11.10*a*) and (11.11) yield

$$N_\theta = -\frac{Etw}{a} \quad (11.19)$$

From Eqs. (11.12) and (11.9*c*),

$$M_x = -D\frac{d^2w}{dx^2} \qquad M_\theta = -\nu D\frac{d^2w}{dx^2} \qquad Q_x = \frac{dM_x}{dx} = -D\frac{d^3w}{dx^3} \quad (11.20)$$

The conditions applicable immediately to the right of the load are

$$Q_x = -D\frac{d^3w}{dx^3} = -\frac{p}{2} \qquad \frac{dw}{dx} = 0 \quad (a)$$

These describe the respective requirements that each half of the cylinder carry one-half the external load and that the slope vanish at the center owing to the symmetry. Introducing Eqs. (*a*) into Eq. (11.18) and setting $x = 0$,

$$C_1 = C_2 = \frac{p}{8\beta^3 D}$$

The displacement is therefore

$$w = \frac{Pe^{-\beta x}}{8\beta^3 D}(\sin \beta x + \cos \beta x) \quad (11.21)$$

This result may be expressed

$$w = \frac{Pe^{-\beta x}}{8\beta^3 D}\left[\sqrt{2}\sin\left(\beta x + \frac{\pi}{4}\right)\right]$$

We observe that the *deflection attenuates with distance as an exponentially damped sine wave* of wave length, for $\nu = 0.3$: $2\pi/\beta \approx 4.89\sqrt{at}$.

Table 11.1

βx	$f_1(\beta x)$	$f_2(\beta x)$	$f_3(\beta x)$	$f_4(\beta x)$	βx	$f_1(\beta x)$	$f_2(\beta x)$	$f_3(\beta x)$	$f_4(\beta x)$
0.0	1.000	0.000	1.000	1.000	3.0	−0.042	0.007	−0.056	−0.049
0.2	0.965	0.163	0.640	0.802	3.2	−0.043	−0.002	−0.038	−0.041
0.4	0.878	0.261	0.356	0.617	3.4	−0.041	−0.009	−0.024	−0.032
0.6	0.763	0.310	0.143	0.453	3.6	−0.037	−0.012	−0.012	−0.024
0.8	0.635	0.322	−0.009	0.313	3.8	−0.031	−0.014	−0.004	−0.018
1.0	0.508	0.310	−0.111	0.199	4.0	−0.026	−0.014	0.002	−0.012
1.2	0.390	0.281	−0.172	0.109	4.2	−0.020	−0.013	−0.006	−0.007
1.4	0.285	0.243	−0.201	0.042	4.4	−0.016	−0.012	0.008	−0.004
1.6	0.196	0.202	−0.208	−0.006	4.6	−0.011	−0.010	0.009	−0.001
1.8	0.123	0.161	−0.199	−0.038	4.8	−0.008	−0.008	0.009	0.001
2.0	0.067	0.123	−0.179	−0.056	5.0	−0.005	−0.007	0.008	0.002
2.2	0.024	0.090	−0.155	−0.065	5.5	0.000	−0.003	0.006	0.003
2.4	−0.006	0.061	−0.128	−0.067	6.0	0.002	−0.001	0.003	0.002
2.6	−0.025	0.038	−0.102	−0.064	6.5	0.002	0.000	0.001	0.001
2.8	−0.037	0.020	−0.078	−0.057	7.0	0.001	0.001	0.000	0.001

The following notations are used to more conveniently represent the expressions for deflection and stress resultants:

$$
\begin{aligned}
f_1(\beta x) &= e^{-\beta x}(\cos \beta x + \sin \beta x) \\
f_2(\beta x) &= e^{-\beta x} \sin \beta x = -\frac{1}{2\beta} f'_1 \\
f_3(\beta x) &= e^{-\beta x}(\cos \beta x - \sin \beta x) = \frac{1}{\beta} f'_2 = -\frac{1}{2\beta^2} f''_1 \\
f_4(\beta x) &= e^{-\beta x} \cos \beta x = -\frac{1}{2\beta} f'_3 = -\frac{1}{2\beta^2} f''_2 = \frac{1}{4\beta^3} f'''_1 \\
f_1(\beta x) &= -\frac{1}{\beta} f'_4
\end{aligned}
\tag{11.22}
$$

Table 11.1 furnishes numerical values of these functions for various values of βx. The term βx is dimensionless and is usually thought of as expressed in radians. Substituting Eq. (11.21) into Eq. (11.19) and (11.20),

$$
\begin{aligned}
w &= \frac{P}{8\beta^3 D} f_1(\beta x) \\
N_\theta &= -\frac{EtP}{8\beta^3 Da} f_1(\beta x) \\
M_x &= \frac{P}{4\beta} f_3(\beta x) \\
M_\theta &= \frac{\nu P}{4\beta} f_3(\beta x) \\
Q_x &= -\frac{P}{2} f_4(\beta x)
\end{aligned}
\tag{11.23}
$$

These expressions are valid for $x \geq 0$. For the left half of the cylinder, one takes x in the opposite direction to that shown in Fig. 11.4. The maximum deflection and moment occur at $x = 0$, found from Eqs. (11.23) to be

$$w_{\max} = \frac{P}{8\beta^3 D} = \frac{Pa^2\beta}{2Et} \qquad M_{\max} = \frac{P}{4\beta} \tag{11.24}$$

The largest values of bending stress are found at $x = 0$ and $z = t/2$, determined by applying Eqs. (11.4) and (11.23):

$$\sigma_{x,\,\max} = \frac{3P}{2\beta t^2} \qquad \sigma_{\theta,\,\max} = \frac{P\beta}{2}\left(-\frac{a}{t} + \frac{3\nu}{\beta^2 t^2}\right) \tag{b}$$

The foregoing are the maximum axial and circumferential stresses in the cylinder, respectively.

Referring to Table 11.1, it is observed that each quantity in Eqs. (11.22) and hence in Eqs. (11.23), decreases with increasing βx. Because of this, in most engineering applications, *the effect of the concentrated loads may be neglected at locations for which* $x > (\pi/\beta)$. Therefore, it is concluded that bending is of a local character. A shell of length $L = 2\pi/\beta$, loaded at midlength, will experience maximum deflection and bending moment nearly identical with those associated with a *long shell.*

Application of Eqs. (11.23) together with the principle of superposition permits the determination of deflection and stress resultants in long cylinders under any other kind of loading.

Example 11.1 A very long cylinder of radius a is subjected to a uniform loading p over L of its length (Fig. 11.5). Derive an expression for the deflection at an arbitrary point O within length L.

SOLUTION Through the use of the first of Eqs. (11.23) the deflection Δw at point O owing to load $P_x = p\,dx$ is

$$\Delta w = \frac{p\,dx}{8\beta^3 D} f_1(\beta x)$$

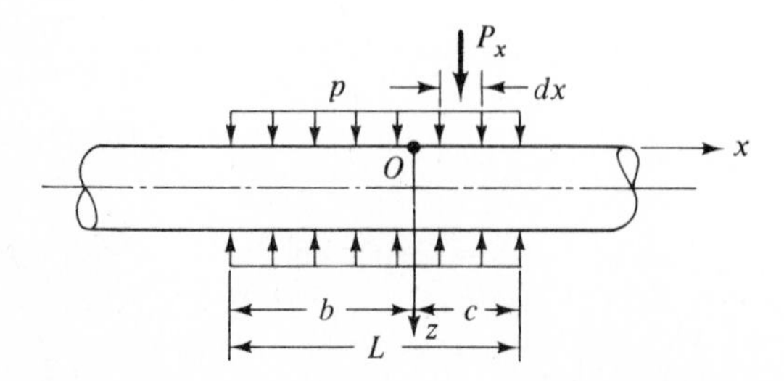

Figure 11.5

The displacement at point O produced by the entire load is then

$$w = \int_0^b \frac{p}{8\beta^3 D} f_1(\beta x)\, dx + \int_0^c \frac{p}{8\beta^3 D} f_1(\beta x)\, dx$$

Inserting into the above $f_1(\beta x)$ from Eqs. (11.22), we obtain after integration

$$w = \frac{pa^2}{2Et}[2 - e^{-\beta b} \cos(\beta b) - e^{\beta c} \cos(\beta c)]$$

or

$$w = \frac{pa^2}{2Et}[2 - f_4(\beta b) - f_4(\beta c)] \qquad (c)$$

The maximum deflection of the cylinder occurs at midlength of the distributed load, at the point at which $b = c$. Note that, *if b and c are large*, the values of $f_4(\beta b)$ and $f_4(\beta c)$ are quite small and the deflection will approximately equal pa^2/Et.

In a like manner, applying the last four of Eqs. (11.23), we can obtain the expressions for the stress resultants at O.

11.8 SHELLS OF REVOLUTION UNDER AXISYMMETRICAL LOADS

Let us consider a body in the *general form of a shell of revolution* subjected to rotationally symmetrical loads. The sphere, cone, and circular cylinder (Sec. 11.6) are typical simple geometries in this category. To begin with, we define the state of stress resultant acting at a point of such shells, represented by an infinitesimal element in Fig. 11.6. Conditions of symmetry dictate that only the resultants Q_ϕ, M_θ, M_ϕ, N_θ, and N_ϕ exist, and that the normal forces N_θ and the bending moments M_θ cannot vary with θ. The notations for the radii of curvature and the angular orientation are identical with those of membrane theory (Fig. 10.2).

The development of the equilibrium equations of shell element $ABCD$ proceeds in a manner similar to that described in Sec. 10.4. The condition that summation of the y-directed forces be equal to zero is fulfilled by

$$\frac{d}{d\phi}(N_\phi r_0\, d\theta)\, d\phi - N_\theta r_1\, d\theta\, d\phi \cos\phi - Q_\phi \cdot r_0\, d\theta \cdot d\phi + p_y r_1\, d\phi r_0\, d\theta = 0 \qquad (a)$$

The first two and the last terms of the above are already specified by Eq. (10.2). The third term is due to the shear forces $Q_\phi \cdot r_0\, d\theta$, on faces AC and BD of the element. These faces form an angle $d\phi$ with one another.

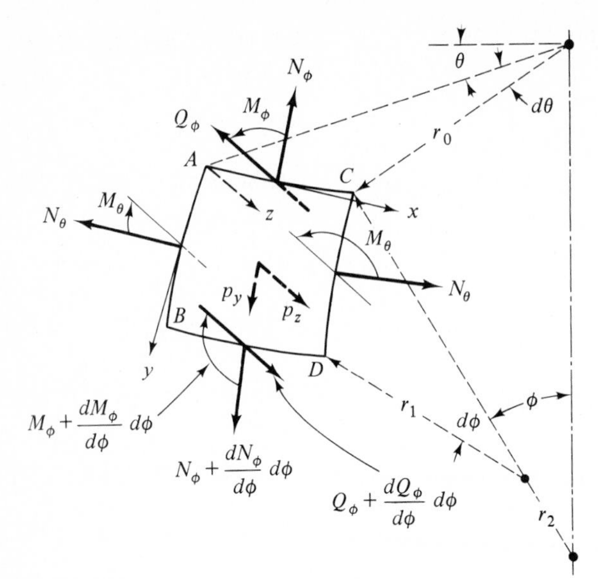

Figure 11.6

The condition of equilibrium in the z direction may readily be obtained by adding the increment of the shear forces $Q_\phi \cdot r_0\, d\theta$ to Eq. (a) of Sec. 10.4. That is,

$$N_\phi r_0\, d\theta\, d\phi + N_\theta r_1\, d\phi\, d\theta \sin\phi + \frac{d}{d\phi}(Q_\phi r_0\, d\theta)\, d\phi + p_z r_1\, d\phi r_0\, d\theta = 0 \quad (b)$$

Finally, one writes the expression

$$\frac{d}{d\phi}(M_\phi \cdot r_0\, d\theta)\, d\phi - Q_\phi \cdot r_0\, d\theta \cdot r_1\, d\phi - M_\theta \cdot r_1\, d\phi \cdot \cos\phi \cdot d\theta = 0 \quad (c)$$

for the equilibrium of the forces around the x axis. The terms of Eq. (c) are described as follows: the first is the increment of the bending moments $M_\phi r_0\, d\theta$; the second represents the moment of the shear forces $Q_\phi r_0\, d\theta$; the third is the resultant of the moments $M_\theta r_1\, d\phi$. Note that the two moment *vectors* $M_\theta r_1\, d\phi$ acting on the side faces AB and CD of the element are not parallel. Their horizontal components $M_\theta r_1\, d\phi \cdot \cos\phi$ form an angle $d\theta$ with one another and thus have a resultant expressed by the last terms.

Dropping the factor $d\theta\, d\phi$ in Eqs. (a) to (c), common to all terms, we obtain the *equations of equilibrium:*

$$\begin{aligned} &\frac{d}{d\phi}(N_\phi r_0) - N_0 r_1 \cos\phi - r_0 Q_\phi + r_0 r_1 p_y = 0 \\ &N_\phi r_0 + N_\theta r_1 \sin\phi + \frac{d(Q_\phi r_0)}{d\phi} + p_z r_1 r_0 = 0 \\ &\frac{d}{d\phi}(M_\phi r_0) - M_\theta r_1 \cos\phi - Q_\phi r_1 r_0 = 0 \end{aligned} \qquad (11.25)$$

The governing equations for the common shells of revolution subjected to axisymmetrical loads may be derived from the above expressions.

Conical shell (Fig. 10.4) For this case it is observed in Sec. 10.5 that $\phi = \text{constant}$ ($r_1 = \infty$). Thus,

$$r_2 = s \cot \phi \qquad r_1\, d\phi = ds \qquad N_\phi = N_s \qquad M_\phi = M_s$$

Employing these, the equations of equilibrium (11.25) assume the form:

$$\begin{aligned} &\frac{d}{ds}(N_s s) - N_\theta = -p_y s \\ &N_\theta + \frac{d}{ds}(Q_s s) \cot \phi = -p_z s \cot \phi \\ &\frac{d}{ds}(M_s s) - Q_s s + M_\theta = 0 \end{aligned} \tag{11.26}$$

Spherical shells Denoting by a the radius of the midsurface of the shell, we have $r_1 = r_2 = a$ and $r_0 = s \sin \phi$. The equilibrium conditions (11.25) then simplify to

$$\begin{aligned} &\frac{d}{d\phi}(N_\phi \sin \phi) - N_\theta \cos \phi - Q_\phi \sin \phi = p_y a \sin \phi \\ &N_\phi \sin \phi + N_\theta \sin \phi + \frac{d}{d\phi}(Q_\phi \sin \phi) = -p_z a \sin \phi \\ &\frac{d}{d\phi}(M_\phi \sin \phi) - M_\theta \cos \phi - a Q_\phi \sin \phi = 0 \end{aligned} \tag{11.27}$$

Cylindrical shells (Fig. 11.3) We can use Eq. (11.26) for the cone, letting $s = x = r_2 \tan \phi$, $\phi = \pi/2$, and $r_2 = a$. By so doing, one obtains expressions which are identical with Eqs. (11.9) of Sec. 11.6.

Interestingly, the first expression of Eqs. (11.26) agrees with the corresponding expression of membrane theory. It is noted that by cancelling the terms involving shear forces and moments, Eqs. (11.26), (11.27), and (11.9) reduce to the conditions of membrane theory of conical, spherical, and cylindrical shells, respectively.

11.9 GOVERNING EQUATIONS FOR AXISYMMETRICAL DISPLACEMENTS

In the preceding section it is observed that three equilibrium conditions (11.25) of an axisymmetrically loaded shell of revolution contain five unknown stress resultants N_ϕ, N_θ, Q_ϕ, M_ϕ, and M_θ. To reduce the number of unknowns to

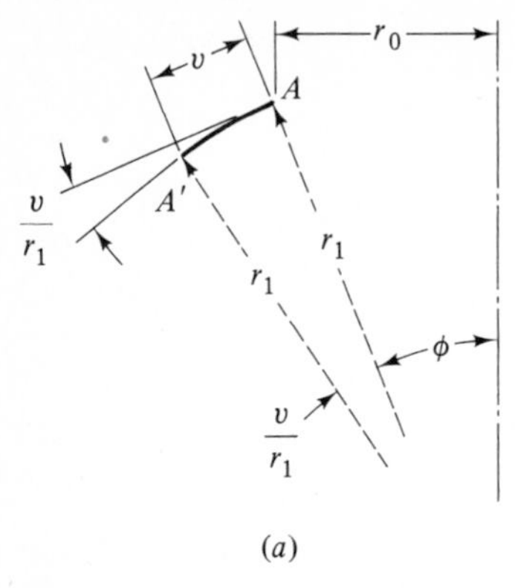

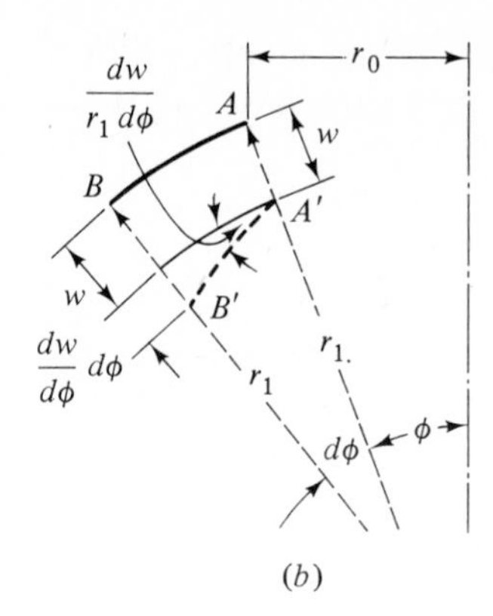

Figure 11.7

three, relationships involving the forces (N_ϕ, N_θ), the moments (M_ϕ, M_θ), and the displacement components (v, w) are developed in the paragraphs which follow.

The membrane strains and the displacements at a point of the midsurface are connected by Eqs. (a) and (b) of Sec. 10.6:

$$\varepsilon_\phi = \frac{1}{r_1}\frac{dv}{d\phi} - \frac{w}{r_1} \qquad \varepsilon_\theta = \frac{v}{r_2}\cot\phi - \frac{w}{r_2} \tag{11.28}$$

The force-resultant strain relations (11.3) then lead to

$$\begin{aligned} N_\phi &= \frac{Et}{1-\nu^2}\left[\frac{1}{r_1}\left(\frac{dv}{d\phi} - w\right) + \frac{\nu}{r_2}(v\cot\phi - w)\right] \\ N_\theta &= \frac{Et}{1-\nu^2}\left[\frac{1}{r_2}(v\cot\phi - w) + \frac{\nu}{r_1}\left(\frac{dv}{d\phi} - w\right)\right] \end{aligned} \tag{11.29}$$

Identical expressions for M_ϕ and M_θ can be obtained by considering the variations in curvature of a shell element (Fig. 11.6). For this purpose we examine the meridional section of the shell element (Fig. 11.7). The rotations of the tangent of the top face AC consist of: a rotation with respect to a perpendicular to the meridian plane by an amount v/r_1 owing to the displacement v of point A to point A' (Fig. 11.7a); a rotation about the same axis by $dw/(r_1\,d\phi)$ produced by the additional displacement of point B with respect to point A (Fig. 11.7b). The total rotation of the upper edge is therefore

$$\frac{v}{r_1} + \frac{dw}{r_1\,d\phi} \tag{a}$$

The top and bottom faces of the element initially makes an angle $d\theta$ with one another. The rotation of the bottom face BD is then

$$\frac{v}{r_1} + \frac{dw}{r_1\,d\phi} + \frac{d}{d\phi}\left(\frac{v}{r_1} + \frac{dw}{r_1\,d\phi}\right)d\phi$$

The variation of curvature of the meridian, the angular variation divided by the length $r_1\,d\phi$ of the arc, is thus

$$\chi_\phi = \frac{1}{r_1}\frac{d}{d\phi}\left(\frac{v}{r_1} + \frac{dw}{r_1\,d\phi}\right) \tag{11.30}$$

It is observed that, owing to the symmetry of deformation, each of the lateral edges AB and CD of the shell element also rotates in its meridian plane by an angle defined by Eq. (a). It may be verified that the unit normal to the right face of a shell element has a y-directed component equal to $-d\phi \cdot \cos\phi$. Thus, the rotation of face CD in its own plane has a component with respect to the y axis given by

$$-\left(\frac{v}{r_1} + \frac{dw}{r_1\,d\phi}\right)\cos\phi\;d\theta$$

Dividing this rotation by length $r_0\,d\theta$, we have the change of curvature:

$$\chi_\theta = \left(\frac{v}{r_1} + \frac{dw}{r_1\,d\phi}\right)\frac{\cos\phi}{r_0} = \left(\frac{v}{r_1} + \frac{dw}{r_1\,d\phi}\right)\frac{\cot\phi}{r_2} \tag{11.31}$$

Finally, inserting Eqs. (11.30) and (11.31) into Eqs. (11.3), the moment-displacement relations are obtained:

$$\begin{aligned} M_\phi &= -D\left[\frac{1}{r_1}\frac{d}{d\phi}\left(\frac{v}{r_1} + \frac{dw}{r_1\,d\phi}\right) + \frac{\nu}{r_2}\left(\frac{v}{r_1} + \frac{dw}{r_1\,d\phi}\right)\cot\phi\right] \\ M_\theta &= -D\left[\left(\frac{v}{r_1} + \frac{dw}{r_1\,d\phi}\right)\frac{\cot\phi}{r_2} + \frac{\nu}{r_1}\frac{d}{d\phi}\left(\frac{v}{r_1} + \frac{dw}{r_1\,d\phi}\right)\right] \end{aligned} \tag{11.32}$$

Now Eqs. (11.25) together with Eqs. (11.29) and (11.32) lead to three expressions in three unknowns: v, w, and Q_ϕ. Furthermore, using the first of the resulting three equations, the shear force Q_ϕ can readily be eliminated in the last two. The expressions (11.25) are thus reduced to two equations in two unknowns: v and w. These *governing equations for displacements*, usually transformed into *new variables*,[32] are employed to treat the shell-bending problem.

An application of the equations derived above is presented in the next section.

11.10 COMPARISON OF BENDING AND MEMBRANE STRESSES

The equations for the stress resultants (11.29) and (11.32) may be used to gauge the accuracy of the membrane analysis discussed in Chap. 10. To determine the bending moments, which were omitted in the membrane theory, the expressions for the displacements developed in Sec. 10.6 are introduced into Eqs. (11.32). The bending stresses are then obtained from Eqs. (11.4). Upon comparison of the stress magnitudes so determined with those of the membrane stresses, conclusions may be drawn with respect to the accuracy of the membrane theory, as in the following example.

Example 11.2 Consider the spherical dome supporting its own weight described in Example 10.1. Assume that the supports are as shown in Fig. 10.5*a*. (*a*) Determine the bending stress, and (*b*) compare the bending and membrane stresses.

SOLUTION (*a*) The displacements v and w, from Eqs. (*g*) and (*c*) of Sec. 10.6 together with Eqs. (10.13), are

$$v = \frac{a^2 p(1+\nu)}{Et}\left(\frac{1}{1+\cos\alpha} - \frac{1}{1+\cos\phi} + \ln\frac{1+\cos\phi}{1+\cos\alpha}\right)\sin\phi$$
$$w = v\cot\phi - \frac{a^2 p}{Et}\left(\frac{1+\nu}{1+\cos\phi} - \cos\phi\right) \tag{a}$$

Introduction of the above into Eqs. (11.32) yields the following expressions for the moments:

$$M_\theta = M_\phi = \frac{pt^2}{12}\frac{2+\nu}{1-\nu}\cos\phi \tag{b}$$

The magnitude of the bending stress σ_b at the surface of the shell is then

$$\sigma_b = \frac{p}{2}\frac{2+\nu}{1-\nu}\cos\phi \tag{c}$$

(b) The value of membrane stress σ from Eqs. (10.13) is

$$\sigma = \frac{ap}{t(1+\cos\phi)}$$

The ratio of the bending stress to the membrane stress,

$$\frac{\sigma_b}{\sigma} = \frac{2+\nu}{2(1-\nu)}\frac{t}{a}(1+\cos\phi)\cos\phi \tag{d}$$

has a maximum value at the top of the shell ($\phi = 0$). For $\nu = 0.3$,

$$\left(\frac{\sigma_b}{\sigma}\right)_{\max} = 3.29\frac{t}{a} \tag{e}$$

For a thin shell ($a > 20t$) the above ratio is small. It is thus seen that membrane theory provides a result of sufficient accuracy.

The values of the forces N_θ and N_ϕ may be determined by introducing Eq. (*b*) into Eq. (11.27). However, it can be verified that (see Prob. 11.8) these closer approximations for the forces will differ little from the results (10.13).

We are led to conclude from the foregoing example that, for thin shells, the stresses (and the displacements) as ascertained from the membrane theory have values of acceptable accuracy. The results will of course be inaccurate if expansion of the shell edge is prevented by a support. In the latter case, the forces exerted by the support on the shell produce bending in the vicinity of the edge. The local stresses caused by concentrated forces discussed in Sec. 11.7, will be treated in considerable detail in Chaps. 12 and 13.

11.11 THE FINITE ELEMENT REPRESENTATIONS OF SHELLS OF GENERAL SHAPE

The factors which complicate the analysis of shell problems may generally be reduced to irregularities in the shape or thickness of the shell and non-uniformity of the applied load. By replacing the actual geometry of the structure and the load configuration with suitable finite element approximations (Sec. 5.7), very little sacrifice in accuracy is encountered.

Consider the case of a shell of variable thickness and general arbitrary shape. There are a number of ways of obtaining an equivalent shell which will not significantly compromise the elastic response. For example, one can replace the actual shell with a series of *curved* or *flat* (straight) triangular elements or finite elements of other form, attached at their edges and corners. Whatever the true load configuration may be, it is then reduced to a series of concentrated or distributed forces applied to each finite element.

When a shell of revolution is subjected to nonuniform load, the usual finite element approach is to replace a shell element with two flat elements, one subjected to direct force resultants and the other to moment resultants. The applied load may be converted to uniform or concentrated forces also acting on the replacement elements. In-plane and bending effects may then be analyzed separately and superposed. Hence, a shell element may be developed as a combination of a *membrane* element and a *plate* element of the same shape. The shell is thus idealized as an assemblage of flat elements.

Curved elements have been proposed to secure an improved approximation of shells, but analysis employing them is more complex than is the case using straight elements. In the general treatment of axisymetrically loaded shells given in the next section, the latter elements are considered.

11.12 THE FINITE ELEMENT SOLUTION OF AXISYMMETRICALLY LOADED SHELLS

An axisymmetrically loaded shell may be represented by a series of *conical frustra* (Fig. 11.8*a*). Each element is thus a ring generated by the *straight line* segment between two parallel circles or "nodes," say i and j (Fig. 11.8*b*). The

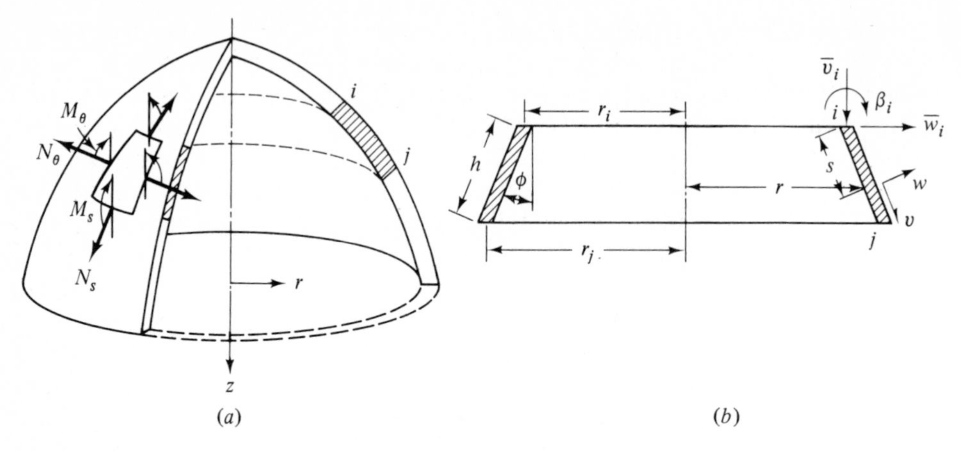

Figure 11.8

thickness may vary from element to element. As before, the displacement of a point in the midsurface is specified by two components v and w in the meridional and normal directions, respectively. Referring to Fig. 11.8b, the "strain"-displacement relations, Eqs. (11.28), (11.30), and (11.31), simplify to

$$\{\varepsilon\} = \begin{Bmatrix} \varepsilon_s \\ \varepsilon_\theta \\ \chi_s \\ \chi_\theta \end{Bmatrix} = \begin{Bmatrix} dv/ds \\ (w \cos\phi + v \sin\phi)/r \\ -d^2w/ds^2 \\ -(dw/ds)\sin\phi/r \end{Bmatrix} \tag{11.33}$$

The stress-resultant strain relations, Eqs. (11.29) and (11.32), are then

$$\begin{Bmatrix} N_s \\ N_\theta \\ M_s \\ M_\theta \end{Bmatrix} = \frac{Et}{1-\nu^2}\begin{bmatrix} 1 & \nu & 0 & 0 \\ \nu & 1 & 0 & 0 \\ 0 & 0 & t^2/2 & \nu t^2/12 \\ 0 & 0 & \nu t^2/12 & t^2/12 \end{bmatrix}\begin{Bmatrix} \varepsilon_s \\ \varepsilon_\theta \\ \chi_s \\ \chi_\theta \end{Bmatrix} \tag{11.34a}$$

or

$$\{N_s, N_\theta, M_s, M_\theta\} = [D]\{\varepsilon\} \tag{11.34b}$$

Here $[D]$ is the elasticity matrix for the isotropic axisymmetrically loaded shell.

Three displacements are chosen at each node (Fig. 11.8b). The element *nodal displacement* matrix is thus

$$\{\delta\}_e = \begin{Bmatrix} \delta_i \\ \delta_j \end{Bmatrix} = \begin{Bmatrix} \bar{v}_i, \bar{w}_i, \beta_i \\ \bar{v}_j, \bar{w}_j, \beta_j \end{Bmatrix} \tag{11.35}$$

where $\bar{v}$, $\bar{w}$, and β represent the axial movement, radial movement, and rotation, respectively. The *displacements within* the element, expressed in standard form, are

$$\{f\} = \begin{Bmatrix} v \\ w \end{Bmatrix} = [N]\{\delta\}_e \tag{11.36}$$

These are to be determined from $\{\delta\}_e$ and position s; slope and displacement continuity are maintained throughout the element. The matrix $[N]$ is a function of position yet to be developed. On evaluating v and w at the nodes i and j, we can relate them to Eqs. (11.35) through use of a transformation matrix. At node i, for example,

$$\begin{Bmatrix} v_i \\ w_j \\ (dw/ds)_i \end{Bmatrix} = \begin{bmatrix} \cos\phi & \sin\phi & 0 \\ -\sin\phi & \cos\phi & 0 \\ 0 & 0 & 1 \end{bmatrix} \begin{Bmatrix} \bar{v}_i \\ \bar{w}_i \\ \beta_i \end{Bmatrix} = [\lambda]\{\delta_i\} \tag{11.37}$$

The following general expressions, employed for $\{f\}$, contain six constants:

$$\begin{aligned} v &= \alpha_1 + \alpha_2 s \\ w &= \alpha_3 + \alpha_4 s + \alpha_5 s^2 + \alpha_6 s^3 \end{aligned} \tag{11.38}$$

To determine the values of the α's, the coordinate s of the nodal points is substituted in the displacement functions (11.38). This will generate six equations in which the only unknowns are the coefficients. By so doing, we can solve for α_1 to α_6 in terms of the nodal displacements $v_i, \ldots, w_i$ and finally obtain[17]

$$\begin{Bmatrix} v \\ w \end{Bmatrix} = \begin{bmatrix} 1-s_1 & 0 & 0 & s_1 & 0 & 0 \\ 0 & 1-3s_1^2+2s_1^3 & s_1(1-2s_1+s_1^2)h & 0 & s_1^2(3-2s_1) & s_1^2(-1+s_1)h \end{bmatrix} \begin{Bmatrix} v_i \\ w_i \\ (dw/ds)_i \\ v_j \\ w_j \\ (dw/ds)_j \end{Bmatrix} \tag{11.39}$$

wherein

$$s_1 = \frac{s}{h} \qquad (0 \le s_1 \le 1) \tag{a}$$

Denoting in Eqs. (11.39) the two-by-six matrix by $[\bar{P}]$, we have

$$\begin{Bmatrix} v \\ w \end{Bmatrix} = [\bar{P}] \begin{bmatrix} [\lambda] & 0 \\ 0 & [\lambda] \end{bmatrix} \{\delta\}_e = \left[[\bar{P}_i][\lambda], [\bar{P}_j][\lambda]\right]\{\delta\}_e = [P]\{\delta\}_e \tag{11.40}$$

Eqs. (11.33) then lead to

$$\{\varepsilon\} = [B]\{\varepsilon\}_e = \left[[B_i][\lambda], [B_j][\lambda]\right]\{\delta\}_e \tag{11.41}$$

in which

$$[B_i] = \begin{bmatrix} -1/h & 0 & 0 \\ (1-s_1)\sin\phi/r & (1-3s_1^2+2s_1^3)\cos\phi/r & hs_1(1-2s_1+s_1^2)\cos\phi/r \\ 0 & 6(1-2s_1)/h^2 & 2(2-3s_1)/h \\ 0 & 6s_1(1-s_1)\sin\phi/rh & (-1+4s_1-3s_1^2)\sin\phi/r \end{bmatrix} \tag{11.42a}$$

$$[B_j] = \begin{bmatrix} 1/h & 0 & 0 \\ s_1\sin\phi/r & s_1^2(3-2s_1)\cos\phi/r & hs_1^2(-1+s_1)\cos\phi/r \\ 0 & 6(-1+2s_1)/h^2 & 2(1-3s_1)/h \\ 0 & 6s_1(-1+s_1)\sin\phi/rh & s_1(2-3s_1)\sin\phi/r \end{bmatrix} \tag{11.42b}$$

The stiffness matrix for the element is given by Eq. (5.23) in the form

$$[k]_e = \int_A [B]^T[D][B]\, dA \tag{b}$$

Here the area of the element equals

$$dA = 2\pi r\, ds = 2\pi rh\, ds_1 \tag{c}$$

Expression (*b*) then appears as

$$[k]_e = 2\pi h \int_0^1 [B]^T[D][B] r\, ds_1 \tag{11.43}$$

Clearly, radius r must be expressed as a function of s prior to integration of the above.

Steps 1 to 3 of the general procedure described in Sec. 5.8 may now be applied to obtain the solution of the shell nodal displacements. We then determine the strains from Eqs. (11.41), the stress resultants from Eqs. (11.34), and the stresses from Eqs. (11.4). In the axisymmetrically loaded shells of revolution, "concentrated" or "nodal" forces are actually loads axisymmetrically distributed around the shells.

Clearly, if only the membrane theory solution is required, the quantities χ_s, χ_θ, β, M_s, M_θ are ignored and the expressions developed in this section are considerably reduced in complexity.

PROBLEMS

Secs. 11.1 to 11.7

11.1 A long steel cylinder 0.060 m in diameter and 2 mm thick is subjected to a uniform line load P distributed over the circumference of the circular cross section at midlength. Employing the energy of distortion theory, predict the value of the maximum load that can be applied to the cylinder without causing the elastic limit to be exceeded. Use $\nu = 0.3$ and $\sigma_{yp} = 210$ MPa.

11.2 If point O is taken to the right of the loaded portion of the cylinder shown in Fig. 11.5, what is the deflection at this point?

11.3 A long circular pipe of diameter $d = 0.5$ m and wall thickness $t = 5$ mm is bent by a load P uniformly distributed along a circular section (Fig. 11.4). The pipe is fabricated of a material of 200 MPa tensile yield strength and 300 MPa compressive yield strength. Let $\nu = 0.3$. Determine the value of P required according to the following theories of failure: (*a*) maximum shear stress and (*b*) maximum principal stress.

11.4 Consider the cylinder loaded as shown in Fig. 11.5. Determine the maximum values of (*a*) axial stress σ_x, (*b*) tangential stress σ_θ, and (*c*) shear stress $\tau_{x\theta}$.

11.5 A long steel pipe of 0.75 m in diameter and 10 mm thickness is subjected to loads P uniformly distributed along two circular sections 0.05 m apart (Fig. 11.4). For the midlength between the loads obtain, by taking $\nu = 0.3$, (*a*) the radial contraction and (*b*) axial and hoop stresses at the outer surface.

Secs. 11.8 to 11.12

11.6 Reduce the differential equations of equilibrium (11.25) to *two* expressions by eliminating shear force Q_ϕ.

11.7 Represent the equations of equilibrium of a conical shell (11.26) by two expressions containing N_s, N_θ, M_s, and M_θ.

11.8 Determine the forces N_θ and N_ϕ in a spherical dome carrying its own weight p using Eq. (*b*) of Sec. 11.10 and Eqs. (11.27). Compare the results with those given by Eqs. (10.13). Take $\nu = 0.3$.

CHAPTER

TWELVE

APPLICATIONS TO PIPES, TANKS, AND PRESSURE VESSELS

12.1 INTRODUCTION

There are many examples of thin shells employed as structural components. In Chap. 10 we observe cases in which, when a cylinder, liquid tank, dome, or tower is subjected to a particular loading, consideration is given the membrane stresses taking place over the entire wall thickness. We now discuss the general analysis of stress and deformation in several typical members, applying equations derived for plates and shells as well as the principle of superposition, as needed.

A degree of caution is necessary when employing formulas for which there is uncertainty as to applicability and restrictions of use. The relatively simple form of many expressions presented in this text result from severely limiting assumptions used in their derivation. Particular cognizance should be taken of the fact that high loadings, extreme temperature, and rigorous performance requirements present difficult design challenges.

The discussions of Secs. 12.2 and 12.3 apply to uniform and reinforced cylinders. A comparison of the bending and membrane stresses for tanks with a vertical axis is made in Sec. 12.4. Sections 12.5 and 12.6 deal with uniform and composite heated shells. All the expressions derived for thin cylindrical vessels (Secs. 12.7 to 12.10) under uniform pressure apply to *internal pressure*. They pertain equally to cases of *external* pressure if the sign of p is changed. However, the stresses thus determined are valid only if the pressure is not significant relative to that which would cause failure by elastic instability. That is, when a vessel such as a vacuum tank is to be constructed to withstand external pressure, the compressive stresses produced by this pressure in the shell must be lower than the critical stresses at which buckling of the walls might occur. Introduced in the last section of this chapter are some relationships employed in the design of plates and shell-like structures.

12.2 PIPES SUBJECTED TO EDGE FORCES AND MOMENTS

The circular *cylinder* or *pipe*, of special significance in engineering, is usually divided into *long* and *short* classifications. A long cylinder loaded at the middle (Fig. 11.4) is defined as of length $L > (2\pi/\beta)$. It is clear from the discussion in Sec. 11.7 that the ends of a long cylinder are not affected appreciably by central loading; therefore, each end can be treated independently. For shorter cylinders, however, the influence at the ends of a central force is substantial. Discussed below is the bending of circular cylinders subject to edge forces and moments.

Long pipes (Fig. 12.1*a*) Consider a long pipe loaded by uniformly distributed forces Q_1 and moments M_1 along edge $x = 0$, of length $L > (\pi/\beta)$. Constants C_1 and C_2 of Eq. (11.18) can be determined by applying the following conditions at the left end of the shell

$$M_x = -D\frac{d^2w}{dx^2} = M_1 \qquad (x = 0)$$
$$Q_x = \frac{dM_x}{dx} = -D\frac{d^3w}{dx^3} = Q_1 \qquad (x = 0) \tag{a}$$

The results are

$$C_1 = -\frac{1}{2\beta^3 D}(Q_1 + \beta M_1) \qquad C_2 = \frac{M_1}{2\beta^2 D} \tag{b}$$

The deflection is now found by substituting C_1 and C_2 into Eq. (11.18):

$$w = \frac{e^{-\beta x}}{2\beta^3 D}[\beta M_1(\sin \beta x - \cos \beta x) - Q_1 \cos \beta x] \tag{12.1}$$

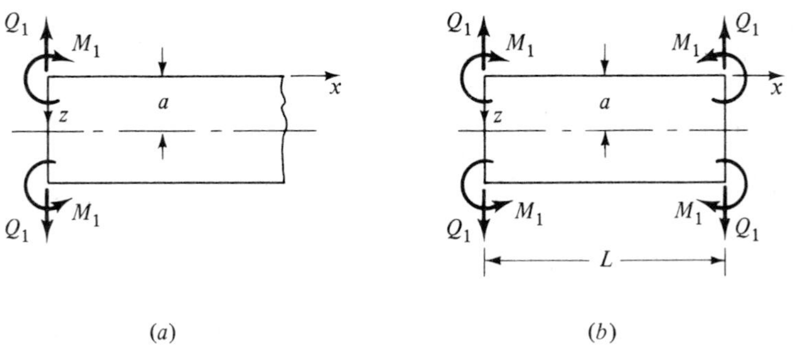

Figure 12.1

The maximum displacement occurs outward at the loaded section and is given by

$$w_{max} = -\frac{1}{2\beta^3 D}(\beta M_1 + Q_1) \qquad (x = 0) \tag{12.2}$$

The deflection is positive radially inward, hence the minus sign. The accompanying slope under the loadings is

$$\frac{dw}{dx} = \frac{1}{2\beta^2 D}(2\beta M_1 + Q_1) \qquad (x = 0) \tag{12.3}$$

Finally, successive differentiations of Eq. (12.1) yield expressions for the derivatives. Altogether, using the notation of Eqs. (11.22), we have

$$\begin{aligned}
w &= -\frac{1}{2\beta^3 D}[\beta M_1 f_3(\beta x) + Q_1 f_4(\beta x)] \\
\frac{dw}{dx} &= \frac{1}{2\beta^2 D}[2\beta M_1 f_4(\beta x) + Q_1 f_1(\beta x)] \\
\frac{d^2 w}{dx^2} &= -\frac{1}{\beta D}[\beta M_1 f_1(\beta x) + Q_1 f_2(\beta x)] \\
\frac{d^3 w}{dx^3} &= \frac{1}{D}[2\beta M_1 f_2(\beta x) - Q_1 f_3(\beta x)]
\end{aligned} \tag{12.4}$$

Upon substituting the above into Eqs. (11.19) and (11.20), one obtains expressions for N_θ, M_x, M_θ, and Q_x. The stresses are then found through the use of the definitions given in Sec. 11.4. As already observed in Sec. 11.7, functions $f_1(\beta x)$ through $f_4(\beta x)$, and thus the results employing them, become negligibly small for $x > (\pi/\beta)$.

Short pipes (Fig. 12.1*b*) The bending of this type of shell loaded along both edges, of length $L < (2\pi/\beta)$, may also be dealt with by applying the general solution (11.17). Now opposite end conditions interact. Expression (11.17) is rewritten in the following alternate form in terms of hyperbolic functions:

$$w = C_1 \sin \beta x \sinh \beta x + C_2 \sin \beta x \cosh \beta x + C_3 \cos \beta x \sinh \beta x + C_4 \cos \beta x \cosh \beta x \tag{12.5}$$

Four constants of integration require evaluation. To accomplish this, two boundary conditions at each end may be applied, leading to four equations with four unknown constants C_1 through C_4. After routine but somewhat lengthy algebraic manipulations, the following expressions for end deflections and end slopes are obtained

$$\begin{aligned}
w &= -\frac{2\beta a^2}{Et}[\beta M_1 h_2(\beta L) + Q_1 h_1(\beta L)] \\
\frac{dw}{dx} &= \pm\frac{2\beta^2 a^2}{Et}[2\beta M_1 h_3(\beta L) + Q_1 h_2(\beta L)]
\end{aligned} \tag{12.6}$$

Table 12.1

βL	0.2	0.4	0.6	1.0	1.4	2.0	3.0	4.0	5.0
$h_1(\beta L)$	5.000	2.502	1.674	1.033	0.803	0.738	0.893	0.005	1.017
$h_2(\beta L)$	0.0068	0.0268	0.0601	0.1670	0.3170	0.6000	0.9770	1.0580	1.0300
$h_3(\beta L)$	0.100	0.200	0.300	0.500	0.689	0.925	1.090	1.050	1.008

where

$$h_1(\beta L) = \frac{\cosh \beta L + \cos \beta L}{\sinh \beta L + \sin \beta L}$$

$$h_2(\beta L) = \frac{\sinh \beta L - \sin \beta L}{\sinh \beta L + \sin \beta L} \tag{12.7}$$

$$h_3(\beta L) = \frac{\cosh \beta L - \cos \beta L}{\sinh \beta L + \sin \beta L}$$

Table 12.1 lists values of $h_1(\beta L)$, $h_2(\beta L)$, and $h_3(\beta L)$ as a function of βL. Interestingly, in the case of long shells $h_1(\beta L) \approx h_2(\beta L) \approx h_3(\beta L) \approx 1$, and Eqs. (12.6), as expected, reduce to Eqs. (12.2) and (12.3).

Example 12.1 A long cylindrical shell is subjected to uniform internal pressure p. Determine the stress and deformation of the pipe for two sets of conditions: (*a*) the ends are built in (Fig. 12.2*a*) and the axial force $N_x = 0$; (*b*) the ends are simply supported.

SOLUTION (*a*) The problem may be treated as the sum of the cases shown in Figs. 12.2*b* and 12.2*c*. The radial expansion owing to p (Fig. 12.2*b*) is, from Eq. (10.11),

$$\delta = \frac{pa^2}{Et} \tag{c}$$

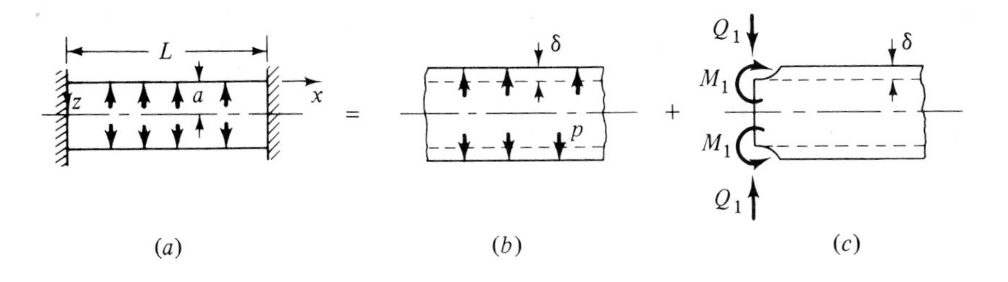

Figure 12.2

The boundary conditions for the edge-loaded pipe (Fig. 12.2c) is represented as follows, using Eqs. (12.2) and (12.3) together with Eq. (c):

$$-\frac{1}{2\beta^3 D}(\beta M_1 + Q_1) = \frac{pa^2}{Et}$$

$$\frac{1}{2\beta^2 D}(2\beta M_1 + Q_1) = 0$$

from which

$$M_1 = \frac{p}{2\beta^2} \qquad Q_1 = -\frac{p}{\beta} \tag{d}$$

In Eqs. (d), the minus sign indicates a negative shear force, acting as shown in Fig. 12.2c.

The expression for deflection, referring to Fig. 12.2c and Eqs. (12.4) and (c), is then

$$w = -\frac{p}{4\beta^4 D}[f_3(\beta x) - 2f_4(\beta x)] - \frac{pa^2}{Et} \tag{12.8}$$

The axial moment, from the first of Eqs. (11.20), is

$$M_x = \frac{p}{2\beta^2}[f_1(\beta x) - 2f_2(\beta x)] \tag{e}$$

and $\sigma_x = 6M_x/t^2$. The circumferential stress,

$$\sigma_\theta = -\frac{Ew}{a} - \nu\sigma_x \tag{f}$$

Clearly, the first and second terms in the above represent the membrane and the bending solutions, respectively.

(b) Since in this case $M_x = 0$ along the edges, we have $M_1 = 0$. Equation (12.2) then yields

$$Q_1 = -2\beta^3 D\, \delta \tag{g}$$

where δ is given by Eq. (c). Upon introduction of Q_1 given by Eq. (g) and $M_1 = 0$ into Eq. (12.4), an expression may be derived for the displacement w.

12.3 REINFORCED CYLINDERS

Cylindrical shells are often stiffened by *rings* or so-called *collars*. Examples are found in pipelines, submarines, and airplane fuselages. We shall discuss the bending of a long pipe reinforced by *equidistant* rings and subjected to a uniform internal pressure of intensity p (Fig. 12.3), applying Eqs. (12.6). The stiffness, size,

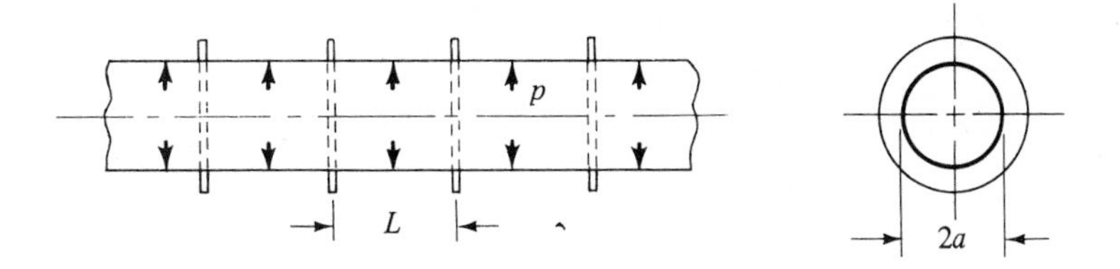

Figure 12.3

and collar spacing are important in the analysis. It is presumed that each ring has *small cross-sectional dimensions* compared with shell radius a and does not resist shell rotation.

Cylinders with collars which prohibit deflection For this case, one assumes the stiffening rings to be very rigid in comparison with the shell. That is, cylinder deflection *is zero* under the collars. If there were no ring, the pressure would produce a constant hoop stress pa/t, given by Eq. (10.10). Referring to Eq. (10.11), the radial displacement would then be

$$\delta_1 = \frac{pa^2}{Et}$$

Inasmuch as the rings are present, reaction forces P per unit circumferential length of pipe are produced between each ring and the shell.

To solve for P, one equates the pipe deflection under the ring, produced by the forces P, to the expansion δ owing to p. The portion of the shell between two adjacent rings may be represented by the cylinder shown in Fig. 12.1*b*. In the problem under consideration, $Q_1 = -P/2$ and M_1 is found by setting the slope $dw/dx = 0$ at an end point. From the second of Eqs. (12.6), we obtain

$$\frac{\beta^2 a^2}{Et}[4\beta M_1 h_3(\beta L) - Ph_2(\beta L)] = 0$$

or

$$M_1 = \frac{Ph_2(\beta L)}{4\beta h_3(\beta L)} \tag{a}$$

Then, applying the first of Eqs. (12.6),

$$\frac{P\beta a^2}{Et}\left[h_1(\beta L) - \frac{h_2^2(\beta L)}{2h_3(\beta L)}\right] = \delta_1 = \frac{pa^2}{Et}$$

from which

$$P\beta\left[h_1(\beta L) - \frac{1}{2}\frac{h_2^2(\beta L)}{h_3(\beta L)}\right] = \frac{\delta_1 Et}{a^2} = p \tag{12.9}$$

Note that when βL is large, the functions $h_1(\beta L)$, $h_2(\beta L)$, and $h_3(\beta L)$ approach unity, and the above reduces to

$$\frac{P\beta a^2}{2Et} = \delta_1$$

This agrees with Eq. (11.24) which was differently derived.

Upon introducing the value of P from Eq. (12.9) into Eq. (a) and simplifying, we obtain

$$M_1 = \frac{p}{2\beta^2} h_2(\beta L) \tag{12.10}$$

Cylinders with collars which resist deflection A second type of reinforced shell is encountered when the rings are relatively *flexible* compared with the flexibility of the shell. In this situation, the deflection of the shell *is not zero* under the collar. The interface forces P increase the inner radius of the ring by

$$\Delta\delta = \left(\frac{Pa}{AE}\right)a = \frac{Pa^2}{AE}$$

in which A and Pa are the area of the ring cross section and the ring tensile force, respectively. Now δ_1 in Eq. (12.9) is replaced by $\delta_2 = \delta_1 - \Delta\delta$ to obtain an equation used to evaluate the force P under a ring:

$$P\beta\left[h_1(\beta L) - \frac{1}{2}\frac{h_2^2(\beta L)}{h_3(\beta L)}\right] = p - \frac{Pt}{A} \tag{12.11}$$

Inserting $p = p - (Pt/A)$ into Eq. (12.10) we also have

$$M_1 = \frac{h_2(\beta L)}{2\beta^2}\left(p - \frac{Pt}{A}\right) \tag{12.12}$$

for the moment under a collar.

Cylinders with closed ends In the two cases discussed above, the ends of the shell are assumed to be open. A variation of the deflection and the stress in the pipe occurs when the ends are closed. The internal pressure acting at the ends creates an axial extension $pa/2Et$ of the pipe. The radial elongation of the shell is thus

$$\delta_2 = \frac{pa^2}{Et}\left(1 - \tfrac{1}{2}\nu\right) \tag{b}$$

For this case, p is replaced by $p(1 - \nu/2)$ in Eqs. (12.10) and (12.11).

12.4 CYLINDRICAL TANKS

Many cylindrical tanks may be treated using a procedure similar to that described in the foregoing sections. Included among these are: tanks with non-rigid or rigid bottom plates resting on the ground; tanks with an elastic roof and

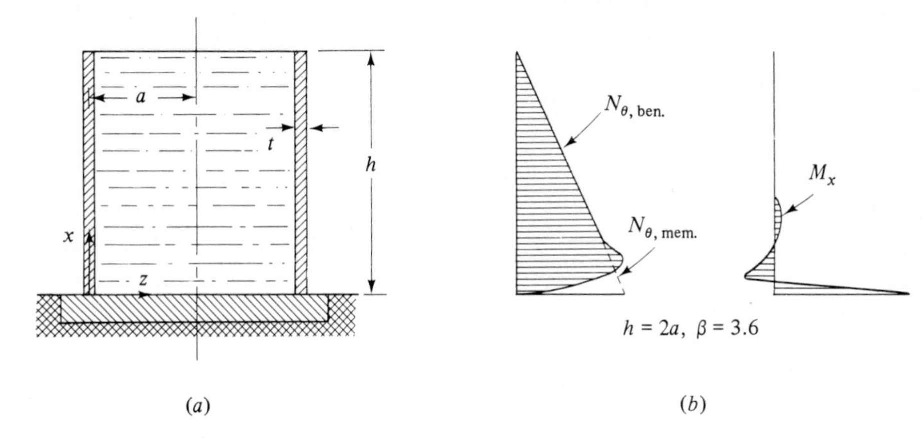

Figure 12.4

bottom; tanks constructed of plates of several different constant thicknesses (Fig. P12.11).

In the case of tanks of variable wall thickness, solution of the problem requires integration of Eq. (11.13*a*). Flexural rigidity D and thickness t must no longer be regarded as constant, but as functions of axial distance x. Thus, one must deal with a linear differential equation of fourth order and variable coefficient, requiring lengthy manipulation. However, in concrete water tanks it is usually common to choose a linear variation of thickness increasing from top to bottom. The solution is thus rendered more manageable.

We here consider a cylindrical tank of *uniform* thickness entirely filled with liquid of specific weight γ (Fig. 12.4*a*). The tank bottom is assumed to be built in and the top is open. The physical conditions indicate that the upper edge is free to deform and that no force exists there ($N_x = 0$). At the lower edge, locating the coordinates in the figure, we have

$$w = 0 \qquad \frac{dw}{dx} = 0 \qquad (x = 0) \tag{a}$$

The differential equation is, from Eq. (11.15*b*),

$$\frac{d^4w}{dx^4} + 4\beta^4 w = -\frac{\gamma(h - x)}{D} \tag{b}$$

where the outward pressure acting at any point on the wall, p_r, is replaced by $-\gamma(h - x)$. The particular solution of Eq. (*b*) is found to be $-\gamma(h - x)a^2/Et$. The general solution is thus

$$w = e^{-\beta x}[C_1 \cos \beta x + C_2 \sin \beta x] + e^{\beta x}[C_3 \cos \beta x + C_4 \sin \beta x] - \frac{\gamma(h - x)a^2}{Et} \tag{c}$$

If t is small relative to a and h, as is usually the case, a longitudinal slice of unit width of the cylinder may be considered infinitely long. It is already observed in Sec. 11.7 that, because w must be finite for all x, $C_3 = C_4 = 0$ in the above expression.

We now satisfy the remaining conditions. Upon substitution of Eqs. (a) into Eq. (c),

$$w = C_1 - \frac{\gamma a^2 h}{Et} = 0$$

$$\frac{dw}{dx} = \beta(C_2 - C_1) + \frac{\gamma a^2}{Et} = 0$$

from which

$$C_1 = \frac{\gamma a^2 h}{Et} \qquad C_2 = \frac{\gamma a^2}{Et}\left(h - \frac{1}{\beta}\right)$$

The radial deflection of the tank is then

$$w = -\frac{\gamma a^2 h}{Et}\left\{1 - \frac{x}{h} - e^{-\beta x}\left[\cos \beta x + \left(1 - \frac{1}{\beta h}\right)\sin \beta x\right]\right\}$$

or

$$w = -\frac{\gamma a^2 h}{Et}\left[1 - \frac{x}{h} - f_4(\beta x) - \left(1 - \frac{1}{\beta h}\right) f_2(\beta x)\right] \tag{12.13}$$

where $f_2(\beta x)$ and $f_4(\beta x)$ are defined by Eq. (11.22). Using Table 11-1, the deflection at any point is easily determined from the above equation. Equation (11.15a), after integration, provides the axial displacement

$$u = \int_0^h \nu \frac{w}{a}\, dx + u_0$$

where the constant, from $u(0) = 0$, is $u_0 = 0$.

We next evaluate the stress resultants from Eqs. (11.11) and (11.12) together with (11.10a) and (12.13):

$$N_\theta = -\frac{Etw}{a} = \gamma a h\left[1 - \frac{x}{h} - f_4(\beta x) - \left(1 - \frac{1}{\beta h}\right) f_2(\beta x)\right]$$

$$M_x = -D\frac{d^2 w}{dx^2} = \frac{2\beta^2 \gamma a^2 D h}{Et}\left[-f_4(\beta x) + \left(1 - \frac{1}{\beta h}\right) f_2(\beta x)\right] \tag{12.14}$$

$$M_\theta = \nu M_x$$

The maximum bending moment occurs at the bottom of the tank, at $x = 0$:

$$M_{x,\,\max} = \left(1 - \frac{1}{\beta h}\right)\frac{2\beta^2 \gamma a^2 D h}{Et} \tag{12.15}$$

The stress resultant N_θ, according to membrane theory, is found from Eq. (10.30*a*) by substituting $p_r = -\gamma(h - x)$:

$$N_\theta = \gamma(h - x)a \tag{d}$$

At $x = 0$, N_θ assumes its maximum value $N_{\theta,\max} = \gamma ha$. Based upon membrane theory, the maximum stress is therefore

$$\sigma_{\theta,\max} = \frac{1}{t}\gamma ha \tag{e}$$

We observe that the force resultant corresponding to the particular solution, $w_p = -\gamma(h - x)a^2/Et$, is from Eq. (12.14):

$$N_\theta = -\frac{Etw_p}{a} = \gamma(h - x)a$$

The above agrees with Eq. (*d*) which was derived by applying membrane theory. This conclusion is also valid for other cases.[33]

According to bending theory, because $N_\theta = 0$ and M_x is a maximum at $x = 0$,

$$\sigma_{x,\max} = \frac{6M_{x,\max}}{t^2} = \frac{12\beta^2\gamma a^2 D h}{Et^3}\left(1 - \frac{1}{\beta h}\right)$$

Introducing $D = Et^3/12(1 - \nu^2)$ and $\beta^2 = \sqrt{3(1 - \nu^2)}/at$ into the above, and treating the terms $1/\beta h$ and ν^2 as small compared with unity, we have

$$\sigma_{x,\max} = \frac{\sqrt{3}}{t}\gamma ha \tag{f}$$

Comparing Eqs. (*e*) and (*f*), it is observed that the *true* maximum stress is in fact $\sqrt{3}$ times larger than that predicated by membrane theory.

A plot of Eqs. (12.14) and (*d*) is presented in Fig. 12.4*b*, where the dashed and solid lines denote the stress-resultant distribution according to the membrane and bending theories, respectively. The figure shows that membrane theory is valid at sections away from fixed end of a long, thin cylinder. For a comparatively thick-walled and short cylinder, however, the differences in the results given by the theories are pronounced in the lower half of the shell. In any case, membrane theory applies reasonably well to the upper portion of the tank.

12.5 THERMAL STRESSES IN CYLINDERS

Thermal stresses are developed whenever the expansion or contraction that would normally be produced by the heating or cooling of a member is restricted. Thermal stress and deformation in thin cylindrical shells is a major factor in the design of structures such as boilers, heat exchangers, pressure vessels, and nuclear piping. In this section, consideration is given two representative examples of temperature fields under which thermal stresses occur in cylindrical shells.

Uniform temperature distribution A uniform temperature change in a cylinder with clamped or simply supported ends causes local bending stresses at the edges. These situations may be analyzed through application of the Eqs. (12.2) and (12.3), following a procedure similar to that described in Example 12.1.

Examine, for example, a long cylinder with *fixed ends*, under a uniform temperature change ΔT (Fig. 12.2*a*). For this case, free radial expansion of the pipe (Fig. 12.2*b*), is described by

$$\delta = a\alpha(\Delta T) \tag{a}$$

Referring to Fig. 12.2*c*, we thus have

$$-\frac{1}{2\beta^3 D}(\beta M_1 + Q_1) = a\alpha(\Delta T)$$

$$\frac{1}{2\beta^2 D}(2\beta M_1 + Q_1) = 0$$

or

$$M_1 = 2a\alpha(\Delta T)\beta^2 D \qquad Q_1 = -4a\alpha(\Delta T)\beta^3 D \tag{b}$$

Expressions for the deflection and the moment are then derived as illustrated in Example 12.1:

$$\begin{aligned} w &= -a\alpha(\Delta T)[f_3(\beta x) - 2f_4(\beta x)] - a\alpha(\Delta T) \\ M_x &= 2a\alpha\beta^2 D(\Delta T)[f_1(\beta x) - f_2(\beta x)] \end{aligned} \tag{12.16}$$

By applying Eqs. (12.16), the stresses at any point can readily be determined as in isothermal bending.

Radial temperature gradient Consider a long cylinder of *arbitrary cross section* subjected to the uniform temperature T_2 at the outer surface and the uniform temperature T_1 at the inner surface (Fig. 12.5). The *temperature gradient* through the thickness is assumed to be linear. Unless otherwise specified, we shall also take $\Delta T = T_1 - T_2$ as positive, i.e., $T_1 > T_2$. In the shell described, *at points*

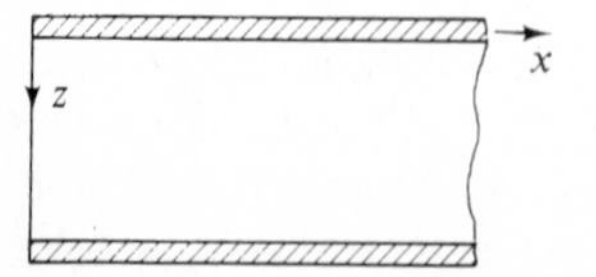

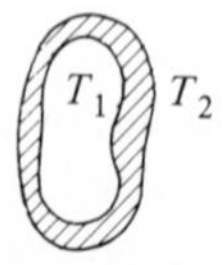

Figure 12.5

remote from the ends, the thermal stress is the same as in a clamped-edge circular plate (Sec. 9.8). The hoop and axial stresses, from Eq. (9.30), are thus

$$\sigma_\theta = \sigma_x = \sigma = \pm \frac{E\alpha(\Delta T)}{2(1 - \nu)} \tag{12.17}$$

Note that the outer surface will be in tension and the inner surface in compression.

Next we discuss the stress distribution *at the edges* of a *circular* cylinder of *uniform thickness* with *free ends*. It is observed from Fig. 12.2*c* that the stresses given by Eq. (12.17) must be balanced near the edge by distributed moments M_1. That is,

$$\sigma_x = -\frac{6M_1}{t^2}$$

from which

$$M_1 = -\frac{E\alpha(\Delta T)t^2}{12(1 - \nu)} \tag{c}$$

Clearly, in order for an edge to be *free*, a moment equal to but opposite in sign to that defined by Eq. (*c*) must be applied there. The deflection from such a moment $(+M_1)$ can be determined by applying Eq. (12.4). By so doing, substituting the resulting expression for w into Eqs. (11.19) and (11.20), one obtains the deflection and stress resultants applicable to the free edge ($x = 0$):

$$\begin{aligned} w &= \frac{M_1}{2\beta^2 D} \\ M_x &= \frac{E\alpha(\Delta T)t^2}{12(1 - \nu)} \\ M_\theta &= \frac{\nu E\alpha(\Delta T)t^2}{12(1 - \nu)} \\ N_\theta &= \frac{Et\alpha(\Delta T)}{2\sqrt{3}(1 - \nu)}\sqrt{1 - \nu^2} \end{aligned} \tag{12.18a–d}$$

Therefore, Eqs. (12.17), (12.18*c*), and (12.18*d*) provide the resultant thermal hoop stress occurring at the free end. Observe that this stress is larger than the axial stress. Thus,

$$\sigma_{\theta,\max} = \frac{E\alpha(\Delta T)}{2\sqrt{3}(1 - \nu)}[\sqrt{3}(1 - \nu) + \sqrt{1 - \nu^2}] \tag{12.19}$$

The stress defined by Eq. (12.19), for $\nu = 0.3$, is 25 percent larger than the stress at points in the pipe away from the ends. This explains why *thermal cracks initiate at the free edge of cylinders.*

Expressions for stress and deformation in cylindrical shells with clamped or simply supported ends, as well as for shells under axial or circumferential temperature gradients, can also be obtained by modifying the cylindrical shell equations appropriate to isothermal bending.[11,34] Generally, used in design computations, is a system of nondimensional curves[35] based upon these solutions or the finite element method.

12.6 THERMAL STRESSES IN COMPOSITE CYLINDERS

The foregoing section and Sec. 9.2 provide the basis for design of composite or compound multishell cylinders, constructed of a number of concentric, thin-walled shells. In the development which follows, each component shell is assumed homogeneous and isotropic. Each may have a different thickness and be subjected to different uniform or variable temperature differentials.[4]

If a free-edged multishell cylinder undergoes a *uniform temperature change* ΔT, the free motion of any shell having different material properties is restricted by components adjacent to it. Only a hoop stress σ_θ and circumferential strain ε_θ are then produced in the cylinder walls. On applying Eqs. (9.2), we thus have

$$\sigma_\theta = E[\varepsilon_\theta - \alpha(\Delta T)] \tag{12.20}$$

In the case of a cylinder subjected to a *temperature gradient*, both axial and hoop stresses take place, and $\sigma_x = \sigma_\theta = \sigma$ (Sec. 12.5). Expressions (9.3) give

$$\sigma = \frac{E}{1-\nu}[\varepsilon_\theta - \alpha(\Delta T)] \tag{12.21}$$

Consider a compound structure comprised of three components, each under different *temperature gradients* (the ΔT's) (Fig. 12.6*a*). Equation (12.21) permits stress determination as follows:

$$\begin{aligned}
\begin{Bmatrix}\sigma_{a1}\\ \sigma_{a2}\end{Bmatrix} &= \frac{E_a}{1-\nu_a}\begin{Bmatrix}[\varepsilon_\theta - \alpha_a(\Delta T_1)]\\ [\varepsilon_\theta - \alpha_a(\Delta T_2)]\end{Bmatrix}\\
\begin{Bmatrix}\sigma_{b2}\\ \sigma_{b3}\end{Bmatrix} &= \frac{E_b}{1-\nu_b}\begin{Bmatrix}[\varepsilon_\theta - \alpha_b(\Delta T_2)]\\ [\varepsilon_\theta - \alpha_b(\Delta T_3)]\end{Bmatrix}\\
\begin{Bmatrix}\sigma_{c3}\\ \sigma_{c4}\end{Bmatrix} &= \frac{E_c}{1-\nu_c}\begin{Bmatrix}[\varepsilon_\theta - \alpha_c(\Delta T_3)]\\ [\varepsilon_\theta - \alpha_c(\Delta T_4)]\end{Bmatrix}
\end{aligned} \tag{12.22a–f}$$

where the subscripts *a*, *b*, and *c* refer to the individual components.

The axial forces corresponding to each layer are next ascertained from the stresses given by Eqs. (12.22)

$$\begin{aligned}
N_a &= \tfrac{1}{2}(\sigma_{a1} + \sigma_{a2})A_a\\
N_b &= \tfrac{1}{2}(\sigma_{b2} + \sigma_{b3})A_b\\
N_c &= \tfrac{1}{2}(\sigma_{c3} + \sigma_{c4})A_c
\end{aligned} \tag{a}$$

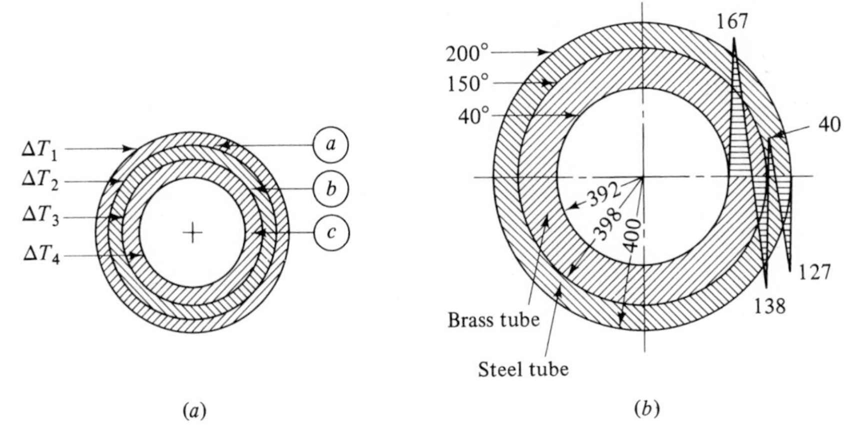

Figure 12.6

Here the A's represent the cross-sectional area of each shell. The condition that the sum of the axial forces be equal to zero is satisfied if

$$N_a + N_b + N_c = 0 \tag{b}$$

From Eqs. (12.22), (a), and (b), assuming $\nu_a = \nu_b = \nu_c$, the following expression for the strain is obtained:

$$\varepsilon_\theta = \frac{A_a E_a \alpha_a (\Delta T)_a + A_b E_b \alpha_b (\Delta T)_b + A_c E_c \alpha_c (\Delta T)_c}{A_a E_a + A_b E_b + A_c E_c} \tag{12.23}$$

In the above, $(\Delta T)_a$, $(\Delta T)_b$, and $(\Delta T)_c$ are the *average* of the temperature differentials at the boundaries of elements a, b, and c, respectively [e.g., $(\Delta T)_a = \frac{1}{2}(\Delta T_1 + \Delta T_2)$]. The stresses at the inner and outer surfaces of each shell may be calculated readily upon inserting Eq. (12.23) into Eqs. (12.22), as illustrated in Examples 12.2 and 12.3.

Note that near the ends there will usually be some bending of the composite cylinder, and the total thermal stresses will be obtained by superimposing upon Eqs. (12.22) such stresses as may be necessary to satisfy the boundary conditions, as shown in the previous section.

Example 12.2 The cylindrical portion of a jet nozzle is made by just slipping a steel shell over a brass shell (Fig. 12.6b). The radii of the tubes are $r_1 = 400$ mm, $r_2 = 398$ mm, and $r_3 = 392$ mm. If the uniform temperature differentials at the boundaries

$$\Delta T_1 = 200°\text{C} \qquad \Delta T_2 = 150°\text{C} \qquad \Delta T_3 = 40°\text{C}$$

must be maintained, what stresses will develop in the two materials? The properties of brass and steel are as follows (Table 1.1):

$$E_b = 103 \text{ GPa} \qquad \alpha_b = 18.9 \times 10^{-6} \text{ per °C} \qquad \nu_b = 0.3$$

$$E_s = 200 \text{ GPa} \qquad \alpha_s = 11.7 \times 10^{-6} \text{ per °C} \qquad \nu_s = 0.3$$

SOLUTION We have $A_b = 2\pi r_2 t_b$ and $A_s = 2\pi r_2 t_s$, where t_b and t_s are the wall thicknesses of the brass and steel tubes, respectively. On applying Eq. (12.23),

$$\varepsilon_\theta = \frac{t_s E_s \alpha_s[\frac{1}{2}(\Delta T_1 + \Delta T_2)] + t_b E_b \alpha_b[\frac{1}{2}(\Delta T_2 + \Delta T_3)]}{t_s E_s + t_b E_b} \tag{12.24}$$

and thus

$$\varepsilon_\theta = \frac{\begin{array}{l}0.002 \times 200 \times 10^9 \times 11.7 \times 10^{-6}(175) \\ \qquad + 0.006 \times 103 \times 10^9 \times 18.9 \times 10^{-6}(95)\end{array}}{0.002 \times 200 \times 10^9 + 0.006 \times 103 \times 10^9} = 1.894 \times 10^{-3}$$

The expansion of the interfacial radius is then $\Delta r_2 = 1.894 \times 10^{-3}(0.398) = 0.754(10^{-3})$ m or 0.754 mm. The radial growth at other locations may be found in a like manner.

Through the use of Eqs. (12.22) and (12.24), the following values are obtained

$$\begin{Bmatrix}\sigma_{s1} \\ \sigma_{s2}\end{Bmatrix} = \frac{200 \times 10^3}{1 - 0.3}\begin{Bmatrix}1894 - 11.7(200) \\ 1894 - 11.7(150)\end{Bmatrix} = \begin{Bmatrix}-127 \\ 40\end{Bmatrix} \text{ MPa}$$

and

$$\begin{Bmatrix}\sigma_{b2} \\ \sigma_{b3}\end{Bmatrix} = \frac{103 \times 10^3}{1 - 0.3}\begin{Bmatrix}1894 - 18.9(150) \\ 1894 - 18.9(40)\end{Bmatrix} = \begin{Bmatrix}-138 \\ 167\end{Bmatrix} \text{ MPa}$$

for the stresses at the outer and inner surfaces of each shell, respectively. It is observed that

$$N_s + N_b = \tfrac{1}{2}(-127 + 40)(0.005) + \tfrac{1}{2}(-138 + 167)(0.015) = 0$$

as required according to Eq. (*b*). A sketch of the stresses throughout the wall thickness is given in Fig. 12.6*b*.

Example 12.3 Reconsider the preceding sample problem for the case in which a brass tube just slips over a steel tube with each shell *uniformly* heated and the temperature raised by ΔT°C. Determine the hoop stress which develops in each component upon *cooling*.

SOLUTION For the case under consideration the composite cylinder is *contracting*. Equation (12.20) then appears as

$$\sigma_\theta = -E[\varepsilon_\theta + \alpha(\Delta T)] \tag{12.25}$$

or

$$\varepsilon_\theta = -\frac{1}{E}\sigma_\theta - \alpha(\Delta T) \tag{12.26}$$

The hoop strain, referring to Eqs. (12.24) and (12.26), is now expressed:

$$\varepsilon_\theta = -\frac{(\Delta T)[t_s E_s \alpha_s + t_b E_b \alpha_b]}{t_s E_s + t_b E_b} \tag{12.27}$$

Hence the hoop stresses, from Eqs. (12.25) and (12.27), after simplification are found to be

$$(\sigma_\theta)_b = -E_b[\varepsilon_\theta + \alpha_b(\Delta T)] = -\frac{E_b(\Delta T)(\alpha_b - \alpha_s)}{1 + (t_b E_b / t_s E_s)} \tag{12.28}$$

$$(\sigma_\theta)_s = -E_s[\varepsilon_\theta + \alpha_s(\Delta T)] = -\frac{E_s(\Delta T)(\alpha_s - \alpha_b)}{1 + (t_b E_b / t_s E_s)} \tag{12.29}$$

Thus the stresses in the brass and steel tubes are tensile and compressive respectively, as $\alpha_b > \alpha_s$ and ΔT is a negative quantity because of cooling.

12.7 DISCONTINUITY STRESSES IN PRESSURE VESSELS

Ever broadening use of vessels for storage, industrial processing, and power generation under unique conditions of temperature, pressure, and environment has given special emphasis to analytical, numerical, and experimental techniques for obtaining appropriate operating stresses. The finite element method (Sec. 5.7) has gained considerable favor in the design of vessels relative to other methods.

A discontinuity of the membrane action in a vessel occurs at all points of external restraint or at the junction of the cylindrical shell and its *head* or *end* possessing different stiffness characteristics. Any incompatibility of deformation at the joint produces bending moments and shearing forces. The stresses owing to this bending and shear are termed *discontinuity stresses*.

It is observed in Sec. 11.7 that the bending is of a local character. Hence, the discontinuity stresses become negligibly small within a short distance. The narrow region at the edge of *spherical*, *elliptical*, and *conical* type vessel heads can be assumed as *nearly cylindrical* in shape. Cylindrical shell equations of Sec. 11.6 can therefore be employed to obtain an approximate solution applicable at the juncture of vessels having spherical, elliptical, or conical ends. In the case of *flat-end* vessels, expressions for circular plates and cylindrical shells are utilized.

Sections 12.8 to 12.10 can provide only a few basic relationships with an active area of contemporary pressure vessel analysis. Additional details and a list of references relevant to stresses in pressure vessels are to be found in a variety of publications.[7,36] The ASME unfired pressure-vessel code furnishes information of practical value for end design (Sec. 12.11).

12.8 CYLINDRICAL VESSEL WITH HEMISPHERICAL HEADS

Consider a cylindrical vessel of radius a, having hemispherical ends, under uniform internal pressure of intensity p, Fig. 12.7*a*. Owing to the action of p the tube and its ends tend to expand by different amounts, as shown in exaggeration by the dashed lines in the figure. Because at some distance away from the joint, membrane theory yields results of sufficient accuracy, for the cylindrical part of the vessel we have, from Eqs. (10.10):

$$N_x = \frac{pa}{2} \qquad N_\theta = pa \tag{a}$$

For the spherical ends, Eq. (10.4) yields

$$N = \frac{pa}{2} \tag{b}$$

The growth in cylinder radius produced by membrane forces N_x and N_θ is expressed by Eq. (10.11):

$$\delta_c = \frac{pa^2}{2Et}(2 - \nu) \tag{12.30}$$

Similarly, extension of the radius of the spherical heads owing to N, through application of Eq. (10.5), is

$$\delta_s = \frac{pa^2}{2Et}(1 - \nu) \tag{12.31}$$

It follows that, if the tube and its ends are disjointed (Fig. 12.7*b*), the differential radial extension owing to the membrane stresses would be $\delta = \delta_c - \delta_s$:

$$\delta = \frac{pa^2}{2Et} \tag{12.32}$$

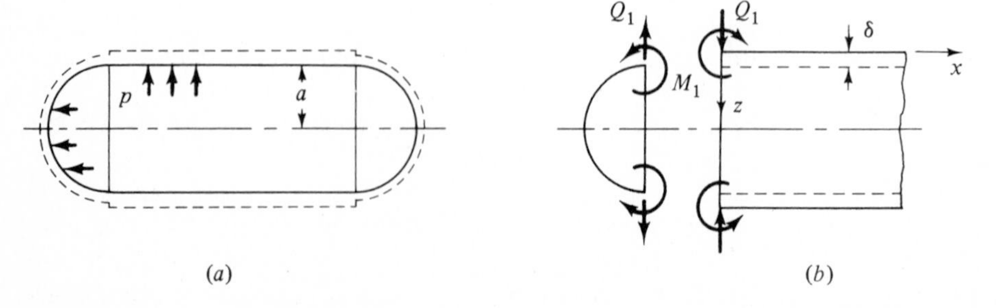

Figure 12.7

In an actual vessel, this discontinuity of displacement is prevented by shearing forces Q_1 and bending moments M_1, both uniformly distributed along the circumference. As already mentioned, these discontinuity effects create bending stresses in the vicinity of the joint.

In practice, the thicknesses of the cylinder and its ends are usually *the same.* In this case, forces Q_1 produce equal deformations at the edges of each component at the joint. Hence, the conditions of continuity at the junction are satisfied if $M_1 = 0$ and Q_1 is of such magnitude as to result in deflections at the edges of the cylinder and its head whose sum is equal to δ. Applying Eqs. (12.2) and (12.32), we thus have

$$\frac{pa^2}{2Et} = \left(\frac{Q_1}{2\beta^3 D}\right)_c + \left(\frac{Q_1}{2\beta^3 D}\right)_h = \frac{Q_1}{\beta^3 D}$$

This, together with definitions (1.11) and (11.14), yields

$$Q_1 = \frac{pa^2\beta^3 D}{2Et} = \frac{p}{8\beta} \tag{c}$$

The displacement, from Eq. (12.4), is then

$$w = \frac{Q_1}{2\beta^3 D} f_4(\beta x) = \frac{p}{16\beta^4 D} f_4(\beta x) \tag{12.33}$$

and the bending moment is therefore

$$M_x = -D\frac{d^2 w}{dx^2} = -\frac{Q_1}{\beta} f_2(\beta x) = -\frac{atp}{8\sqrt{3(1-\nu^2)}} f_2(\beta x) \tag{12.34}$$

The maximum moment takes place at $x = \pi/4\beta$, as $dM_x/dx = 0$ at this point.

The highest stress occurs on the outer face at $x = \pi/4\beta$ of the tube. By means of Eqs. (*a*), (12.34), and Eq. (11.4), for $\nu = 0.3$, it is found that

$$\sigma_{x,\,\max} = \frac{ap}{2t} + \frac{3}{4}\frac{ap}{t\sqrt{3(1-\nu^2)}} f_2\left(\frac{\pi}{4}\right) = 1.293\frac{ap}{2t} \tag{12.35}$$

Hoop stress also occurs at the outer surface of the cylinder. This stress is composed of the membrane solution component pa/t, a circumferential stress produced by deflection w (see Sec. 11.6) and a bending stress resulting from the moment $M_\theta = -\nu M_x$:

$$\sigma_\theta = \frac{ap}{t} - \frac{Ew}{a} - \frac{6\nu}{t^2} M_x$$

Substituting Eqs. (12.33) and (12.34) into the above, we have

$$\sigma_\theta = \frac{ap}{t}\left[1 - \tfrac{1}{4} f_3(\beta x) + \frac{3\nu}{4\sqrt{3(1-\nu^2)}} f_4(\beta x)\right] \tag{d}$$

For $\nu = 0.3$, the maximum stress, referring to Table 11.1, is found to be

$$\sigma_{\theta,\,\max} = 1.032 \frac{ap}{t} \qquad (\beta x = 1.85) \tag{12.36}$$

Inasmuch as the membrane stress is lower in the heads than in the tube, the maximum stress in the spherical ends is always smaller than the value given above. We are led to conclude therefore that the *hoop stress* [Eq. (12.36)], *is most critical* in the design of the vessel.

12.9 CYLINDRICAL VESSEL WITH ELLIPSOIDAL HEADS

The general procedure for the determination of the stress in a vessel with ellipsoidal ends (Fig. 12.8) follows the same pattern as employed in the foregoing section. The radial increment δ_c in the cylindrical portion is given by Eq. (12.30). Referring to Eqs. (10.20) the increment at the edge of the ellipsoidal member is found to be

$$\delta_e = \frac{a}{E}(\sigma_\theta - \nu\sigma_\phi) = \frac{pa^2}{Et}\left(1 - \frac{a^2}{2b^2} - \frac{\nu}{2}\right) \tag{a}$$

The radial extension $\delta = \delta_c - \delta_e$ of the two components is thus

$$\delta = \frac{pa^2}{2Et}\left(\frac{a^2}{b^2}\right) \tag{12.37}$$

A comparison of Eqs. (12.37) and (12.32) reveals that the extension at the joint of an ellipsoidal-ended vessel is greater by the ratio a^2/b^2 than that for a vessel with hemispherical heads. Hence, the shearing force and the discontinuity stresses also increase in the identical proportions.

In the particular case of a vessel for which $a = 2b$, Eqs. (12.35), (12.36), and (*d*) of Sec. 12.8 yield the following expressions for maximum axial and the hoop stresses in the vessel:

$$\begin{aligned}\sigma_{x,\,\max} &= \frac{ap}{2t} + \frac{3ap}{t\sqrt{3(1-\nu^2)}} f_2\left(\frac{\pi}{4}\right) = 2.172\frac{ap}{2t} \\ \sigma_{\theta,\,\max} &= 1.128\frac{ap}{t}\end{aligned} \tag{12.38}$$

Note that the last stress, as before, has the largest magnitude.

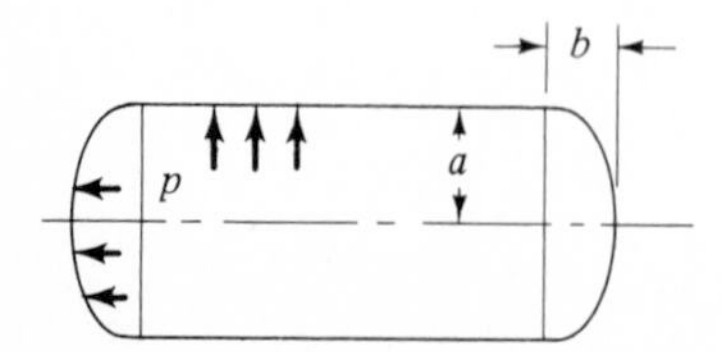

Figure 12.8

12.10 CYLINDRICAL VESSEL WITH FLAT HEADS

Cylindrical vessels are often constructed with flat ends as shown in Fig. 12.9. In this case, each head is a circular plate. A solution may be obtained by assuming that the heads bend into spherical surfaces subsequent to the application of the uniformly distributed loading p. The internal pressure will produce extensions in the tube and its heads as observed in the cases of the vessels previously discussed. Therefore, at the joint there must act shearing forces Q_1 and bending moments M_1 to make the displacements compatible.

For a circular plate head the edge slopes α owing to p and the edge moment M_1, are, from Eqs. (2.17) and (2.27), respectively,

$$\alpha_h' = \frac{pa^3}{8D(1+\nu)} \qquad \alpha_h'' = \frac{aM_1}{D(1+\nu)} \tag{a}$$

Similarly, the slope α_c due to Q_1 and M_1 of the cylindrical portion at the joint is, referring to Eq. (12.3):

$$\alpha_c = \frac{1}{2\beta^2 D}(2\beta M_1 + Q_1) \tag{b}$$

The condition that the edges of the two parts rotate the same amount is satisfied by

$$\alpha_h' - \alpha_h'' = \alpha_c \tag{12.39}$$

We next consider the radial displacement. The extension of the cylinder radius owing to p is, according to Eq. (12.30):

$$\delta_c' = \frac{pa^2}{2Et_c}(2-\nu) \tag{c}$$

It can be verified that the radial displacement of the circular plate head owing to Q_1 is expressed by

$$\delta_h = -\frac{aQ_1}{Et_h}(1-\nu) \tag{d}$$

Equation (12.2) provides the increment in the cylinder radius at the joint:

$$\delta_c'' = -\frac{1}{2\beta^3 D}(\beta M_1 + Q_1) \tag{e}$$

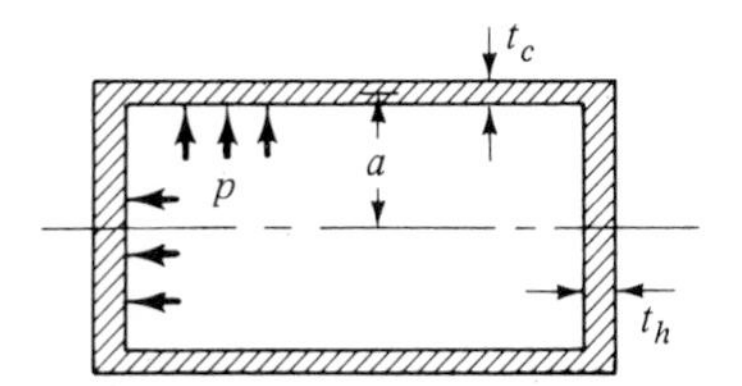

Figure 12.9

The compatibility of the radial displacements of the edges of the two parts requires that

$$\delta_c' = \delta_h + \delta_c'' \tag{12.40}$$

Finally, the solution is completed by determining Q_1 and M_1 using Eqs. (12.39) and (12.40). The results are presented in the following form:

$$\begin{aligned} Q_1 &= -\frac{\rho^3k^3 + 2(2-\nu)\rho^3k + 2(2-\nu)(1+\nu)}{2\rho^3k^2 + [(1-\nu)\rho^4 + (1+\nu)]k + (1-\nu^2)\rho}\,\frac{pa}{4} \\ M_1 &= \frac{2\rho^3k^3 + (1-\nu)\rho^4k^2 + 2(2-\nu)(1+\nu)}{2\rho^3k^2 + [(1-\nu)\rho^4 + (1+\nu)]k + (1-\nu^2)\rho}\,\frac{pa^2}{8k} \end{aligned} \tag{12.41}$$

where

$$\rho = \frac{t_c}{t_h} \qquad k = a\beta = \sqrt[4]{\frac{3(1-\nu^2)a^2}{t_c^2}} \tag{12.42}$$

The displacement and the discontinuity stresses may now be obtained by following a similar procedure to that described in Sec. 12.8.

12.11 DESIGN FORMULAS FOR CONVENTIONAL PRESSURE VESSELS

It is observed in Secs. 12.8 and 12.9 that the discontinuity stresses at the joint of a cylindrical shell and its head is quite a bit larger than the membrane stress in either portion. When a vessel is properly designed and constructed, however, these stresses are reduced greatly and it becomes *unnecessary* to consider them. The ASME *code for unfired pressure-vessels*[37] furnishes formulas for calculating the required minimum thickness of the shells and the ends. All the formulas, except those associated with the case of flat heads, are based solely upon the membrane stresses, but the shape and proportions of the end and the manner of attachment to the shell are specified so as to avoid high discontinuity stresses.

The following factors and a host of others contributing to an ideal vessel design are described by the code: approved techniques for joining the end to the shell; formulas for computing the thickness of shell and end; materials in combination; temperature ranges; maximum allowable stress values; corrosion; types of closures. For some of the shells and the end types discussed in Chaps. 10 through 12, Table 12.2 lists the minimum required thicknesses. The following symbols are employed in the table:

p = internal pressure (lb/in^2)
r = inside radius of shell or hemispherical head (in)
σ = maximum allowable stress (lb/in^2)
e = lowest joint efficiency
D = inside diameter of conical head or inside length of the major axis of an ellipsoidal head (in)

Table 12.2

Geometry	Required thickness (in)
Cylindrical shell	$t = \dfrac{pr}{\sigma e - 0.6p}$
Hemispherical head	$t = \dfrac{pr}{2\sigma e - 0.2p}$
Conical head	$t = \dfrac{pD}{2\cos\alpha(\sigma e - 0.6p)}$
Flat head	$t = d\sqrt{\dfrac{cp}{\sigma}}$
Ellipsoidal head	$t = \dfrac{pD}{2\sigma e - 0.2p}$
Wholly spherical shell	$t = \dfrac{pr}{2\sigma e - 0.2p}$

α = half the apex angle of the conical head
d = diameter of flat head (in)
c = a numerical coefficient depending upon the method of attachment of head, e.g., $c = 0.5$ for circular plates welded to the end of the shell

It is mentioned that the required wall thicknesses for *tubes* and *pipes* under internal pressure are determined in accordance with the rules for shell in the code. In its present status, the code is applicable when the pressure *does not exceed* 3000 lb/in^2. Pressures in excess of this amount may require special attention in the design and construction of the vessels, closures, and branch connections of piping systems.[38]

Note that the preceding is only a *partial* description of the code specifications which, if complied with, provide assurance that discontinuity stresses may be neglected. For the complete requirements, reference should be made to the current edition of the code.

PROBLEMS

Secs. 12.1 to 12.4

12.1 A long steel pipe is subjected to a circumferential load Q_1 at its end (Fig. 12.1*a*). The radius and the thickness of the pipe are $a = 0.08$ m and $t = 3$ mm. Determine: (*a*) the least distance from the end at which no deflection occurs; (*b*) the allowable value of Q_1 if diametrical expansion at the end is limited to 15×10^{-5} m. Assume $E = 200$ GPa and $\nu = 0.3$.

12.2 A steel pipe of 0.5-m radius and 10-mm wall thickness is subjected to an internal pressure of 3.6 MPa. The pipe is joined to a pressure vessel. The juncture of the two parts is assumed to be rigid (i.e. $w = dw/dx = 0$ at the end of the pipe). Find, for practical considerations: (*a*) the minimum distance from the joint at which the pipe diameter attains its completely expanded value; (*b*) the maximum bending stress. Use $E = 200$ GPa and $\nu = 0.3$.

12.3 A long steel boiler tube of 0.05-m radius and 3-mm wall thickness is filled with hot water under 4 MPa uniform pressure. The tube is attached to a "header" in a simply supported manner at its end. Determine the deflection and the hoop stress at a distance of $L = 0.05$ m from the end. Let $E = 200$ GPa and $\nu = 0.3$.

12.4 A short cylindrical shell subjected to uniform internal pressure is simply supported at both ends (Fig. P12.4). Derive the following expression, setting $\alpha = \beta L/2$:

$$w_{\max} = -\frac{pL^4}{64D\alpha^4}\left(1 - \frac{2\cos\alpha\cosh\alpha}{\cosh\alpha + \cosh 2\alpha}\right) \tag{P12.4}$$

The foregoing represents the maximum displacement, occurring at the midlength of the shell. Employ Eqs. (12.5) as the homogeneous solution.

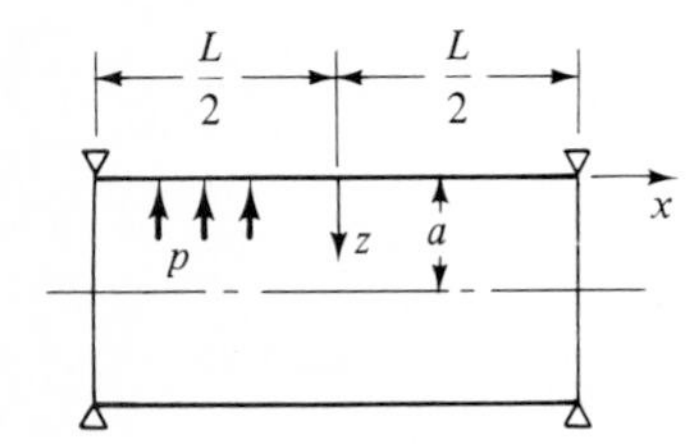

Figure P12.4

12.5 When the cylinder described in Prob. 12.4 is long, show that Eq. (P12.4) reduces to Eq. (*c*) of Sec. 12.2. What conclusion can be drawn from this result?

12.6 A circular steel pipe is reinforced by collars spaced L apart (Fig. 12.3). The cylinder is under an internal pressure p. The cross-sectional area of the collar is 0.025 m^2. What is the maximum bending stress in the pipe if the collars are constructed of a very rigid material. Use $L = 1.3$ m, $a = 0.6$ m, $t = 10$ mm, $p = 1.4$ MPa, and $\nu = 0.3$.

12.7 A long thin-walled bronze pipe of radius a and wall thickness t has a bronze collar shrunk over it at its midlength. Verify that if the cross-sectional area of the collar is $A = 1.56\sqrt{at^3}$, the decrease of shell diameter equals the expansion of collar diameter at the time of the shrinkage. Let $\nu = 0.3$.

12.8 A narrow ring is given a shrink fit onto a long pipe of radius a and thickness t. The cross-sectional area of the ring is A. If the outer radius of the pipe is greater by δ than the inner radius of the ring, what is the shrink-fit bending stress in the pipe?

12.9 Redo Prob. 12.6 for a ring made of a relatively flexible material as compared with the flexibility of the shell.

12.10 Compute the maximum bending and membrane stresses in a cylindrical tank wall with clamped base (Fig. 12.4*a*). The tank is filled to the top with water (specific weight 9.81 kN/m^3). The dimensions are $a = 2.7$ m, $h = 3.7$ m, and $t = 20$ mm. Use $\nu = 0.3$.

12.11 Figure P12.11 shows a water tower of radius a and height h is filled to capacity with a liquid of specific weight γ. The upper half portion of the tank is constructed of sheet steel of thickness t_2 and the lower half of sheet steel of thickness t_1. Determine the values of the discontinuity moments and shearing forces along the joint at mn.

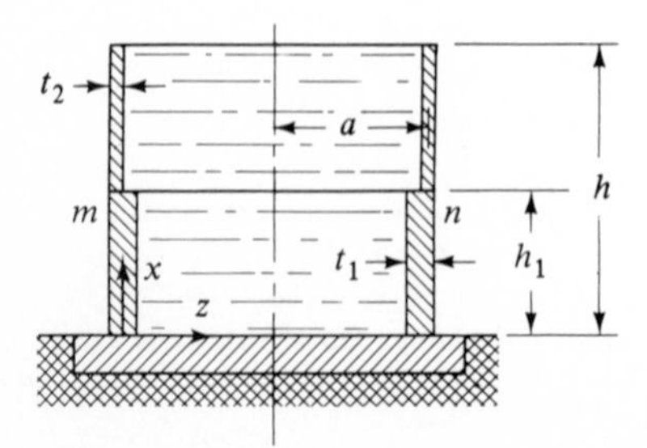

Figure P12.11

Secs. 12.5 to 12.11

12.12 A long copper tube of mean diameter 0.60 m and thickness $t = 5$ mm is heated in oil to 120°C above room temperature 20°C. Assume that the ends of the tube are simply supported and that

$$E_c = 120 \text{ GPa} \qquad \alpha_c = 17 \times 10^{-6} \text{ per °C} \qquad \nu_c = 0.3$$

What are the maximum thermal stresses.

12.13 A long brass cylinder of mean radius $a = 0.5$ m and thickness $t = 10$ mm has the uniform temperatures $T_1 = 100$°C and $T_2 = 300$°C at the inner and the outer surfaces, respectively. Calculate the maximum stress if the ends of the shell are assumed to be free. Use $E_b = 100$ GPa, $\alpha_b = 19 \times 10^{-6}$ per °C, and $\nu_b = 0.3$.

12.14 Compute the stresses in the composite cylinder described in Example 12.2 (Fig. 12.6*b*) for uniform temperature differentials equal to $\Delta T_1 = \Delta T_2 = \Delta T_3 = 134$°C.

12.15 Determine the maximum discontinuity and the membrane stresses in the vessel shown in Fig. 12.7 for $p = 1$ MPa, $a = 0.6$ m, $t = 10$ mm, and $\nu = 0.3$.

12.16 Determine the maximum discontinuity and the membrane stresses in the vessel of Fig. 12.8 for $p = 1$ MPa, $a = 0.6$ m, $b = 0.3$ m, $t = 10$ mm, and $\nu = 0.3$.

12.17 Obtain the maximum discontinuity and membrane stresses in the flat-head vessel (Fig. 12.9) for $p = 1$ MPa, $a = 0.6$ m, $t_c = t_h = 10$ mm, and $\nu = 0.3$.

12.18 A spherical steel vessel is rated to operate at an internal pressure of up to 2 MPa. The dimensions and strength of the shell are $r = 0.5$ m, $e = 0.8$, $\sigma = 100$ MPa. Determine the minimum required thickness of vessel according to: (*a*) the membrane theory; (*b*) the ASME unfired pressure-vessel code formula.

CHAPTER

THIRTEEN

CYLINDRICAL SHELLS UNDER GENERAL LOADS

13.1 INTRODUCTION

This chapter is devoted to methods for determining the deflection and stress in circular cylindrical shells under arbitrary radial, circumferential, or axial loads. In order to develop the relationships involving the applied-loading, cross-sectional, and material properties of a shell, internal stress resultants, and deformations, the approach applied earlier in Chap. 11 is again employed. This requires first that the equations governing the stress variation within the shell be developed (Sec. 13.2); second, that the deformation causing strain be related to the stress resultants through the appropriate stress-strain relationships as well as to the loading (Secs. 13.3 and 13.4); and finally, that the governing equations for deflection be solved by satisfying the edge conditions (Sec. 13.5). Special cases of the above-described *general procedure* form the basis of the theories of plates and beams.

The *inextensional*-shell theory and the *shallow-shell* theory are *simplified* shell theories.[11,39] The former is valid when no midsurface straining of the plate occurs owing to a particular type of loading (Sec. 13.6). The latter applies to thin shells of very large radius or small depth.

To complete the analysis of shell deformation and stress, the critical loads must also be considered. *Buckling* action underscores the difference in physical behavior of a thin shell under compression and tension. The critical load is particularly significant under axial compression, described in the last two sections, because axial buckling is synonymous with collapse of the shell.

13.2 DIFFERENTIAL EQUATIONS OF EQUILIBRIUM

To establish the differential equations of equilibrium for a circular cylindrical shell, we proceed as in Sec. 11.6, this time taking into account *all* stress resultant and surface-loading components. An element separated from a cylinder of radius a will now have acting on it the internal and surface-force resultants (Fig. 13.1*a*) and moment resultants (Fig. 13.1*b*). Clearly, the force and moment intensities vary across the elements. To simplify the diagrams, the notation N_x^+ is employed to denote $N_x + (\partial N_x/\partial x)\,dx$, etc.

Equilibrium of forces in the x, y, and z directions, after dividing by $dx\,d\theta$, now requires that

$$
\begin{aligned}
a\frac{\partial N_x}{\partial x} + \frac{\partial N_{\theta x}}{\partial \theta} + p_x a &= 0 \\
\frac{\partial N_\theta}{\partial \theta} + a\frac{\partial N_{x\theta}}{\partial x} - Q_\theta + p_y a &= 0 \\
\frac{\partial Q_\theta}{\partial \theta} + a\frac{\partial Q_x}{\partial x} + N_\theta + p_r a &= 0
\end{aligned}
\tag{13.1a-c}
$$

In a like manner, summation of moments, relative to the x and y coordinate directions, respectively, yields the expressions

$$
\begin{aligned}
\frac{\partial M_\theta}{\partial \theta} + a\frac{\partial M_{x\theta}}{\partial x} - aQ_\theta &= 0 \\
a\frac{\partial M_x}{\partial x} + \frac{\partial M_{\theta x}}{\partial \theta} - aQ_x &= 0
\end{aligned}
\tag{13.1d,e}
$$

The sixth equilibrium equation, $\sum M_z = 0$, leads to an identity and provides no new information.

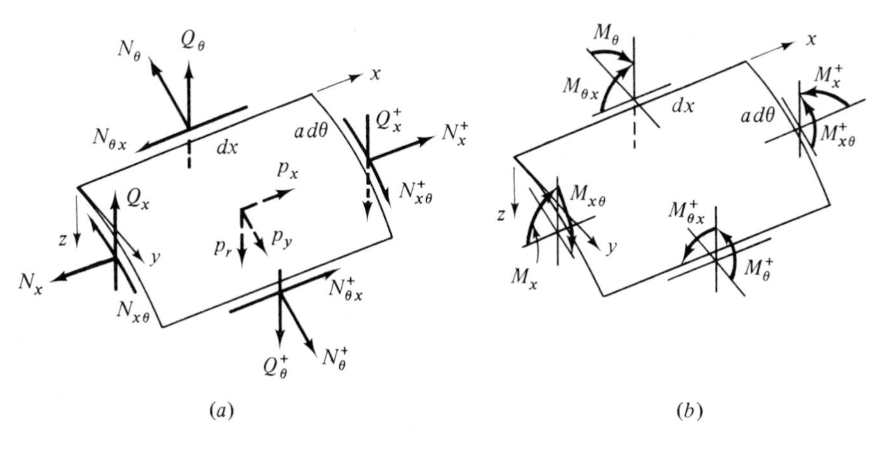

Figure 13.1

The transverse shears Q_x and Q_θ may be eliminated from the above expressions. As already noted in Sec. 11.2, for thin shells: $N_{x\theta} = N_{\theta x}$ and $M_{x\theta} = M_{\theta x}$. Collection of the *equilibrium equations* and rearrangement then yields

$$
\begin{aligned}
a\frac{\partial N_x}{\partial x} + \frac{\partial N_{x\theta}}{\partial \theta} + p_x a &= 0 \\
\frac{\partial N_\theta}{\partial \theta} + a\frac{\partial N_{x\theta}}{\partial x} - \frac{1}{a}\frac{\partial M_\theta}{\partial \theta} - \frac{\partial M_{x\theta}}{\partial x} + p_y a &= 0 \qquad (13.2a\text{–}c) \\
\frac{1}{a}\frac{\partial^2 M_\theta}{\partial \theta^2} + 2\frac{\partial^2 M_{x\theta}}{\partial x\,\partial \theta} + a\frac{\partial^2 M_x}{\partial x^2} + N_\theta + p_r a &= 0
\end{aligned}
$$

These three expressions contain six unknown stress resultants. It is therefore necessary to identify three additional relationships, based upon the deformation of the shell as discussed in the next section.

We note that in the derivation of Eqs. (13.2), *stretching* of the midsurface and change in curvature of the shell element were neglected. Thus, the results obtained hold *only* if the forces N_x, N_y, and N_{xy} are *small* in comparison with their *critical* values, at which buckling of shell may take place.

13.3 KINEMATIC RELATIONSHIPS

In studying the *geometry* of strains and curvatures of a nonsymmetrically deformed cylindrical shell, we may begin as in the analysis of plates and shells of revolution. Kinematic expressions relating the *midsurface* strains to the displacements are thus:

$$
\varepsilon_x = \frac{\partial u}{\partial x} \qquad \varepsilon_\theta = \frac{1}{a}\frac{\partial v}{\partial \theta} - \frac{w}{a} \qquad \gamma_{x\theta} = \frac{1}{a}\frac{\partial u}{\partial \theta} + \frac{\partial v}{\partial x} \qquad (13.3a\text{–}c)
$$

The components ε_x and $\gamma_{x\theta}$ are the same as in the case of the plates except that $a d\theta$ is replaced by dy. Circumferential strain ε_θ is obtained from Eq. (*a*) of Sec. 10.6 by setting $r_1 = a$.

Changes in curvatures, χ_x and χ_θ, and twist $\chi_{x\theta}$ are expressed by

$$
\chi_x = \frac{\partial^2 w}{\partial x^2} \qquad \chi_\theta = \frac{1}{a^2}\left(\frac{\partial v}{\partial \theta} + \frac{\partial^2 w}{\partial \theta^2}\right) \qquad \chi_{x\theta} = \frac{1}{a}\left(\frac{\partial v}{\partial x} + \frac{\partial^2 w}{\partial x\,\partial \theta}\right) \qquad (13.3d\text{–}f)
$$

The terms of χ_θ are determined by considering the deformation of a circumferential element (Fig. 13.2*a*). Solid line *AB* of length *ds* represents the side view of the midsurface prior to deformation, having a curvature in the circumferential direction: $\partial\theta/\partial s = \partial\theta/a\,\partial\theta = 1/a$. It is then displaced to position *A′B′*, which is now *ds′* long. Its curvature then becomes

$$
\frac{\partial \theta'}{\partial s'} \approx \frac{d\theta + (\partial^2 w/\partial s^2)\,ds}{(a - w)\,d\theta}
$$

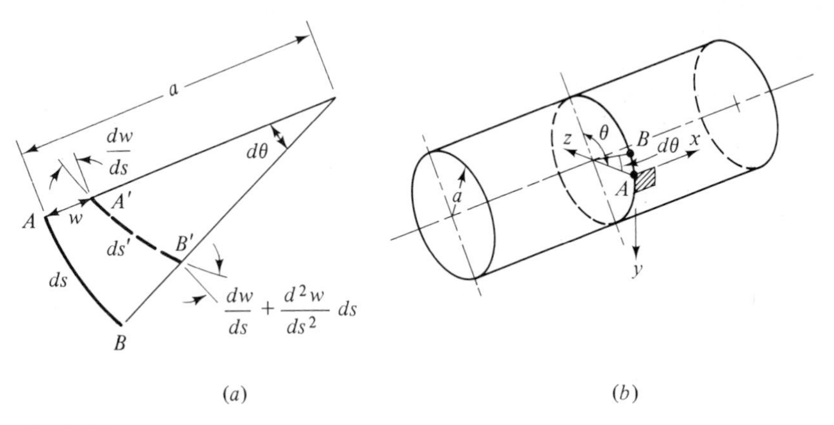

Figure 13.2

Hence,

$$\chi_\theta = \frac{\partial \theta'}{\partial s'} - \frac{1}{a} \approx \frac{1}{a^2}\left(w + \frac{\partial^2 w}{\partial \theta^2}\right)$$

Note that an increase in circumferential length dv owing to the radial displacement w is only $w\, d\theta$. Neglecting the effect of circumferential midsurface straining, we can therefore replace w in the above expression by $\partial v/\partial \theta$ to obtain Eq. (13.3*e*).

With respect to the terms of the $\chi_{x\theta}$ in Eqs. (13.3*f*), representing the twist of each midsurface shell element, as indicated at point A in Fig. 13.2*b*, additional explanation is required. During deformation, this element rotates through an angle $-\partial w/\partial x$ about the y axis and through an angle $\partial v/\partial x$ about the z axis. Here the sign convention used is based upon the right-hand-screw rule. Now consider a similar element at point B. Owing to displacement w, the second element experiences the following angular displacement about the y axis:

$$-\frac{\partial w}{\partial x} - \frac{\partial^2 w}{\partial \theta\, \partial x}\, d\theta \tag{a}$$

The same element also experiences the following angular displacement in the plane tangent to the shell

$$\frac{\partial v}{\partial x} + \frac{\partial(\partial v/\partial x)}{\partial \theta}\, d\theta$$

Neglecting a higher-order infinitesimal term, it is found that the latter rotation, owing to angle $d\theta$, has a component with respect to the y axis

$$-\frac{\partial v}{\partial x}\, d\theta \tag{b}$$

We observe from Eqs. (*a*) and (*b*) that the total angle of twist between the element at A and B is equal to

$$-\chi_{x\theta} a\, d\theta = -\left(\frac{\partial^2 w}{\partial\theta\,\partial x} + \frac{\partial v}{\partial x}\right) d\theta$$

This expression yields the result given by Eq. (13.3*f*).

13.4 THE GOVERNING EQUATIONS FOR DEFLECTIONS

The expressions governing the deformation of cylindrical shells subjected to direct and bending forces can now be developed. This is accomplished by first introducing Eqs. (13.3) into the constitutive equations (11.2). In so doing, we have the following stress-resultant displacement relations

$$\begin{aligned}
N_x &= \frac{Et}{1-\nu^2}\left[\frac{\partial u}{\partial x} + \nu\left(\frac{1}{a}\frac{\partial v}{\partial\theta} - \frac{w}{a}\right)\right] \\
N_\theta &= \frac{Et}{1-\nu^2}\left[\frac{1}{a}\frac{\partial v}{\partial\theta} - \frac{w}{a} + \nu\frac{\partial u}{\partial x}\right] \\
N_{x\theta} &= \frac{Et}{2(1+\nu)}\left(\frac{\partial v}{\partial x} + \frac{1}{a}\frac{\partial u}{\partial\theta}\right) \\
M_x &= -D\left[\frac{\partial^2 w}{\partial x^2} + \frac{\nu}{a^2}\left(\frac{\partial v}{\partial\theta} + \frac{\partial^2 w}{\partial\theta^2}\right)\right] \\
M_\theta &= -D\left[\frac{1}{a^2}\left(\frac{\partial v}{\partial\theta} + \frac{\partial^2 w}{\partial\theta^2}\right) + \nu\frac{\partial^2 w}{\partial x^2}\right] \\
M_{x\theta} &= -D(1-\nu)\frac{1}{a}\left(\frac{\partial v}{\partial x} + \frac{\partial^2 w}{\partial x\,\partial\theta}\right)
\end{aligned} \qquad (13.4a\text{–}f)$$

Then, upon substituting Eqs. (13.4) into Eqs. (13.2), there is derived a set of three expressions in three displacements u, v, and w:

$$\begin{aligned}
&\frac{\partial^2 u}{\partial x^2} + \frac{1-\nu}{2a^2}\frac{\partial^2 u}{\partial\theta^2} + \frac{1+\nu}{2a}\frac{\partial^2 v}{\partial x\,\partial\theta} - \frac{\nu}{a}\frac{\partial w}{\partial x} = -\frac{p_x(1-\nu^2)}{Et} \\
&\frac{1+\nu}{2a}\frac{\partial^2 u}{\partial x\,\partial\theta} + \frac{1-\nu}{2}\frac{\partial^2 v}{\partial x^2} + \frac{1}{a^2}\frac{\partial^2 v}{\partial\theta^2} - \frac{1}{a^2}\frac{\partial w}{\partial\theta} + \frac{t^2}{12a^2}\left(\frac{\partial^3 w}{\partial x^2\,\partial\theta}\right. \\
&\qquad \left. + \frac{1}{a^2}\frac{\partial^3 w}{\partial\theta^3}\right) + \frac{t^2}{12a^2}\left[(1-\nu)\frac{\partial^2 v}{\partial x^2} + \frac{1}{a^2}\frac{\partial^2 v}{\partial\theta^2}\right] = -\frac{p_y(1-\nu^2)}{Et} \qquad (13.5a\text{–}c) \\
&\frac{\nu}{a}\frac{\partial u}{\partial x} + \frac{1}{a^2}\frac{\partial v}{\partial\theta} - \frac{w}{a^2} - \frac{t^2}{12}\left(\frac{\partial^4 w}{\partial x^4} + \frac{2}{a^2}\frac{\partial^4 w}{\partial x^2\,\partial\theta^2} + \frac{1}{a^4}\frac{\partial^4 w}{\partial\theta^4}\right) \\
&\qquad - \frac{t^2}{12a^2}\left(\frac{\partial^3 v}{\partial x^2\,\partial\theta} + \frac{1}{a^2}\frac{\partial^3 v}{\partial\theta^3}\right) = -\frac{p_r(1-\nu^2)}{Et}
\end{aligned}$$

These are the *governing equations for the displacements* in thin-walled circular cylindrical shells under *general* loading.

Approximate relations Equations (13.5) contain a number of terms, which in many applications, yield numerical results of negligible practical importance.[11,32] The last two terms on the left-hand side of Eq. (13.5*b*) and the last term on the left-hand side of Eq. (13.5*c*) are of this type. Such quantities are of the same order as those which have already been neglected in Secs. 11.3 and 13.2 for the thin shells ($t \ll a$). By dropping these terms of small magnitude, much time and effort is saved in the solving for the displacements. By so doing, the following *simplified* set of *governing equations* for displacements of circular cylindrical shells are obtained

$$\begin{aligned}
\frac{\partial^2 u}{\partial x^2} + \frac{1-\nu}{2a^2}\frac{\partial^2 u}{\partial \theta^2} + \frac{1+\nu}{2a}\frac{\partial^2 v}{\partial x\,\partial \theta} - \frac{\nu}{a}\frac{\partial w}{\partial x} &= -\frac{p_x(1-\nu^2)}{Et} \\
\frac{1+\nu}{2a}\frac{\partial^2 u}{\partial x\,\partial \theta} + \frac{1-\nu}{2}\frac{\partial^2 v}{\partial x^2} + \frac{1}{a^2}\frac{\partial^2 v}{\partial \theta^2} - \frac{1}{a^2}\frac{\partial w}{\partial \theta} &= -\frac{p_y(1-\nu^2)}{Et} \\
\frac{\nu}{a}\frac{\partial u}{\partial x} + \frac{1}{a^2}\frac{\partial v}{\partial \theta} - \frac{w}{a^2} - \frac{t^2}{12}\left(\frac{\partial^4 w}{\partial x^4} + \frac{2}{a^2}\frac{\partial^4 w}{\partial x^2\,\partial \theta^2} + \frac{1}{a^4}\frac{\partial^4 w}{\partial \theta^4}\right) &= -\frac{p_r(1-\nu^2)}{Et}
\end{aligned} \tag{13.6}$$

When a circular cylindrical shell is loaded symmetrically with respect to its axis, as expected the solution (13.5) or (13.6) can readily be reduced to the form given by Eqs. (11.9). Several examples of such a case are given in Sec. 11.6 and Chapter 12.

Based upon the simplifications which lead to Eqs. (13.6), the effect of displacements u and v on the bending and twisting moments must be regarded as negligible. From Eqs. (13.4), the *simplified elastic law* is then

$$\begin{aligned}
N_x &= \frac{Et}{1-\nu^2}\left[\frac{\partial u}{\partial x} + \nu\left(\frac{1}{\nu}\frac{\partial v}{\partial \theta} - \frac{w}{a}\right)\right] \\
N_\theta &= \frac{Et}{1-\nu^2}\left(\frac{1}{a}\frac{\partial v}{\partial \theta} - \frac{w}{a} + \nu\frac{\partial u}{\partial x}\right) \\
N_{x\theta} &= \frac{Et}{2(1+\nu)}\left(\frac{\partial v}{\partial x} + \frac{1}{a}\frac{\partial u}{\partial \theta}\right) \\
M_x &= -D\left(\frac{\partial^2 w}{\partial x^2} + \frac{\nu}{a^2}\frac{\partial^2 w}{\partial \theta^2}\right) \\
M_\theta &= -D\left(\frac{1}{a^2}\frac{\partial^2 w}{\partial \theta^2} + \nu\frac{\partial^2 w}{\partial x^2}\right) \\
M_{x\theta} &= -D(1-\nu)\frac{1}{a}\frac{\partial^2 w}{\partial x\,\partial \theta}
\end{aligned} \tag{13.7}$$

We observe that the problem of circular cylindrical shells reduces, in each particular case, to the solution of a set of three differential equations (13.5) or (13.6). The latter set is often preferred for practical applications. A particular example of such a solution is illustrated in the next section.

13.5 A TYPICAL CASE OF ASYMMETRICAL LOADING

A pipe supported on saddles at intervals or at the edges and filled partially with liquid is a typical case of axially *unsymmetrically* loaded cylindrical shells.

Consider a circular cylindrical tank filled to a given level with a liquid of specific weight γ (Fig. 13.3*a*). Both ends of the tank may be taken as supported by end plates so that, at $x = 0$ and $x = L$:

$$v = 0 \qquad w = 0 \qquad N_x = 0 \qquad M_x = 0 \tag{a}$$

It is observed that the expressions (*a*) and the conditions of symmetry of deformation are satisfied by assuming the following displacements

$$\begin{aligned} u &= \sum_{m=1}^{\infty} \sum_{n=0}^{\infty} a_{mn} \cos n\theta \cos \frac{m\pi x}{L} \\ v &= \sum_{m=1}^{\infty} \sum_{n=0}^{\infty} b_{mn} \sin n\theta \sin \frac{m\pi x}{L} \\ w &= \sum_{m=1}^{\infty} \sum_{n=0}^{\infty} c_{mn} \cos n\theta \sin \frac{m\pi x}{L} \end{aligned} \tag{13.8}$$

in which the angle θ is measured as shown in Fig. 13.3*a*. The upper part $(\alpha \le \theta \le \pi)$ of the tank carries no load. For the lower part $(0 \le \theta \le \alpha)$,

$$p_r = p = -\gamma a(\cos\theta - \cos\alpha) \qquad p_x = p_y = 0 \tag{13.9}$$

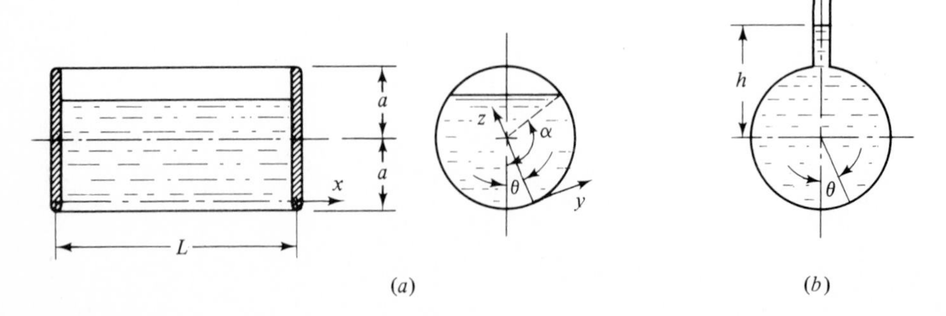

Figure 13.3

Here the angle α defines the liquid level. It is useful to expand the foregoing load into a Fourier series:

$$p = -\sum_{m}^{\infty}\sum_{n}^{\infty} p_{mn} \cos n\theta \sin \frac{m\pi x}{L} \tag{13.10}$$

The coefficients p_{mn}, determined in the usual manner (Sec. A.3), are of the form

$$p_{mn} = \frac{8\gamma a(\cos\alpha \sin n\alpha - n\cos n\alpha \sin\alpha)}{mn\pi^2(n^2-1)} \qquad \begin{pmatrix} m = 1, 3, \ldots \\ n = 2, 3, \ldots \end{pmatrix} \tag{13.11a}$$

$$p_{m0} = \frac{4\gamma a}{m\pi^2}(\sin\alpha - \alpha\cos\alpha) \qquad p_{m1} = \frac{2\gamma a}{m\pi}(2\alpha - \sin 2\alpha) \tag{13.11b}$$

If the cylinder is *filled to capacity* (Fig. 13.3*b*), the loading is expressed

$$p = -\gamma(h + a\cos\theta)$$

and the coefficients are

$$p_{mn} = 0 \qquad p_{m0} = \frac{4\gamma h}{m\pi} \qquad p_{m1} = \frac{4\gamma a}{m\pi} \qquad (m = 1, 3, \ldots) \tag{b}$$

We now introduce the notation

$$\lambda = \frac{L}{a} \qquad \eta = \frac{t}{2L}$$

The governing equations (13.6), upon substituting (13.8) and (13.10) and using the above notation are, therefore:

$$\begin{aligned} a_{mn}[2m^2\pi^2 + (1-\nu)\lambda^2 n^2] - b_{mn}(1+\nu)\lambda mn\pi + c_{mn}2\nu\lambda m\pi &= 0 \\ a_{mn}(1+\nu)\lambda mn\pi - b_{mn}[(1-\nu)m^2\pi^2 + 2\lambda^2 n^2] - c_{mn}2\lambda^2 n &= 0 \\ a_{mn}3\nu\lambda m\pi - b_{mn}3\lambda^2 n + c_{mn}[3\lambda^2 + \eta^2(m^2\pi^2 + \lambda^2 n^2)^2] &= -\frac{p_{mn}L^2t^2}{4D} \end{aligned} \tag{13.12}$$

Inasmuch as p_{mn} is given by expressions (13.11) or (*b*), the parameters a_{mn}, b_{mn}, c_{mn} can be determined in each particular case upon application of Eqs. (13.12) for any m and n. The stress resultant and displacement at any cylinder location may then be ascertained using Eqs. (13.7) and (13.8).

Example 13.1 Calculate the maximum values of lateral deflection and stress in the tank filled with liquid ($\alpha = \pi$), shown in Fig. 13.3*b*. Use: $a = h = 50$ cm; $L = 25$ cm; $t = 7$ cm; $\nu = 0.3$.

SOLUTION As $a = h$, by setting

$$Z = \frac{2\gamma a L^2 t}{\pi^2 D}$$

Table 13.1

m	$a_{m0}\dfrac{2(10^3)}{Zt}$	$c_{m0}\dfrac{2(10^3)}{Zt}$	$a_{m1}\dfrac{2(10^3)}{Zt}$	$b_{m1}\dfrac{2(10^3)}{Zt}$	$c_{m1}\dfrac{2(10^3)}{Zt}$
1	57.88	−1212	49.18	−66.26	−1183
3	0.1073	−6.742	0.1051	−0.0432	−6.704
5	0.00503	−0.526	0.00499	−0.00122	−0.525

for $n = 0$, $m = 1, 3, \ldots$, Eqs. (13.12) and (b) yield

$$c_{m0} = -\frac{m\pi}{\lambda\nu}a_{m0} \qquad b_{m0} = 0 \qquad a_{m0} = \frac{\nu Z\lambda t}{2m^2[3\lambda^2(1-\nu^2)+\eta^2 m^4\pi^4]} \tag{c}$$

For $n = 1$, the expressions for the coefficients become more complicated. The numerical values[40] of the coefficients are furnished in Table 13.1 for $m = 1, 3, 5$ and $n = 0, 1$.

The largest values of the lateral deflection, force, and moment are found at $x = L/2$, $\theta = 0$. These will be computed by taking $m = 1, 3, 5$ and $n = 0, 1$. On applying the third of Eqs. (13.8), together with the given data, we have

$$w_{\max} = c_{10} + c_{11} - c_{30} - c_{31} + c_{50} + c_{51} = -11{,}768.7\,\frac{\gamma}{E} \quad \text{cm}$$

In a like manner, from Eqs. (13.8) and (13.7):

$$N_{x,\max} = -2481.2\gamma \quad \text{N/cm}$$

$$N_{\theta,\max} = 1607.7\gamma \quad \text{N/cm}$$

$$M_{x,\max} = -5657.5\gamma \quad \text{N}$$

$$M_{\theta,\max} = -1763.7\gamma \quad \text{N}$$

Through the use of Eqs. (11.4), we then obtain

$$\sigma_{x,\max} = -\frac{2481.2\gamma}{7} - \frac{6(5657.5\gamma)}{49} = 1047.2\gamma \quad \text{N/cm}^2$$

$$\sigma_{\theta,\max} = \frac{1607.7\gamma}{7} - \frac{6(1763.7\gamma)}{49} = 13.7\gamma \quad \text{N/cm}^2$$

as the maximum axial and tangential stresses in the tank.

It is observed in the foregoing example that the coefficients a_{mn}, b_{mn}, and c_{mn} diminish very rapidly. We conclude therefore that only a few terms in the series (13.8) suffice to yield fairly accurate results. However, in the case of longer and thinner shells a_{mn}, b_{mn} and c_{mn} diminish rather slowly, and it is necessary to calculate a somewhat large number of coefficients in order to obtain the deformation and stress resultants with sufficient accuracy.

13.6 INEXTENSIONAL DEFORMATIONS

Associated with the *inextensional deformations* of shells, a *simplified* shell theory is useful under certain conditions. The *inextensional shell* theory is applicable to a variety of shell forms. However, we shall deal only with the inextensional deformation of circular cylindrical shells.

The inextensional theory is often preferred when shell structures resist loading principally through bending action. Such cases include: a cylinder subjected to loads without axial symmetry and confined to a small circumferential portion (Fig. 13.4*a*); a cylinder with free ends under variable pressure $p(x, \theta)$ where $\int_0^{2\pi} p(x, \theta)\, d\theta \approx 0$ (Fig. 13.4*b*). In both situations shortening of the vertical diameter along which P or $p(x, \theta)$ act and lengthening of the horizontal diameter occur. Hence, there is considerable bending caused by the changes in curvature, but *no* stretching of midsurface length. Deformations of these types are thus described as inextensional.

In inextensional-shell theory, the midsurface in-plane strain components given by Eqs. (13.3*a–c*) are taken to be zero. The *conditions of inextensibility* are therefore

$$\frac{\partial u}{\partial x} = 0 \qquad \frac{1}{a}\frac{\partial v}{\partial \theta} - \frac{w}{a} = 0 \qquad \frac{1}{a}\frac{\partial u}{\partial \theta} + \frac{\partial v}{\partial x} = 0 \qquad (13.13a\text{–}c)$$

Upon introduction of the above in Eqs. (13.4*a–c*), we obtain

$$N_x = N_\theta = N_{x\theta} = 0 \qquad (a)$$

We also observe from Eqs. (13.13*a*) that u depends on θ only, and Eqs. (13.13*b*) leads to

$$w = \frac{\partial v}{\partial \theta} \qquad (b)$$

The conditions (13.13) are thus fulfilled by assuming the displacements as follows:[11]

$$u = u_1 + u_2 \qquad v = u_1 + v_2 \qquad w = w_1 + w_2 \qquad (13.14)$$

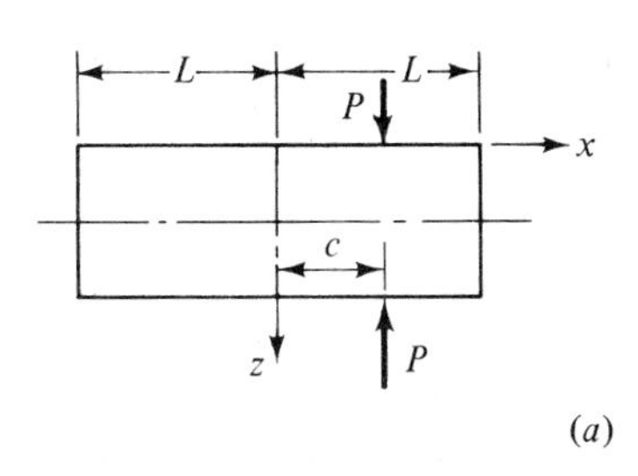

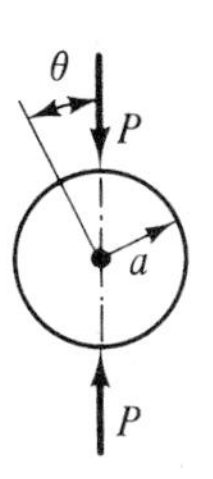

(*a*)

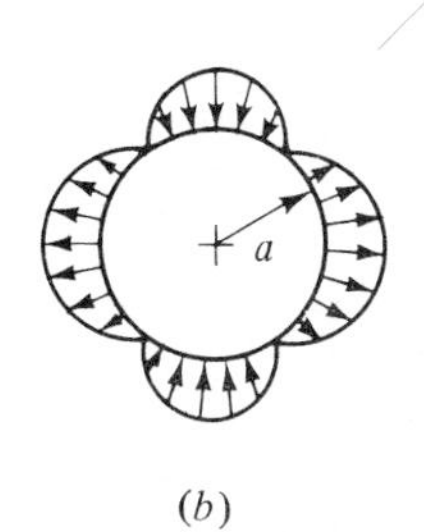

(*b*)

Figure 13.4

with

$$u_1 = 0$$

$$v_1 = a \sum_{n=1}^{\infty} (a_n \cos n\theta - \bar{a}_n \sin \theta) \tag{13.15a–c}$$

$$w_1 = -a \sum_{n=1}^{\infty} n(a_n \sin \theta + \bar{a}_n \cos \theta)$$

and

$$u_2 = -a \sum_{n=1}^{\infty} \frac{1}{n}(b_n \sin n\theta + \bar{b}_n \cos n\theta)$$

$$v_2 = x \sum_{n=1}^{\infty} (b_n \cos n\theta - \bar{b}_n \sin n\theta) \tag{13.16a–c}$$

$$w_2 = -x \sum_{n=1}^{\infty} n(b_n \sin n\theta + \bar{b}_n \cos n\theta)$$

In the above, a represents the radius of cylinder. The constants a_n, $\bar{a}_n$, b_n, and $\bar{b}_n$ are determined for each loading situation. According to expressions (13.15), all cross sections of the cylinder deform alike. The displacements varying along the length of the shell are given by the last two equations of (13.16).

A shell experiences inextensional deformations under a prescribed load in such a manner that a minimum strain energy is produced. Since expressions (13.14) yield $\partial^2 w/\partial x^2 = \chi_x = 0$, the required expression for the bending-strain energy of the shell is given by (Sec. 11.5):

$$U = -\frac{D}{2} \iint [\chi_\theta^2 + 2(1-\nu)\chi_{x\theta}^2] a \, d\theta \, dx \tag{c}$$

The strain energy of a cylindrical shell undergoing inextensional deformation, upon substituting Eqs. (13.3) into Eq. (*c*), is then

$$U = \frac{D}{2a^4} \iint \left[\left(\frac{\partial v}{\partial \theta} + \frac{\partial^2 w}{\partial \theta^2} \right)^2 + 2(1-\nu)a^2 \left(\frac{\partial^2 w}{\partial x \, \partial \theta} + \frac{\partial v}{\partial x} \right)^2 \right] a \, d\theta \, dx \tag{13.17}$$

Inserting the displacements w and v from Eqs. (13.14) for a cylindrical shell of length $2L$ (Fig. 13.4*a*), Eq. (13.17) takes the form

$$U = \pi D L \sum_{n=2}^{\infty} \frac{(n^2-1)^2}{a^3} \{a^2 n^2 [a_n^2 + (\bar{a}_n)^2] + \tfrac{1}{3} L^2 n^2 [b_n^2 + (\bar{b}_n)^2] + 2(1-\nu)a^2 [b_n^2 + (b_n)^2]\} \tag{13.18}$$

Omission of the term $n = 1$ in the above expression is explained as follows. For $n = 1$, one has

$$v_1 = a(a_1 \cos \theta - \bar{a}_1 \sin \theta)$$

$$w_1 = -a(a_1 \sin \theta + \bar{a} \cos \theta)$$

The foregoing describes the rigid-body displacement of a circle in its plane and hence does not contribute to the strain energy of the shell. The same conclusion holds for dispacements u_2, v_2, and w_2.

Equation (13.18) is useful in determining the components of shell displacement as is next observed.

Example 13.2 A circular cylinder of length $2L$ is loaded by two equal and opposite concentrated forces a distance c from the midspan (Fig. 13.4a). The ends of the shell are free. Derive the equations describing the elastic surface.

SOLUTION The radial displacements at $x = c$, in the direction of the loads, are $w(0)$ and $w(\pi)$. It is clear that the terms involving a_n and b_n in Eqs. (13.15c) and (13.16c) become zero at these points. The remaining terms with the coefficients $\bar{a}_n$ and $\bar{b}_n$ in w contribute *only* to the work produced by the forces. The potential energy $\Pi = U - W$ is therefore

$$\Pi = \pi DL \sum_{n=2}^{\infty} \frac{(n^2-1)^2}{a^2} \{n^2[a^2(\bar{a}_n)^2 + \tfrac{1}{3}L^2(\bar{b}_n)^2] + 2(1-\nu)a^2(\bar{b}_n)^2\}$$
$$+ \left\{a \sum_{n=1}^{\infty} n\bar{a}_n[1 + \cos n\pi] + c \sum_{n=1}^{\infty} n\bar{b}_n[1 + \cos n\pi]\right\} P$$

Applying Eqs. (1.43), from conditions $\partial\Pi/\bar{a}_n = 0$ and $\partial\Pi/\bar{b}_n = 0$, coefficients $\bar{a}_n$ and $\bar{b}_n$ are readily evaluated:

$$\bar{a}_n = -\frac{a^2(1 + \cos n\pi)P}{2n\pi DL(n^2-1)^2}$$

$$\bar{b}_n = -\frac{nca^3(1 + \cos n\pi)P}{2\pi DL(n^2-1)^2[\tfrac{1}{3}n^2L^2 + 2(1-\nu)a^2]}$$

or

$$\bar{a}_n = -\frac{a^2P}{n(n^2-1)^2\pi DL} \qquad (n = 2, 4, \ldots)$$

$$\bar{b}_n = -\frac{nca^3P}{(n^2-1)^2\pi DL[\tfrac{1}{3}n^2L^2 + 2(1-\nu)a^2]} \qquad (n = 2, 4, \ldots)$$

and $\bar{a}_n = \bar{b}_n = 0$ for $n = 1, 3, \ldots$. Equations (13.14) become, in this case,

$$u = \frac{Pa^3}{\pi DL} \sum_n^{\infty} \frac{ac \cos n\theta}{(n^2-1)^2[\tfrac{1}{3}n^2L^2 + 2(1-\nu)a^2]}$$

$$v = \frac{Pa^3}{\pi DL} \sum_n^{\infty} \left\{\frac{1}{n(n^2-1)^2} + \frac{ncx}{(n^2-1)^2[\tfrac{1}{3}n^2L^2 + 2(1-\nu)a^2]}\right\} \sin n\theta \qquad (13.19)$$

$$w = \frac{Pa^3}{\pi DL} \sum_n^{\infty} \left\{\frac{1}{n(n^2-1)^2} + \frac{n^2cx}{(n^2-1)^2[\tfrac{1}{3}n^2L^2 + 2(1-\nu)a^2]}\right\} \cos n\theta$$

where $n = 2, 4, \ldots$. The expressions for the moments can be obtained from the general solution (13.19) by applying Eqs. (13.4). The stresses are then determined from Eqs. (11.4).

When the loads are applied at midlength ($c = 0$), the *vertical reduction* of the cross section at the center, from $w(0) + w(\pi)$, is

$$w_v = \frac{2Pa^3}{\pi DL} \sum_n^\infty \frac{1}{(n^2 - 1)^2} = 0.149 \frac{Pa^3}{2DL} \tag{d}$$

At the center cross section the *horizontal increase*, as ascertained from $w(\pi/2) + w(3\pi/2)$, equals

$$w_h = \frac{2Pa^3}{\pi DL} \sum_n^\infty \frac{(-1)^{(n/2)+1}}{(n^2 - 1)^2} = 0.137 \frac{Pa^3}{2DL} \tag{e}$$

The displacements at any section of the cylinder may be found similarly.

It is observed in the above solution that all coefficients a_n, b_n, $\bar{a}_n$, $\bar{b}_n$ are uniquely determined by the applied load. Hence, in inextensional-shell theory, *no* boundary conditions can be fulfilled at the free edges of the shell. However, the edge bending due to displacements (13.19) is of local character and does not have a pronounced effect on the accuracy of the results given by Eqs. (*d*) and (*e*).

13.7 SYMMETRICAL BUCKLING UNDER UNIFORM AXIAL PRESSURE

In numerous examples of the preceding chapters, it is observed that thin shell structures are often subjected to compressive stresses at particular areas. To check the stability of the elastic equilibrium of compressed shells, the methods described in Secs. 1.8 and 1.9 may be applied. In this section, we describe the fundamental concepts of the energy approach by considering the *symmetrical buckling* of a thin-walled circular cylinder under uniform axial compression. We thus provide only an introduction to shell instability,[24,32] a critically important area of engineering design. In the section following, a more general case of buckling is discussed which employs the equilibrium approach.

When a circular cylindrical shell is under uniform axial compression, axisymmetrical buckling of the shell may take place at a particular value of compressive load (Fig. 13.5). We shall determine the critical value of compressive forces N per unit circumferential length. At the start of buckling, the shell-strain energy is increased by the following: midsurface straining in the circumferential direction, bending, and axial compression. At the critical value of load, this increase in energy must be equal to the work done by the compressive load owing to axial straining and bending as the cylinder deflects due to buckling action. Thus, the critical load is that for which

$$\delta U = \delta W \tag{a}$$

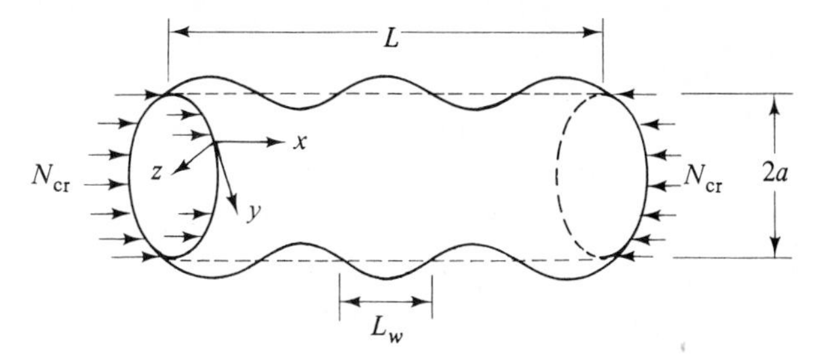

Figure 13.5

This is the principle of virtual work (Sec. 1.9). It is therefore assumed that the axial compressive force *does not change during the buckling.*

If the cylinder is *simply supported* at both ends, the radial deflection satisfying the boundary conditions is

$$w = -c_0 \sin \frac{m\pi x}{L} \tag{b}$$

Here c_0 is a constant, L is the length of the shell, and m is the number of half-sine waves in the axial direction. As the cylinder deflects, it becomes shorter in length. The axial midsurface strain *before* buckling is given by

$$\varepsilon_{x1} = -\frac{N}{Et}$$

where t is the thickness of the shell. From Hooke's law,

$$\varepsilon_{x1} E = \frac{E}{1-\nu^2}(\varepsilon_{x2} + \nu\varepsilon_{\theta 2}) \tag{c}$$

or

$$\varepsilon_{x2} + \nu\varepsilon_{\theta 2} = (1-\nu^2)\varepsilon_{x1} \tag{d}$$

in which ε_{x2} and $\varepsilon_{\theta 2}$ are the axial and circumferential midsurface strains *after* buckling. Referring to Sec. 11.6, the latter strain is expressed

$$\varepsilon_{\theta 2} = -\nu\varepsilon_{x1} - \frac{w}{a} = -\nu\varepsilon_{x1} + \frac{c_0}{a}\sin\frac{m\pi x}{L} \tag{e}$$

Inserting the above into Eq. (*d*),

$$\varepsilon_{x2} = \varepsilon_{x1} - \nu\frac{c_0}{a}\sin\frac{m\pi x}{L} \tag{f}$$

Owing to the axisymmetrical deformation,

$$\gamma_{xy} = \chi_y = \chi_{xy} = 0 \tag{g}$$

and the change in curvature in the axial plane is

$$\chi_x = c_0 \frac{m^2\pi^2}{L^2} \sin \frac{m\pi x}{L} \tag{h}$$

As N remains constant during buckling, the work done is

$$\delta W = 2\pi N \left[\nu \int_0^L c_0 \sin \frac{m\pi x}{L} dx + \frac{a}{2} \int_0^L \left(c_0 \frac{m\pi}{L} \sin \frac{m\pi x}{L} \right)^2 dx \right] \tag{i}$$

The components of this work are associated with the change of the axial strain $(\varepsilon_{x2} - \varepsilon_{x1})$ and the bending of the generators defined by Eq. (*b*). Next, the increase in strain energy during buckling must be determined. For this purpose, Eqs. (*e*) to (*h*), are introduced into Eqs. (11.7) and (11.8). After addition of δU_b and δU_m, we have

$$\delta U = -2\pi t E \nu \varepsilon_{x1} \int_0^L c_0 \sin \frac{m\pi x}{L} dx + \frac{\pi^2 c_0^2 EtL}{2a} + c_0^2 \frac{\pi^4 m^4}{2L^4} \pi a L D \tag{j}$$

The criterion for buckling, $\delta W = \delta U$, together with Eqs. (*i*) and (*j*), is applied to find the critical value of N. It follows that

$$\left(\frac{\pi EtL}{2a} + \frac{\pi^5 am^4 D}{2L^3} - \frac{\pi^3 am^2 N}{2L} \right) c_0^2 = 0$$

For the above to be valid for any c_0, it is required that

$$N_{\text{cr}} = D\left(\frac{m^2\pi^2}{L^2} + \frac{EtL^2}{Da^2m^2\pi^2} \right) \tag{13.20}$$

As for columns and plates, for each value of m there is a unique buckling-mode shape and a unique buckling load. The lowest critical load is of greatest interest and is found by setting the derivative of N_{cr} with respect to L equal to zero for $m = 1$. We thus determine, for $\nu = 0.3$, the length at which the minimum buckling load has effect:

$$L_w = \pi \sqrt[4]{\frac{a^2t^2}{12(1-\nu^2)}} \approx 1.72\sqrt{at} \tag{13.21}$$

The above is the *length of half-sine waves* into which the shell buckles (Fig. 13.5). The corresponding *minimum buckling load* is

$$N_{\text{cr}} = \frac{Et^2}{a\sqrt{3(1/\nu^2)}} = 0.605 \frac{Et^2}{a} \tag{13.22}$$

We therefore have the relationship

$$\sigma_{cr} = \frac{Et}{a\sqrt{3(1-\nu^2)}} = 0.605\frac{Et}{a} \tag{13.23}$$

for the *critical stress* for axially symmetric buckling of a simply supported circular cylindrical shell under uniform axial loading.

When the compression load N exceeds N_{cr}, given by Eq. (13.22), the system is *unstable* as the work done by N is greater than the increase in strain energy. We are led to conclude that the straight column is unstable when $N > N_{cr}$. If $N = N_{cr}$ the cylinder is in *neutral equilibrium* for small displacements. If $N < N_{cr}$ the straight cylinder is in *stable equilibrium*. The solution $c_0 = 0$ therefore represents an unstable or a stable configuration according to whether N is greater or smaller than N_{cr}.

13.8 NONSYMMETRICAL BUCKLING UNDER UNIFORM AXIAL COMPRESSION

The preceding section concerned simple buckling of an axially loaded cylinder. However, owing to *initial imperfections*,[41] i.e., small, unintentional deviations from the assumed initial state of the structure, cylindrical shells often buckle into axially nonsymmetrical forms. This case of instability is considered here through application of the equilibrium method.

We shall first modify the governing equations (13.1) for shell displacements to include axial load effects, as is done for plates and columns. This approach is based upon the assumption that all surface forces, except N_x, are very small. Thus, the product formed of these forces and the derivatives of displacement are neglected. Following a procedure similar to that given in Sec. 7.2, it is found that the force N_x yields a y component $N_x(\partial^2 v/\partial^2 x)\,dx \cdot a\,d\theta$ in the expression $\sum F_y = 0$ and a z component $N_x(\partial^2 w/\partial^2 x)\,dx \cdot a\,d\theta$ in the expression $\sum F_z = 0$. These components, after division by $dx\,d\theta$, are added to Eqs. (13.1). The equations of equilibrium, assuming $p_x = p_y = p_z = 0$, are then

$$\begin{gathered}
a\frac{\partial N_x}{\partial x} + \frac{\partial N_{x\theta}}{\partial \theta} = 0 \\
\frac{\partial N_\theta}{\partial \theta} + a\frac{\partial N_{x\theta}}{\partial x} + aN_x\frac{\partial^2 v}{\partial x^2} - Q_\theta = 0 \\
a\frac{\partial Q_x}{\partial x} + \frac{\partial Q_\theta}{\partial \theta} + aN_x\frac{\partial^2 w}{\partial x^2} + N_\theta = 0 \\
\frac{\partial M_\theta}{\partial \theta} + a\frac{\partial M_{x\theta}}{\partial \theta} - aQ_\theta = 0 \\
a\frac{\partial M_x}{\partial x} + \frac{\partial M_{x\theta}}{\partial \theta} - aQ_x = 0
\end{gathered} \tag{13.24}$$

The shear forces Q_x and Q_θ may be eliminated from the above equations. In so doing, we have

$$a\frac{\partial N_x}{\partial x} + \frac{\partial N_{x\theta}}{\partial \theta} = 0$$

$$\frac{\partial N_\theta}{\partial \theta} + a\frac{\partial N_{x\theta}}{\partial x} + aN_x\frac{\partial^2 v}{\partial x^2} - \frac{\partial M_{x\theta}}{\partial x} - \frac{1}{a}\frac{\partial M_\theta}{\partial \theta} = 0 \tag{13.25}$$

$$aN_x\frac{\partial^2 w}{\partial x^2} + N_\theta + a\frac{\partial^2 M_x}{\partial x^2} + 2\frac{\partial^2 M_{x\theta}}{\partial x\,\partial \theta} + \frac{1}{a}\frac{\partial^2 M_\theta}{\partial \theta^2} = 0$$

We now employ the elastic law (13.4) to express all stress resultants in terms of u, v, w and their derivatives. Taking the compressive force $N_x = N$ to be positive and introducing the dimensionless parameters

$$\alpha = \frac{t^2}{12a^2} \qquad q = \frac{N(1-\nu^2)}{Et} \tag{13.26}$$

the differential equations of the buckling problem are expressed as follows:

$$\frac{\partial^2 u}{\partial x^2} + \frac{1+\nu}{2a}\frac{\partial^2 v}{\partial x\,\partial \theta} - \frac{\nu}{a}\frac{\partial w}{\partial x} + \frac{1-\nu}{2}\frac{1}{a^2}\frac{\partial^2 u}{\partial \theta^2} = 0$$

$$\frac{1+\nu}{2a}\frac{\partial^2 u}{\partial x\,\partial \theta} + \frac{1-\nu}{2}\frac{\partial^2 v}{\partial x^2} + \frac{1}{a^2}\frac{\partial^2 v}{\partial \theta^2} - \frac{1}{a^2}\frac{\partial w}{\partial \theta}$$

$$+\alpha\left(\frac{1}{a^2}\frac{\partial^2 v}{\partial \theta^2} + \frac{1}{a^3}\frac{\partial^3 w}{\partial \theta^3} + \frac{\partial^3 w}{\partial x^2\,\partial \theta} + \frac{1-\nu}{2}\frac{\partial^2 v}{\partial x^2}\right) - q\frac{\partial^2 v}{\partial x^2} = 0 \tag{13.27}$$

$$-aq\frac{\partial^2 w}{\partial x^2} + \nu\frac{\partial u}{\partial x} + \frac{1}{a}\frac{\partial v}{\partial \theta} - \frac{w}{a}$$

$$-\alpha\left(\frac{1}{a}\frac{\partial^3 v}{\partial \theta^3} + a\frac{\partial^3 v}{\partial x^2\,\partial \theta} + a^3\frac{\partial^4 w}{\partial x^4} + \frac{1}{a}\frac{\partial^4 w}{\partial \theta^4} + 2a\frac{\partial^4 w}{\partial x^2\,\partial \theta^2}\right) = 0$$

It is noted that Eqs. (13.27) are satisfied for the particular case in which the solution is expressed in terms of constants c_1 and c_2,

$$u = \frac{c_1}{\nu a}x + c_2 \qquad v = 0 \qquad w = c_1$$

This represents the cylindrical form of equilibrium wherein the compressed cylinder uniformly expands laterally. Also, when one assumes $v = 0$, and u and w to be functions of x only, a solution is obtained for the axisymmetrical buckling treated in Sec. 13.7.

The general solution of Eqs. (13.27), if the origin of coordinates is placed at one end of the shell, can be expressed by the series

$$
\begin{aligned}
u &= \frac{c_1}{\nu a} x + c_2 + \sum_m^\infty \sum_n^\infty a_{mn} \sin n\theta \cos \frac{m\pi x}{L} \\
v &= \sum_m^\infty \sum_n^\infty b_{mn} \cos n\theta \sin \frac{m\pi x}{L} \qquad (13.28) \\
w &= c_1 + \sum_m^\infty \sum_n^\infty c_{mn} \sin n\theta \sin \frac{m\pi x}{L}
\end{aligned}
$$

For *long cylinders*, the results obtained from Eqs. (13.28) can be used *irrespective of the type of edge supports*. This is attributable to the fact that the edge conditions have only a minor influence on the magnitude of the critical load provided that the shell length is not small (say $L > 2a$).

When we introduce the notation

$$\lambda = \frac{m\pi a}{L} \qquad (13.29)$$

and substitute the solution (13.28) into Eqs. (13.27), the trigonometric functions drop out entirely, and we find that

$$
\begin{aligned}
&a_{mn}\left(\lambda^2 + \frac{1-\nu}{2} n^2\right) + b_{mn} \frac{n(1-\nu)\lambda}{2} + c_{mn}\, \nu\lambda = 0 \\
&a_{mn} \frac{n(1+\nu)\lambda}{2} + b_{mn}\left[\frac{(1-\nu)(1+\alpha)\lambda^2}{2} + (1+\alpha)n^2 - q\lambda^2\right] \qquad (13.30) \\
&\qquad + c_{mn}[n + \alpha n(n^2 + \lambda^2)] = 0 \\
&a_{mn}\, \nu\lambda + b_{mn}[1 + \alpha(n^2 + \lambda^2)] + c_{mn}[1 - q\lambda^2 + \alpha(\lambda^2 + n^2)^2] = 0
\end{aligned}
$$

The foregoing represent three linear equations with *buckling amplitudes* a_{mn}, b_{mn}, and c_{mn} as unknowns. The nontrivial solution, the buckling condition of the shell, is determined by setting equal to zero the determinant of coefficients. Usually α and q defined by Eqs. (13.26) are much smaller than unity, and we shall therefore neglect the terms containing the square of these quantities.

Observing that the minimum value of q takes place when λ^2 and n^2 are large numbers, the expanded determinantal equation then results in

$$q = \frac{N(1-\nu^2)}{Et} = \alpha \frac{(n^2 + \lambda^2)^2}{\lambda^2} + \frac{(1-\nu)\lambda^2}{(n^2 + \lambda^2)^2} \qquad (13.31)$$

We see that for $n = 0$, this expression reduces to Eq. (13.20), the result obtained for symmetrical buckling.

To determine the minimum value of q, we let $\eta = (n^2 + \lambda^2)^2/\lambda^2$. Equation (13.31) becomes $q = \alpha\eta + (1-\nu)/\eta$. The minimizing condition, $dq/d\eta = 0$, leads to $\eta = \sqrt{(1-\nu^2)/\alpha}$. Hence,

$$q_{\min} = 2\sqrt{\alpha(1-\nu^2)}$$

The corresponding critical stress is therefore

$$\sigma_{cr} = \frac{N_{cr}}{t} = \frac{Et}{a\sqrt{3(1-\nu^2)}} \tag{13.32}$$

This coincides with Eq. (13.23), which was derived in a different way for axisymmetrical buckling. Note that the critical stress *depends* upon the material properties, thickness, and radius, while it is *independent* of cylinder length.

The value of critical stress defined by Eq. (13.32), which is based upon the small-displacement theory, often does *not* agree with experimental data. This discrepancy is explained by applying the large-deflection theory of buckling, which takes into account the squares of the derivatives of the deflection w, initial imperfections, and a host of additional factors.

To relate the theoretical value obtained in this section to actual test data, it is necessary to incorporate an *empirical factor* in Eq. (13.32). For example, based upon the coefficients[42]

$$K = 1 - 0.901(1 - e^{-\psi}) \qquad \psi = \frac{1}{16}\sqrt{\frac{a}{t}}$$

Equation (13.32), for $\nu = 0.3$, becomes

$$\sigma_{cr} = 0.605K\frac{Et}{a} \tag{13.33}$$

The result, Eq. (13.33), is in satisfactory agreement with the tests for cylinders having $L/a < 5$.

There are many other kinds of shell-buckling problems, many of which are difficult to analyze. This is especially true, as in the cases of plates, for shell structures with end restraints other than those of simple support.

PROBLEMS

Secs. 13.1 to 13.8

13.1 Determine the maximum deflection w and stress in the tank described in Example 13.1. Use $t = 5$ mm. Assume that all the other data are unchanged.

13.2 During a stage of firing, a long, thin cylindrical missile casing of 1 m radius and 25 mm thickness, is loaded in axial compression. If $E = 200$ GPa, $\nu = 0.3$, and $\sigma_{yp} = 300$ MPa, what is the critical stress?

13.3 A 2.5-m-long steel pipe of 1.2 m diameter and 12 mm thickness is used in a structure as a column. What axial load can be tolerated without causing the shell to buckle? Use $E = 200$ GPa.

13.4 A water tower support, constructed of long steel piping of 0.8-m diameter, is to carry an axial compression load of 450 kN. What should be the minimum thickness of the shell? Assume that $\nu = 0.3$, $E = 210$ PGa, and $\sigma_{yp} = 400$ MPa.

APPENDIX

A

FOURIER SERIES EXPANSIONS

A.1 SINGLE FOURIER SERIES

The Fourier series are indispensable aids in the analytical treatment of many problems in the field of applied mechanics. The representation of periodic functions using trigonometric series is commonly called the *Fourier series expansion.* A function $f(x)$ defined in the interval $(-L, L)$ and determined outside of this interval by $f(x + 2L) = f(x)$ is said to be *periodic*, of period $2L$. Here L is a nonzero constant; for example, for $\sin x$ there are periods 2π, -2π, 4π, ... The Fourier expansion corresponding to $f(x)$ is of the form:[10]

$$f(x) = \frac{a_0}{2} + \sum_{n=1}^{\infty} \left(a_n \cos \frac{n\pi x}{L} + b_n \sin \frac{n\pi x}{L} \right) \tag{A.1}$$

in which a_0, a_n, and b_n are the *Fourier coefficients.*

To determine the coefficients a_n, we multiply both sides of Eq. (A.1) by $\cos(m\pi x/L)$, integrate over the interval of length $2L$, and make use of the *orthogonality relations:*

$$\begin{aligned}
&\int_{-L}^{L} \cos \frac{m\pi x}{L} \cos \frac{n\pi x}{L}\, dx = \begin{cases} 0 & m \neq n \\ L & m = n \end{cases} \\
&\int_{-L}^{L} \cos \frac{m\pi x}{L} \sin \frac{n\pi x}{L}\, dx = 0 \qquad \text{for all } m, n \\
&\int_{-L}^{L} \sin \frac{n\pi x}{L} \sin \frac{m\pi x}{L}\, dx = \begin{cases} 0 & m \neq n \\ L & m = n \end{cases}
\end{aligned} \tag{A.2}$$

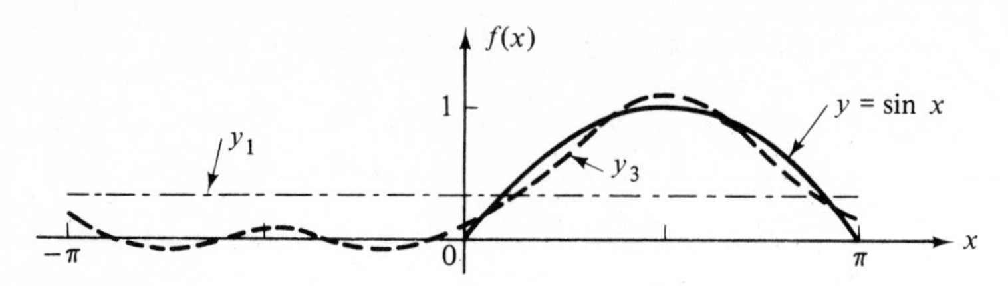

Figure A.1

In so doing, we obtain

$$a_n = \frac{1}{L} \int_{-L}^{L} f(x) \cos \frac{n\pi x}{L} \, dx \qquad (n = 0, 1, \ldots) \tag{A.3a}$$

Similarly, multiplication of both sides of Eq. (A.1) by $\sin (m\pi x/L)$ and integration over the interval $(-L, L)$ results in

$$b_n = \frac{1}{L} \int_{-L}^{L} f(x) \sin \frac{n\pi x}{L} \, dx \qquad (n = 1, 2, \ldots) \tag{A.3b}$$

Example A.1 Develop the Fourier expansion of the periodic function defined by

$$f(x) = 0 \quad \text{if} \quad -\pi < x < 0 \qquad f(x) = \sin x \quad \text{if} \quad 0 < x < \pi$$

SOLUTION The function is shown by *solid* line in Fig. A.1. The period $2L = 2\pi$ and $L = \pi$. Upon application of Eq. (A.3),

$$\begin{aligned} a_n &= \frac{1}{\pi} \int_{-\pi}^{0} (0) \cos nx \, dx + \frac{1}{\pi} \int_{0}^{\pi} \sin x \cos nx \, dx \\ &= -\frac{1}{2\pi} \left(\frac{-\cos n\pi}{1 - n} + \frac{-\cos n\pi}{1 + n} - \frac{2}{1 - n^2} \right) \\ &= \frac{1 + \cos n\pi}{\pi(1 - n^2)} \qquad (n \neq 1) \\ a_1 &= \frac{1}{\pi} \int_{0}^{\pi} \sin x \cos x \, dx = 0 \end{aligned}$$

and

$$b_n = \frac{1}{\pi} \int_{-\pi}^{0} (0) \sin nx \, dx + \frac{1}{\pi} \int_{0}^{\pi} \sin x \sin nx \, dx = 0 \qquad (n \neq 1)$$

$$b_1 = \frac{1}{\pi} \int_{0}^{\pi} \sin^2 x \, dx = \frac{1}{2}$$

The required Fourier series, for $n = 0, 1, 2, \ldots$, is thus

$$f(x) = \frac{1}{\pi} + \frac{\sin x}{2} - \frac{2}{\pi}\left(\frac{\cos 2x}{3} + \frac{\cos 4x}{15} + \frac{\cos 6x}{35} + \frac{\cos 8x}{63} + \cdots\right)$$

The first and the first three terms

$$y_1 = \frac{1}{\pi} \qquad y_3 = \frac{1}{\pi} + \frac{\sin x}{2} - \frac{2 \cos 2x}{3\pi}$$

are sketched by *dotted* and *dashed* lines, respectively, in the figure. We observe that as the number of terms increases, the approximating curves approach the graph of $f(x)$.

A.2 HALF-RANGE EXPANSIONS

In numerous applications, the evaluation of the Fourier coefficients can be simplified by employing a *half-range* series over the interval $(0, L)$.

Assume first that $f(x)$ is an *even* function, i.e.,

$$f(-x) = f(x)$$

for all x. Viewed geometrically, a graph of even function is symmetric with respect to the vertical, y axis. For example, x^2 and $\cos x$ are even functions. We can rewrite Eq. (A.3*a*) as

$$a_n = \frac{1}{L}\int_{L}^{0} f(x) \cos \frac{n\pi x}{L}\, dx + \frac{1}{L}\int_{0}^{L} f(x) \cos \frac{n\pi x}{L}\, dx \tag{a}$$

In the first integral above, we substitute $x = -u$ and hence $dx = -du$. Inasmuch as $x = -L$ implies $u = L$, and $x = 0$ implies $u = 0$, this integral is therefore

$$\frac{1}{L}\int_{L}^{0} f(-u) \cos \left(\frac{-n\pi u}{L}\right)(-du) \tag{b}$$

Because

$$f(-u) = f(u) \qquad \cos \left(\frac{-n\pi u}{L}\right) = \cos \frac{n\pi u}{L}$$

expression (*b*) becomes

$$\frac{1}{L}\int_{0}^{L} f(u) \cos \frac{n\pi u}{L}\, du \tag{c}$$

Here, the dummy variable of integration u can be replaced by any other symbol, and in particular x. Substitution of Eq. (*c*) into Eq. (*a*) then yields, for an even, periodic function $f(x)$, the coefficients a_n:

$$a_n = \frac{2}{L}\int_{0}^{L} f(x) \cos \frac{n\pi x}{L}\, dx \qquad (n = 0, 1, 2, \ldots) \tag{A.4}$$

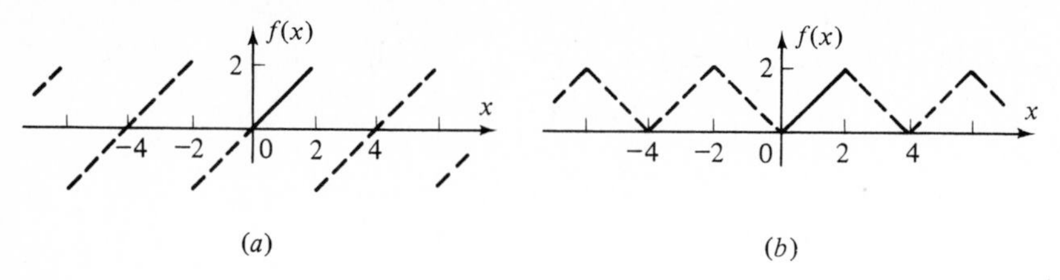

Figure A.2

By using Eq. (A.3*b*), it is readily verified that $b_n = 0$.

In the case of an *odd* function,

$$f(-x) = -f(x)$$

for all values of x. A procedure identical with that described above yields

$$b_n = \frac{2}{L} \int_0^L f(x) \sin \frac{n\pi x}{L}\, dx \qquad (n = 1, 2, \ldots) \tag{A.5}$$

and $a_n = 0$. The graph of an odd function (e.g., x and $\sin x$) is *skew-symmetric*.

The half-range single series expansion of various common loadings is listed in Table 3.2.

Example A.2 Determine the half-range expansion of the function

$$f(x) = x \qquad (0 < x < 2)$$

in (*a*) sine series and (*b*) cosine series.

SOLUTION The function is indicated by the *solid* line in Fig. A.2.

(*a*) We shall treat $f(x)$ as an odd function of period $2L = 4$. The *extended function* is shown in Fig. A.2*a* by the *dashed* lines. For the sine series $a_n = 0$, and Eq. (A.5) gives

$$b_n = \frac{2}{2} \int_0^2 x \sin \frac{n\pi x}{2}\, dx = -\frac{4}{n\pi} \cos n\pi$$

Hence

$$f(x) = \sum_{n=1}^{\infty} \frac{-4}{n\pi} \cos n\pi \sin \frac{n\pi x}{2}$$

$$= \frac{4}{\pi}\left(\sin \frac{\pi x}{2} - \frac{1}{2} \sin \frac{2\pi x}{2} + \frac{1}{3} \sin \frac{3\pi x}{2} - \cdots\right) \tag{d}$$

It is observed that the odd function is discontinuous $x = 2$ and $x = -2$, and the sine series (*d*) converges to zero, the *mean value* of the function, at the points of discontinuity.

(b) Now the definition of $f(x)$ is *extended* to that of an even function of period $2L = 4$ (Fig. A.2*b*). As $f(x)$ is even, $b_n = 0$ and

$$a_n = \frac{2}{2}\int_0^2 x \cos\frac{n\pi x}{2}\,dx = -\frac{4}{n^2\pi^2}(\cos n\pi - 1) \qquad (n \neq 0)$$

When $n = 0$

$$a_0 = \int_0^2 x\,dx = 2$$

It follows that

$$f(x) = 1 + \sum_{n=1}^{\infty} \frac{4}{n^2\pi^2}(\cos n\pi - 1)\cos\frac{n\pi x}{2}$$

$$= 1 - \frac{8}{\pi^2}\left(\cos\frac{\pi x}{2} + \frac{1}{3^2}\cos\frac{3\pi x}{2} + \frac{1}{5^2}\cos\frac{5\pi x}{2} + \cdots\right) \qquad (e)$$

The series (d) and (e) represent the extended function shown in Figs. A.2*a* and A.2*b*, respectively; however, our concern is only with the interval $0 < x < 2$. Note that the latter representation converges more rapidly than the former.

A.3 DOUBLE FOURIER SERIES

The idea of a Fourier series expansion for a function of a single variable can be extended to the case of functions of two or more variables. For instance, we can expand $p(x, y)$ into a *double Fourier sine series*, Eq. (3.1*a*):

$$p(x, y) = \sum_{m=1}^{\infty}\sum_{n=1}^{\infty} p_{mn} \sin\frac{m\pi x}{a} \sin\frac{n\pi y}{b}$$

The above represents a half-range sine series in x, multiplied by a half-range sine series in y, using for the period of expansions $2a$ and $2b$, respectively. That is

$$p(x, y) = \sum_{m=1}^{\infty} p_m(y) \sin\frac{m\pi x}{a} \qquad (A.6)$$

with

$$p_m(y) = \sum_{n=1}^{\infty} p_{mn} \sin\frac{n\pi y}{b} \qquad (a)$$

Treating Eq. (A.6) as a Fourier series wherein y is kept constant, Eq. (A.5) is applied to yield

$$p_m(y) = \frac{2}{a}\int_0^a p(x, y) \sin\frac{m\pi x}{a}\,dx \qquad (A.7)$$

Similarly, for Eq. (*a*):

$$p_{mn} = \frac{2}{b} \int_0^b p_m(y) \sin \frac{n\pi y}{b} \, dy \tag{b}$$

Equation (*b*), together with (A.7), then leads to

$$p_{mn} = \frac{4}{ab} \int_0^b \int_0^a p(x, y) \sin \frac{m\pi x}{a} \sin \frac{n\pi y}{b} \, dx \, dy \tag{c}$$

This agrees with Eq. (3.3) of Sec. 3.2 which was derived differently.

Similarly, the results can be obtained for cosine series or for series having both sines and cosines.

The double-sine series expansion of some typical loadings is illustrated in Sec. 3.3.

APPENDIX

B

SOLUTION OF SIMULTANEOUS LINEAR EQUATIONS

B.1 INTRODUCTION

Simultaneous algebraic equations are encountered frequently in the numerical treatment of plate and shell problems. A system of linear equations

$$
\begin{aligned}
a_{11}x_1 + a_{12}x_2 + \cdots + a_{1n}x_n &= c_1 \\
a_{21}x_1 + a_{22}x_2 + \cdots + a_{2n}x_n &= c_2 \\
\cdots\cdots\cdots\cdots\cdots\cdots\cdots\cdots \\
a_{n1}x_1 + a_{n2}x_2 + \cdots + a_{nn}x_n &= c_n
\end{aligned}
\tag{B.1}
$$

can be written in the matrix form

$$
\begin{bmatrix} a_{11} & a_{12} & \cdots & a_{1n} \\ a_{21} & a_{22} & \cdots & a_{2n} \\ \cdots & \cdots & \cdots & \cdots \\ a_{n1} & a_{n2} & \cdots & a_{nn} \end{bmatrix}
\begin{Bmatrix} x_1 \\ x_2 \\ \cdots \\ x_n \end{Bmatrix}
=
\begin{Bmatrix} c_1 \\ c_2 \\ \cdots \\ c_n \end{Bmatrix}
\tag{B.2}
$$

or concisely

$$[A]\{X\} = \{C\}$$

wherein the rule for the *matrix multiplication* is evident from the original equations. In order that Eqs. (B.2) equal to Eqs. (B.1), the terms of each *row* of the coefficient matrix $[A]$ must be multiplied by the unknown terms of the *column* matrix $\{X\}$. The quantities a_{ij} and c_i $(i, j = 1, 2, \ldots, n)$ are treated as constants.

There are many techniques[16,43] applied to facilitate the solution of simultaneous equations by means of digital and analog computers and desk calculators. Most methods require a knowledge of the basic theory, operation, and the behavior of matrices.

In the next section is described a common elimination approach. The development is based upon the assumption that the determinant of the coefficient matrix is nonzero, $|A| \neq 0$.

B.2 THE GAUSS REDUCTION METHOD

The object of the *Gauss reduction* or *Gauss' method* is to eliminate the unknowns systematically from the simultaneous equations. This is the simplest practical procedure, involving a minimum number of computational steps. Numerous modified forms of the Gauss reduction process exist; some are well adopted for hand computations while others are more suitable for electronic computers.

To describe the method, consider an $n \times n$ system of equations (B.2). Note that the *same matrix row operations* that are performed on $[A]$ must be performed on $\{C\}$. When these operations have been properly completed, the unit matrix $[I]$ will appear on the left, and the column matrix of the *solution* $\{S\}$ on the right.[10] That is,

$$\begin{bmatrix} 1 & 0 & \dots & 0 \\ 0 & 1 & \dots & 0 \\ \dots & \dots & \dots & \dots \\ 0 & 0 & \dots & 1 \end{bmatrix} \begin{Bmatrix} x_1 \\ x_2 \\ \dots \\ x_n \end{Bmatrix} = \begin{Bmatrix} s_1 \\ s_2 \\ \dots \\ s_n \end{Bmatrix} \tag{B.3}$$

or $x_1 = s_1$, $x_2 = s_2$, ..., and $x_n = s_n$.

As a first example, consider the problem of determining the unknowns in a 3×3 set of equations:

$$\begin{bmatrix} 20 & -14 & 3 \\ -2 & 1 & 0 \\ 5 & -4 & 1 \end{bmatrix} \begin{Bmatrix} x_1 \\ x_2 \\ x_3 \end{Bmatrix} = \begin{Bmatrix} 14 \\ -3 \\ 2 \end{Bmatrix} \tag{a}$$

The Gauss reduction can be carried out in the following steps:

Step 1 Transform the first colum to the unit matrix. To accomplish this

divide the first row by (*pivot* element a_{11}) 20
subtract -2 times the first row from the second row
subtract five times the first row from the third row

The result of performing these three operations on both sides is

$$\begin{bmatrix} 1 & -7/10 & 3/20 \\ 0 & -2/5 & 3/10 \\ 0 & -1/2 & 1/4 \end{bmatrix} \begin{Bmatrix} x_1 \\ x_2 \\ x_3 \end{Bmatrix} = \begin{Bmatrix} 7/10 \\ -8/5 \\ -3/2 \end{Bmatrix}$$

Step 2 Similar operations are performed with the second row. Now the second element of the second row is the pivot element. Thus, this element must be transformed to unity, while the other elements of the second column are transformed to zero. For this purpose

divide the second row by $-2/5$
subtract $-7/10$ times the second row from the first row
subtract $-1/2$ times the second row from the third row

Carrying out the above operations on both sides, we obtain

$$\begin{bmatrix} 1 & 0 & -3/8 \\ 0 & 1 & -3/4 \\ 0 & 0 & -1/8 \end{bmatrix} \begin{Bmatrix} x_1 \\ x_2 \\ x_3 \end{Bmatrix} = \begin{Bmatrix} 7/2 \\ 4 \\ 1/2 \end{Bmatrix}$$

Step 3 In this case, the third element of the third row is the pivot element and the third column must be transformed to a unit matrix. Thus, on both sides we must

divide the third row by $-1/8$
subtract $-3/8$ times the third row from the first row
subtract $-3/4$ times the third row from the second row

In so doing, one has

$$\begin{bmatrix} 1 & 0 & 0 \\ 0 & 1 & 0 \\ 0 & 0 & 1 \end{bmatrix} \begin{Bmatrix} x_1 \\ x_2 \\ x_3 \end{Bmatrix} = \begin{Bmatrix} 2 \\ 1 \\ -4 \end{Bmatrix}$$

The successive transformations have now been completed. Inasmuch as the until matrix appears on the left, the matrix on the right is the solution. We thus have $x_1 = 2$, $x_2 = 1$, and $x_3 = -4$.

One observes that a value for x_3 having been found at step 2 by the described *forward-elimination* process, values of x_2 and x_1 can be ascertained (without applying the operations of step 3) by *back substitution*, each calculation involving only one unknown.

Consider, as a second example, the case of a 4×4 set:

$$\begin{bmatrix} 1 & 2 & 3 & -1 \\ 1 & 2 & 4 & 3 \\ 3 & 5 & 6 & -2 \\ 4 & 2 & 1 & 0 \end{bmatrix} \begin{Bmatrix} x_1 \\ x_2 \\ x_3 \\ x_4 \end{Bmatrix} = \begin{Bmatrix} 5 \\ 1 \\ 13 \\ 10 \end{Bmatrix} \qquad (b)$$

The first step of the operations, described in preceding example, yields

$$\begin{bmatrix} 1 & 2 & 3 & -1 \\ 0 & 0 & 1 & 4 \\ 0 & -1 & -3 & 1 \\ 0 & -6 & -11 & 4 \end{bmatrix} \begin{Bmatrix} x_1 \\ x_2 \\ x_3 \\ x_4 \end{Bmatrix} = \begin{Bmatrix} 5 \\ -4 \\ -2 \\ -10 \end{Bmatrix}$$

Next, the second column must be reduced to unity. This is *impossible* because a_{22} (current pivot) is *zero*. However, interchanging the second and the third equations, one can continue the elimination:

$$\begin{bmatrix} 1 & 2 & 3 & -1 \\ 0 & -1 & -3 & 1 \\ 0 & 0 & 1 & 4 \\ 0 & -6 & -11 & 4 \end{bmatrix} \begin{Bmatrix} x_1 \\ x_2 \\ x_3 \\ x_4 \end{Bmatrix} = \begin{Bmatrix} 5 \\ -2 \\ -4 \\ -10 \end{Bmatrix}$$

It is noted that at the beginning of *any* step, the interchange can always occur between the equation with zero pivot element and an equation *below* it.

Then, successive transformations to unity of the second, third, and fourth columns will develop

$$\begin{bmatrix} 1 & 0 & -3 & 1 \\ 0 & 1 & 3 & -1 \\ 0 & 0 & 1 & 4 \\ 0 & 0 & 7 & -2 \end{bmatrix} \begin{Bmatrix} x_1 \\ x_2 \\ x_3 \\ x_4 \end{Bmatrix} = \begin{Bmatrix} 1 \\ 2 \\ -4 \\ 2 \end{Bmatrix}$$

$$\begin{bmatrix} 1 & 0 & 0 & 13 \\ 0 & 1 & 0 & -13 \\ 0 & 0 & 1 & 4 \\ 0 & 0 & 0 & -30 \end{bmatrix} \begin{Bmatrix} x_1 \\ x_2 \\ x_3 \\ x_4 \end{Bmatrix} = \begin{Bmatrix} -11 \\ 14 \\ -4 \\ 30 \end{Bmatrix}$$

$$\begin{bmatrix} 1 & 0 & 0 & 0 \\ 0 & 1 & 0 & 0 \\ 0 & 0 & 1 & 0 \\ 0 & 0 & 0 & 1 \end{bmatrix} \begin{Bmatrix} x_1 \\ x_2 \\ x_3 \\ x_4 \end{Bmatrix} = \begin{Bmatrix} 2 \\ 1 \\ 0 \\ -1 \end{Bmatrix}$$

We thus have $x_1 = 2$, $x_2 = 1$, $x_3 = 0$, and $x_4 = -1$.

REFERENCES

1. A. C. Ugural and S. K. Fenster, *Advanced Strength and Applied Elasticity*, Elsevier, 2d ed., 1981.
2. S. Marcus, *Die Theorie elastischer Gewebe*, Berlin, 1932.
3. H. L. Langhaar, *Energy Methods in Applied Mechanics*, Wiley, 1962.
4. L. H. Abraham, *Structural Design of Missiles and Spacecraft*, McGraw-Hill, 1962.
5. D. F. Miner and J. B. Seastone (eds.), *Handbook of Engineering Materials*, Wiley, 1955.
6. D. McFarland, B. L. Smith, and W. D. Bernhart, *Analysis of Plates*, Spartan Books, 1972.
7. J. R. Roark and W. C. Young, *Formulas for Stress and Strain*, McGraw-Hill, 1975.
8. W. Flügge (ed.), *Handbook of Engineering Mechanics*, McGraw-Hill, 1968.
9. T. W. Gawain and E. C. Durry, "Stresses in Laterally Loaded Disks of Nonuniform Thickness," *Prod. Eng.*, December, 1949, pp. 130–134.
10. I. S. Sokolnikoff and R. M. Redheffer, *Mathematics of Physics and Modern Engineering*, McGraw-Hill, 1966.
11. S. P. Timoshenko and S. Woinowsky-Krieger, *Theory of Places and Shells*, McGraw-Hill, 1959.
12. R. Szilard, *Theory and Analysis of Plates—Classical and Numerical Methods*, Prentice-Hall, 1974.
13. A. Nadai, *Die elastischen Platten*, Springer, 1925.
14. R. E. Peterson, *Stress Concentration Design Factors*, Wiley, 1974.
15. M. G. Salvadori and M. L. Baron, *Numerical Methods in Engineering*, Prentice-Hall, 1967.
16. R. L. Ketter and S. P. Prawell, Jr., *Modern Methods of Engineering Computation*, McGraw-Hill, 1969.
17. D. C. Zienkiewitcz, *The Finite Element Method in Engineering Science*, McGraw-Hill, 1971.
18. *Proceedings of the Second Conference on Matrix Methods in Structural Mechanics*, Air Force Flight Dynamics Lab., Wright-Patterson AFB, AFFDL-TR-68-150, 1969.
19. S. G. Lekhnitskii, *Anisotropic Plates*, Gordon & Breach, 1968.
20. J. R. Vinson and T. W. Chou, *Composite Materials and Their use in Structures*, Wiley, 1975.
21. H. J. Huffington, "Theoretical Determination of Rigidity Properties of Orthotropic Stiffened Plates," *J. Appl. Mech.*, **23**, 1956.
22. M. S. Troitsky, *Stiffened Plates—Bending, Stability, and Vibrations*, Elsevier, 1976.
23. K. S. Pister and S. B. Dong, "Elastic Bending of Layered Plates," *Proc. ASCE*, **84**, *J. Eng. Mech. Div.*, October, 1950, 1–10.
24. S. P. Timoshenko and J. M. Gere, *Theory of Elastic Stability*, McGraw-Hill, 1961.
25. B. Aalami and D. G. Williams, *Thin Plate Design for Transverse Loading*, Wiley, 1975.
26. B. A. Boley and J. H. Weiner, *Theory of Thermal Stresses*, Wiley, 1960.
27. D. J. Johns, *Thermal Stress Analysis*, Pergamon, 1965.

28. J. N. Goodier, "Thermal Stresses," pp. 74–77 of *Design Data and Methods, ASME*, 1953.
29. H. S. Tsien, "Similarity Laws for Stressing Heated Wings," *J. Aero. Sci.*, **20**(1), January, 1953, pp. 1–11.
30. J. Heyman, *Equilibrium of Shell Structures*, Oxford University Press, 1977.
31. P. L. Gould, *Static Analysis of Shells*, Lexington Books, 1977.
32. W. Flügge, *Stresses in Shells*, Springer, 1962.
33. A. Pflüger, *Elementary Statics of Shells*, McGraw-Hill, 1961.
34. J. H. Faupel, *Engineering Design*, Wiley, 1964.
35. H. D. Tabakman and Y. J. Lin, "Quick Way to Calculate Thermal Stresses in Cylindrical Shells," *Mach. Des.*, September 21, 1978.
36. J. F. Harvey, *Theory and Design of Modern Pressure Vessels*, Van Nostrand Reinhold, 1974.
37. ASME *Boiler and Pressure Vessel Code*, Section VIII.
38. The M.W. Kellogg Co., *Design of Piping Systems*, Wiley, 1956.
39. L. H. Donnell, "Stability of Thin Walled Tubes Under Torsion," *NACA Rep. 479*, 1933.
40. I. A. Wojtaszak, "Deformation of Thin Cylindrical Shells Subjected to Internal Loading," *Phil. Mag.*, ser. 7, **18,** pp. 1099–1116, 1934.
41. D. O. Brush and B. O. Almroth, *Buckling of Bars, Plates, and Shells*, McGraw-Hill, 1975.
42. J. R. Vinson, *Structural Mechanics: The Behavior of Plates and Shells*, Wiley, 1974.
43. T. E. Shoup, *A Practical Guide to Computer Methods for Engineers*, Prentice-Hall, 1979.

ANSWERS TO SELECTED PROBLEMS

CHAPTER 1

1.5 $\sigma_x = \sigma_y = Etc_0/32(1-\nu)a^2 \qquad \tau_{xy} = 0$

1.6 $w = -[M_0/D(1-\nu)][xy - (a/2)^2]$

1.7(*a*) $M_x = \frac{1}{3}\left(2M - x\dfrac{\partial M}{\partial x}\right) \qquad \sigma_{x,\,\max} = 4P/\pi t^2$

1.9 $U = \dfrac{Dc_0^2}{ab}\left(0.1625\dfrac{a^4+b^4}{a^2b^2} + 0.0929\right)$

CHAPTER 2

2.1 $t = 1.28$ mm

2.3 $N = 3.9$

2.12 $w_{\max} = 0.0237M_1 a^2/Et^3$

2.15 $w_{\max} = p_0 a^2/48(\ln a - 1.3)$

2.16 $w_{\max} = Pa^2/8\pi D(\ln a - 1.3)$

2.22 $w_c = (Pb^2/16\pi D)[2\ln(b/a) + (a/b)^2 - 1]$

CHAPTER 3

3.1(*a*) $p_0 = 126.396$ kPa **(*b*)** $w_{\max} = 11.59$ mm

3.4(*a*) $w_{\max} = 0.00328\ Pa^2/D$ **(*b*)** $\sigma_{y,\,\max} = 0.836\ P/t^2$

3.6 $w_c = 0.00697\ a^2(P_1 + P_2)/D$

3.17 $a = 679$ mm $\quad w_{max} = 1.03$ mm

3.21 $w_{1,\,max} = 0.0055\ p_0 a^4/D \quad w_{2,\,max} = -0.0102\ p_0 a^4/D$

3.24 $\sigma_{max} = 0.1986\ p_0(a/t)^2$

3.25(*a*) $t = 0.496a(Np_0/\sigma_{yp})^{1/2}$ **(*b*)** $t = 0.480a(Np_0/\sigma_{yp})^{1/2}$

3.28 $k = 1.011$ MPa/m

3.31 $w_{max} = 0.0112\ p_0 a^4/D$

CHAPTER 4

4.2 $M_{AC} = (M_0/2a)(\sqrt{3}x - y) \quad R = -2M_0/\sqrt{3}$

4.3(*a*) $p_0 = 56.26$ kPa **(*b*)** $p_0 = 78.68$ kPa

4.6 $w_{max,\,e}/w_{max,\,c} = 2.271 \quad M_{max,\,e}/M_{max,\,c} = 1.838$

CHAPTER 5

5.4 $\sigma_{max} = 2.14\ p_0(a/t)^2$

5.5 $w_3 = w_{max} = 0.003211\ p_0 a^4/D$

5.7 $w_5 = w_{max} = 0.04506\ p_0 a^4/D$

5.10 $w_5 = w_{max} = 0.00283\ p_0 a^4/D$

5.12 $w_2 = w_{max} = 0.03335\ p_0 a^4/D$

CHAPTER 6

6.1 $D_x = H = 146.5$ kN·m $\quad D_y = 1346.5$ kN·m

6.2 $D_x = 19.36$ kN·m $\quad D_y = 110.25$ kN·m $\quad D_{xy} = 0$
$H = 17.43$ kN·m

6.6 $p_0 = 0.107/a^4$ MPa

6.9 $w_{max} = 0.483 \times 10^{-6} p_0$ m

6.10 $w_{max} = 6.706 \times 10^{-6} p_0$ m

CHAPTER 7

7.3 $N_{cr} = 2\pi^2 D/a^2$

7.7 $N_{cr} = 4.86\pi^2 D/a^2$

7.9 $N_{cr} = 6.08\pi^2 D/a^2$

7.10 $w_{max} = a_0/[1 + (Na^2/2\pi^2 D)]$

CHAPTER 9

9.1 $M^* = 146.16$ N $\quad N^* = 380$ kN/m
$N_x = N_y = 542.86$ kN/m $\quad M_x = M_y = -208.8$ N

9.3 $w_{max} = 16042 \times 10^{-6} \alpha A a^2 t^2 \quad M_{x,\,max} = 19605.6 \times 10^{-6} \alpha A E t^5$

9.9 $w = -0.0045(1 + \nu)\alpha B a^4 t^4$

CHAPTER 10

10.1 $N = 40.2$

10.2 $\Delta p = 896$ kPa

10.3(*a*) 2.815 m (***b***) 1.407 m

10.14 $\sigma_\theta = pa/t \quad \sigma_x = \nu pa/t$

10.18 $N = 2.4$

CHAPTER 11

11.1 $P_{max} = 81.08$ kN/m

11.3(*a*) $P = 96.893$ kN/m (***b***) $P = 121.189$ kN/m

11.5(*a*) $2w = 218 \times 10^{-7} P/D$

11.5(*b*) $\sigma_x = 308.195P$ Pa $\quad \sigma_\theta = -543.965P$ Pa

CHAPTER 12

12.1(*a*) $L = 18.93$ mm (***b***) $Q_1 = 42.371$ kN/m

12.3 $w = 0.0447 \times 10^{-6}$ m $\quad \sigma_\theta = -347.555$ kPa

12.6 $\sigma_{\theta,\,max} = -129.75$ MPa

12.8 $\sigma_{\max} = 1.5(P/t^2)[a^2t^2/3(1 - \nu^2)]^{1/4}$

12.13 $\sigma_{\theta,\max} = 339.5$ MPa

12.16 $\sigma_{\theta,b} = 67.68$ MPa $\quad \sigma_{\theta,m} = 60$ MPa

12.18(*a*) $t = 5$ mm $\quad$ **(*b*)** $t = 6.27$ mm

CHAPTER 13

13.1 $w_{\max} = -164{,}761.2\gamma/E$ cm $\quad \sigma_{x,\max} = -5655.12\gamma$ N/cm^2

13.3 $P_{cr} = 74{,}246$ kN

INDEX